Skyrmions and Hall Transport

Skyrmions and Hall Transport

Bom Soo Kim

Jenny Stanford
PUBLISHING

Published by

Jenny Stanford Publishing Pte. Ltd.
101 Thomson Road
#06-01, United Square
Singapore 307591

Email: editorial@jennystanford.com
Web: www.jennystanford.com

British Library Cataloguing-in-Publication Data
A catalogue record for this book is available from the British Library.

Skyrmions and Hall Transport

Copyright © 2023 Jenny Stanford Publishing Pte. Ltd.

All rights reserved. This book, or parts thereof, may not be reproduced in any form or by any means, electronic or mechanical, including photocopying, recording or any information storage and retrieval system now known or to be invented, without written permission from the publisher.

For photocopying of material in this volume, please pay a copying fee through the Copyright Clearance Center, Inc., 222 Rosewood Drive, Danvers, MA 01923, USA. In this case permission to photocopy is not required from the publisher.

ISBN 978-981-4968-34-8 (Hardcover)
ISBN 978-1-003-37253-0 (eBook)

To Myung-Duk, Young-Kyung, and Dong-Yon

Contents

Preface   xi

Acknowledgements   xv

Introduction   xvii

1 Symmetries of Magnetic Skyrmions   1
    1.1 Symmetry   1
    1.2 Action formalisms for localized spins   6
       1.2.1 Landau-Lifshitz equation   6
       1.2.2 Landau-Lifshitz revisited   11
       1.2.3 Spin Lagrangian   15
    1.3 Stability of magnetic skyrmions in chiral magnet   19
       1.3.1 Helicoidal state in Nematic liquid crystals   19
       1.3.2 Stability of spin spiral state   23
       1.3.3 Stability of skyrmion crystals   29
       1.3.4 Topological nature and skyrmion charge   37
    1.4 Symmetries of skyrmion action   40
       1.4.1 Stress energy tensor   41
       1.4.2 Translation symmetry   44
       1.4.3 Rotation symmetry   47
       1.4.4 (Angular) Momentum of the skyrmion   51
    1.5 Mutually compatible observables   53

2 Hydrodynamics   57
    2.1 Introduction: Relativistic hydrodynamics   57
       2.1.1 Ideal hydrodynamics   58
       2.1.2 First derivative order   60
    2.2 Hydrodynamics: New developments   62
       2.2.1 Hydrodynamics with broken parity   62

		2.2.2	Hydrodynamics with broken boost	64
	2.3	Hall viscosity: Introduction		66
		2.3.1	Another view of stress tensor	66
		2.3.2	Hall viscosity: A geometric picture	71
	2.4	Charged hydrodynamics		74
		2.4.1	With broken parity	78
		2.4.2	With broken boost	81
	2.5	Green's function and Kubo formula		83
		2.5.1	Linear response theory	83
		2.5.2	Background field methods	88
		2.5.3	Thermodynamic response functions	92
	2.6	Irreversible thermodynamics		94
		2.6.1	Onsager's reciprocal relation	96
		2.6.2	Seebeck, Peltier, and Thompson effects	100
		2.6.3	Thermo-electromagnetic effects	104
3	**Hall Viscosity**			**109**
	3.1	Charged particles in electromagnetic fields		109
		3.1.1	Quantum Hall fluid	113
		3.1.2	Wave function	115
	3.2	Landau Hamiltonian on torus		120
		3.2.1	Flat torus	120
		3.2.2	Deformed torus	124
	3.3	Berry phase and Hall viscosity		128
		3.3.1	Adiabatic process and Berry phase	128
		3.3.2	Kubo formula for Hall viscosity	130
		3.3.3	Hall viscosity of quantum Hall states	132
	3.4	Quantum Hall systems and Hall conductivity		135
		3.4.1	Conductivity for a single particle	137
		3.4.2	Kubo formula for Hall conductivity	141
		3.4.3	Topology and Hall conductivity	144
		3.4.4	Hall conductivity with momentum dependence	148
	3.5	Hall conductivity and Hall viscosity		148
		3.5.1	Non-homogeneous electric field	148
		3.5.2	Hall viscosity in terms of Hall conductivity	151
4	**Spin Dynamics**			**155**
	4.1	Landau-Lifshitz-Gilbert equation		156
		4.1.1	Ferromagnetism	156

		4.1.2	Basic understanding of spin torque	159
		4.1.3	Domain Wall illustration of LLG equation	162
	4.2	Spin torques	164
		4.2.1	Spin transfer torque	165
		4.2.2	Domain wall illustration of STT	168
		4.2.3	Spin-orbit torque	173
		4.2.4	Domain Wall illustration of SOT	177
		4.2.5	Spin Hall torque	180
		4.2.6	Domain Wall illustration of SHT	185
		4.2.7	Emergent electromagnetic field	187
	4.3	Generalized LLG equation	187
	4.4	Thiele equation	189
		4.4.1	Thiele's original derivation	189
		4.4.2	Generalizations with various spin torques	192
		4.4.3	Generalization with transverse velocity	196
	4.5	Spin effects on transport	197
		4.5.1	Spin Seebeck effect	200
		4.5.2	Spin Peltier effect	203
5	**Ward Identity**		**207**
	5.1	Symmetries and Ward identity	208
		5.1.1	Symmetry and equation of motion	208
		5.1.2	Geometric understanding of Ward identities	210
	5.2	An example	214
		5.2.1	Ward identity for global currents	214
		5.2.2	Ward identity in momentum space	217
	5.3	Quantum field theory Ward identities	220
		5.3.1	Ward identities for stress energy tensor	221
		5.3.2	Ward identities based on symmetries	223
		5.3.3	Ward identity with conserved current	234
		5.3.4	Galilean invariant Ward identities	243
		5.3.5	Ward identities with dissipative terms	248
	5.4	Hall viscosity from Ward identities	250
		5.4.1	Spectral representation	253
		5.4.2	Fermions in magnetic field	256
	5.5	Ward identity with topological charge	262
		5.5.1	Topological charge as a central extension	262
		5.5.2	Topological Ward identity	267

6	**Skyrmion Transport in Magnets**		**273**
	6.1 Conducting magnets		273
		6.1.1 Topological Hall effect	275
		6.1.2 Skyrmion Hall effect	280
	6.2 Ward identity for conducting magnets		288
		6.2.1 With magnetic field and electric current	288
		6.2.2 Modeling the Hund's rule coupling	291
		6.2.3 Hall viscosity in conducting materials	293
	6.3 Insulating magnets		295
		6.3.1 Magnon Hall effect	296
		6.3.2 Skyrmion Seebeck effect	301
		6.3.3 Rotational motion of skyrmions	312
	6.4 Ward identity for insulating magnets		318
		6.4.1 A way to measure Hall viscosity	318
	6.5 Outlook: Measuring Hall viscosity?		321
7	**Modeling Hall Viscosity**		**323**
	7.1 Ferro-, antiferro- and ferrimagnets		323
		7.1.1 Domain wall velocity in ferrimagnets	324
		7.1.2 Theory for ferrimagnet dynamics	326
	7.2 Hall angle data of skyrmion and antiskyrmion		332
		7.2.1 For a ferromagnet	333
		7.2.2 For a ferrimagnet	335
	7.3 Direction of Hall viscosity		336
		7.3.1 Hall viscosity is independent of skyrmion charge	338
	7.4 Modeling Hall viscosity in Thiele equation		340
		7.4.1 Generalized skyrmion Hall angle	343
		7.4.2 Hall viscosity from data	343
		7.4.3 Contributions from spin torques?	345
	7.5 Outlook		350
	Bibliography		353
	Index		365

Preface

About a decade ago, inspired by the study of broken parity symmetry in holographic fluid dynamics, a surprising development in hydrodynamics filled a missing gap in Landau and Lifshitz's seminal work. One of the fruitful outcomes of the community's efforts is the appreciation of a universal and mysterious quantity known as "Hall viscosity." Although a part of the viscosity tensor, it is dissipationless and does not create entropy. Hall viscosity continues to be extensively investigated in quantum Hall systems but is yet to be empirically measured.

Around 2009, another exciting event took place in the context of broken parity symmetry—the discovery of chiral magnetic skyrmions in what is called the A-phase of the material MnSi. Since then, the subject has attracted much scientific attention, both in experiment and in theory. Early experiments revealed that skyrmions are stable with topological protection, small to the nanoscale, and highly energy efficient in their creations, annihilations, and movements. Skyrmions, with these properties, are fascinating candidates in the search to construct a next-generation storage device. To understand their motion, the community has relied on phenomenological approaches based on the Landau-Lifshitz-Gilbert equation due to their complexity and extended nature.

This book aims to understand how Hall viscosity plays an important role in the transport properties of skyrmions. It is no coincidence that both phenomena can only exist without parity symmetry. Though symmetries can be mathematically formulated through Ward identities, a truly systematic investigation of skyrmions within broken parity symmetry would require something more sophisticated than the conventional Ward identity.

In developing a more generalized topological Ward identity (covered in §5.5), this book introduces a new mechanism that empowers a broader treatment of topological objects. It also enables us to include Hall viscosity as an explanans for skyrmion motion.

A closer look at the systematic experimental data published in Nature journals reveals that skyrmion Hall angles are 10% larger than antiskyrmion Hall angles. According to the Thiele equation, which describes the motion of the skyrmion center, the skyrmion and the antiskyrmion have the same Hall angles with opposite signs. The current theoretical explanation does not account for this discrepancy. Yet this 10% can be explained by incorporating the effect of Hall viscosity as another contribution to the skyrmion Hall effect by generalizing the Thiele equation with a transverse collective coordinate, introduced in §4.4. With this theoretical foundation, we next turn to experimental verification. Verifying the existence of Hall viscosity through focused investigation of precise data on both skyrmion and antiskyrmion motion is more than feasible given advanced experimental and the vast number of ongoing experiments on chiral magnetic skyrmions. It is discussed in Chapters 6 and 7.

Though this book aspires to comprehensively introduce the subject to the reader, each subtopic has been designed to be modular and self-contained. Each chapter begins with the context in the form of an introduction, a background, or a relevant example. The connections between and within sections are explicitly drawn. Furthermore, most mathematical derivations and their intermediate steps have been provided for the readers to verify themselves. For clarity, I present these derivations in both index and vector notation.

Bom Soo Kim
Mount Pleasant, Wisconsin
October 2022

Acknowledgements

This book is a result of more than two decades of learning physics and research collaborations at various institutes around the globe. I am grateful for my teachers and supervisors for their inspiration and encouragement: Jae Hyung Yee at Yonsei University, Ori Ganor and Petr Hořava at University of California Berkeley, Elias Kiritsis and Christos Panagopoulos at the University of Crete, Yaron Oz at Tel Aviv University, and Al Shapere and Sumit Das at the University of Kentucky. Engaging in numerous rewarding (often lengthy) discussions with them at offices, seminar rooms, cafés, corridors, and even near entrance doors has constituted my most rigorous research experiences as well as many delightful and heartwarming memories.

I am also thankful to Korkut Bardakci, Dah-Wei Chiou, Andrea Erdas, Mitsutoshi Fujita, Blaise Goutéraux, Jung Hoon Han, Sean Hartnoll, Jelle Hartong, Aki Hashimoto, Chris Herzog, Yoon Pyo Hong, Carlos Hoyos, Seungjoon Hyun, Ioannis Iatrakis, Nissan Itzhaki, Matti Järvinen, Eunhwa Jeong, Jaehoon Jeong, Bum-Hoon Lee, Hyukjae Lee, Kimyeong Lee, Matthew Lippert, Paul Mohazzabi, Takeshi Morita, Hitoshi Murayama, Ganpathy Murthy, Rene Myer, Andy O'Bannon, Cheol-Hwan Park, Tassos Petkou, Sumiran Pujari, Mukund Rangamani, Shinsei Ryu, Subir Sachdev, Ambrose Seo, Yunseok Seo, Sang-Jin Sin, Cobi Sonnenschein, Shigeki Sugimoto, Mahiko Suzuki, Marika Taylor, Theodore Tomaras, David Tong, Ashvin Vishwanath, Daiske Yamada, Amos Yarom, and Piljin Yi for their intellectual influences on me through collaborations, discussions, and/or correspondences.

My family, Myung-Duk, Young-Kyung, and Dong-Yon, has journeyed together with me from Korea to California, Greece, Israel, Kentucky, Maryland, and finally Wisconsin. I am indebted to their sincere and heartfelt love and care that have helped me carry on studies and research for all these years. I am grateful for my parents and parents-in-law for their kind and generous support. Finally, I am truly grateful to God Almighty who reveals the wonderful secrets of nature to mankind!

Introduction

While emphases are given to the first principle methods using symmetries and conservation equations throughout this book, phenomenological approaches remain pivotal due to the complexity of skyrmion systems. It is reflected in the topics chosen for this book. It starts with the celebrated Landau-Lifshitz-Gilbert (LLG) equation (in §1.2) that has been extensively used to understand the skyrmion motion. A simple action that accommodates skyrmions is assembled in §1.3 by adding minimal ingredients. The corresponding symmetries are thoroughly studied in §1.4. Chapter 1 serves as an introduction to skyrmion physics. Individual chapters cover separate topics with their own introductions. They are self-contained and can be read independently. Here the contents of the chapters are highlighted so that the readers can taste the range of topics covered in this book.

Hydrodynamics describes a system's universal properties based on symmetries. Chapter 2 covers the recent developments of the hydrodynamics without parity symmetry, 'parity-breaking hydrodynamics,' along with the hydrodynamics with broken boost symmetry. The chapter provides the original Landau-Lifshitz hydrodynamics as a background material in §2.1 and the parity-breaking generalizations in §2.2. Basic properties for Hall viscosity are given in §2.3 with an intuitive geometric illustration. §2.4 explains charged hydrodynamics with a $U(1)$ symmetry that can be directly applied for conducting materials with an electric current. §2.5 covers the linear response theory, Green's functions, and Kubo formula, which are useful for the rest of this book. When more than one deriving mechanisms exist, such as temperature gradient and electric voltage, they influence each other. This subject is covered in §2.6, Irreversible Thermodynamics, along with a further generalization in the presence of an external magnetic field in §2.6.3.

Parity-breaking hydrodynamics predicts the existence of Hall viscosity. The latter has been extensively studied in quantum Hall systems. Interestingly, Hall viscosity is directly related to the angular momentum for a gapped system as discussed in §3.2 & §3.3. It is also surprising to find that the same Hall viscosity is also related to Hall conductivity, which is covered in §3.4 & §3.5. These two different relations for Hall viscosity are the consequences of two inequivalent, mutually compatible sets of quantum operators, which is explained in §1.5. Accordingly, there are also two inequivalent sets of Ward identities as discussed in Chapter 5. Chapter 3 starts with an elementary background content, the motion of a charged particle in a magnetic field in §3.1.

Spin dynamics, introduced in Chapter 4, is crucial for understanding the skyrmion motion. It starts again with the LLG equation for ferromagnets in §4.1. §4.2 & §4.3 generalize the LLG equation with various spin torques such as spin transfer torque (STT), spin orbit torque (SOT), spin Hall torque (SHT), and also emergent electromagnetic fields. Effects of these torques on spins are illustrated with the domain wall motion in detail. The chapter presents two basic (and equivalent) mechanisms: (i) Hund rule coupling approach and (ii) effective field direction approach. In the second half of the chapter, the well-known Thiele equation from the LLG equation is derived in §4.4. The Thiele equation is also extended by including the spin torques, SST, SOT, SHT, and the electromagnetic fields. In §4.5, the spin Seebeck effect and the spin Peltier effect, which are connected to the irreversible thermodynamics presented in §2.6, have been explained.

Chapter 5 comes back to the symmetry considerations to discuss the Ward identity. §5.1 presents a basic idea of the Ward identity. Fortunately, there is an illuminating geometric picture that directly appeals to our intuition. In §5.2, the Ward identity for a global current has been explained. It serves as a simple, yet full-fledged, example. The quantum field theory Ward identity using the stress energy tensors is explained in §5.3 along with its generalization with a conserved current. §5.4 revisits the Hall viscosity and its connections to angular momentum as a consequence of the identity. Finally, the Ward identity is generalized to include topological objects in §5.5. This is a crucial step for applying the Ward identity to the skyrmion systems.

Chapter 6 contains some well-established experimental aspects of skyrmion motion in magnets and the corresponding theoretical developments through the topological Ward identities. The latter reveals that Hall viscosity can play important roles in skyrmion motion, which was not appreciated by other approaches. Application of Ward identity to skyrmion systems is under development. §6.1 focuses on the experimental facts about skyrmions in conducting magnets: the topological Hall effects on the conduction electrons and the skyrmion Hall effects on the skyrmions. The Ward identity approach for skyrmions in conducting magnets is presented in §6.2 along with the role of Hall viscosity. The next two sections are devoted to insulating magnets and skyrmions. The experimental aspects of the magnon Hall effects, skyrmion Seebeck effect, and rotation motion are covered in §6.3. In the following section, the topological Ward identity for insulating magnets and the ways to measure Hall viscosity are discussed.

The last chapter models Hall viscosity by generalizing the Thiele equation with the skyrmion's transverse collective coordinate to find surprising predictions. It starts with the recent progresses of domain wall motion in ferromagnets, antiferromagnets, and ferrimagnets in §7.1. It also introduces a theory of ferrimagnet dynamics and presents the systematic experimental Hall angle data for skyrmions and antiskyrmions for a ferromagnet and a ferrimagnet (across the so-called angular momentum compensation point) in §7.2. It turns out that there are unexpected large differences between the skyrmion & antiskyrmion Hall angles. The conventional Thiele equation cannot describe their differences. §7.3 shows that the direction of Hall viscosity is the same for both skyrmions and antiskyrmions. Finally, the role of Hall viscosity in skyrmion and antiskyrmion Hall motions is discussed in §7.4 using the generalized Thiele equation.

Chapter 1

Symmetries of Magnetic Skyrmions

One of the overarching themes of the book across the chapters is various symmetries that exist and absent in the systems that accommodate the skyrmions. In particular, absence of the parity or mirror symmetry plays important roles in the existence of the skyrmions and the corresponding dynamics, especially the transverse skyrmion motion. We use first principle methods as well as phenomenological approaches to understand skyrmion dynamics. As we see below the first principle method we adapt provides new perspectives on skyrmion physics that have not been available through phenomenological approaches.

1.1 Symmetry

Symmetries have been a central theme of various exciting advancements in physics. From the high energy physics point of view, unification of existing symmetries is one underlying theme of its development, while effective field theories based on symmetry principles is a powerful tool for low energy physics.

Let us start with a simple example that is useful for constructing a skyrmion action and for appreciating the power of the symmetry in general. Consider a particle of mass m constrained to move on a two dimensional unit sphere, which serves as an illustrative example for quantizing constrained mechanical systems by Dirac. The action is

$$S = \frac{m}{2} \int dt \, \dot{x}_i^2 \,, \tag{1.1}$$

with a constraint $x_i^2 = 1$, where we sum over the repeated indices, $i = 1, 2, 3$. The Lagrangian has several different symmetries. Without the constraint, the simplest two derivative Lagrangian (1.1) has the translational symmetries along the three (real) spatial directions $x_i \to x_i + c_i$ with constants c_i, rotational symmetries $x_i \to O_{ij} x_j$ with 3×3 orthogonal matrix O, such that $O^T O = 1$, where T is the transpose of the matrix. There are also discrete symmetries, the parity $x_i \to -x_i$ and time reversal $t \to -t$. Once the constraint is imposed, translation symmetry is broken as $x_i^2 = 1$ is not appreciated by $x_i \to x_i + c_i$ for any non-zero c_i, while the rotation symmetry along with the two discrete symmetries remain valid.

In the context of relativistic quantum mechanics and quantum field theory, the continuous symmetries can be extended to Poincaré symmetry where time and space are combined to be in the same footing. Poincaré group includes the inhomogeneous space-time translations in addition to the homogeneous Lorentz boost and spatial rotations.

Back to the discussion of the constraint, it is more convenient to use the spherical polar coordinates (r, θ, ϕ) given by

$$(x_1, x_2, x_3) = (r \sin \theta \cos \phi, r \sin \theta \sin \phi, r \cos \theta) \,. \tag{1.2}$$

The constraint becomes $r^2 = x_1^2 + x_2^2 + x_3^2 = 1$ and the Lagrangian $S = \int dt \, L$ has the form

$$S = \frac{m}{2} \int dt \, (\dot{\theta}^2 + \sin^2 \theta \, \dot{\phi}^2) \,. \tag{1.3}$$

The Euler equation of motion, $\partial L/\partial q_i - (d/dt)(\partial L/\partial \dot{q}_i) = 0$ for $q_i = \theta, \phi$, can be evaluated as

$$\ddot{\theta} - \sin\theta\cos\theta\dot{\phi}^2 = 0,$$
$$\sin\theta\ddot{\phi} + 2\cos\theta\dot{\theta}\dot{\phi} = 0. \qquad (1.4)$$

The second equation can be integrated to have $\sin^2\theta\dot{\phi} = c_1 = $ const., which can be recognized from the Euler equation $(d/dt)(\partial L/\partial\dot{\phi}) = 0$ as the Lagrnagian (1.3) is independent of the coordinate ϕ. $\dot{\phi}$ can be plugged into the first equation to find $\ddot{\theta} - c_1^2\cos\theta/\sin^3\theta = 0$. This equation can be converted into a solvable first order differential equation $\dot{\theta}^2 + c_1^2\cot^2\theta = c_2$ by multiplying $\dot{\theta}$ on both terms and recognizing $(d/dt)(1/\tan\theta) = -\dot{\theta}/\sin^2\theta$. Our focus here is not to solve this classical equations of motion.

To discuss the symmetries in the system, it is more convenient to write the equations of motion in the Cartesian coordinate

$$m\ddot{x}_i + mx_i(\sum_k \dot{x}_k^2) = 0. \qquad (1.5)$$

One can check the equations for x_1 and x_2 are combined into the first equation in (1.4), while that of x_3 provides the second equation. The equations of motion appreciate the symmetries of the Lagrangian and is invariant under both the time reversal transformation $t \to -t$ and the parity transformation $x_i \to -x_i$ separately.

We ask a slightly different question [1] that leads to a useful spin action. We are interested in the equations that are only invariant under the combined operations $t \to -t$ and $x_i \to -x_i$ from the view point of low energy effective theory. At low energy, the terms with the small number of derivatives dominate, and thus the simplest choice is to add a term $\epsilon_{ijk}x_j\dot{x}_k$ to the equation of motion.

$$m\ddot{x}_i + mx_i(\sum_k \dot{x}_k^2) = \alpha\epsilon_{ijk}x_j\dot{x}_k, \qquad (1.6)$$

where α is a constant and ϵ_{ijk} is the totally antisymmetric tensor with $\epsilon_{123} = 1$. One of the difficulties is that there is no obvious Lagrangian, whose variation produces the new term.

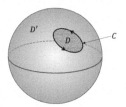

Figure 1.1 A closed orbit C on a spherical surface and two disks D and D', whose boundaries coincide with C.

It turns out that this has a well known solution. The added term is the Lorentz force for an electric charge interacting a magnetic monopole located at the center of the sphere. $\vec{F} \propto \dot{\vec{x}} \times \vec{B} = \dot{\vec{x}} \times \vec{x}$ as $\nabla \cdot \vec{B} = 4\pi\delta^{(3)}(\vec{x})$ and $\vec{B} = \nabla \times \vec{A} = \hat{x}/|\vec{x}|^2 = \hat{x} = \hat{r}$. The modified action has the form

$$S = \int dt \left(\frac{m}{2} \dot{x}_i^2 + \alpha A_i \dot{x}_i \right) . \tag{1.7}$$

This new Lagrangian yields (1.6) as the new term provides two extra contributions to the Euler-Lagrange equation

$$\alpha \frac{\partial A_j}{\partial x_i} \dot{x}_j - \frac{d}{dt}(\alpha A_i) = \alpha \dot{x}_j \left(\frac{\partial A_j}{\partial x_i} - \frac{\partial A_i}{\partial x_j} \right) = -\alpha \epsilon_{ijk} \dot{x}_j B_k , \tag{1.8}$$

which is $-\alpha \epsilon_{ijk} \dot{x}_j x_k$ for $|\vec{x}| = 1$. Here we use the fact that the vector potential only depends on the spatial coordinates as $A_i(\vec{x})$, independent of the time t and their time derivatives $\dot{x}_i$.

The system (1.7) has a Dirac string that stretches from the origin to the spatial infinity and poses difficulties. To explore the problem quantum mechanically, consider the Feynman path integral, $\mathcal{Z} = \text{Tr} e^{-\beta H} = \oint dx_i e^{-(1/\hbar) \int_0^\beta L d\tau}$ with the periodic boundary condition $x_i(\beta) = x_i(0)$ for Euclidean time $\tau = it$. Here β is the inverse temperature and H the Hamiltonian. The new term is

$$\exp \left(i \oint_C \alpha A_i dx^i \right) , \tag{1.9}$$

with the integration goes over the closed orbit C, a boundary of a certain disk. Now there is an ambiguity of choosing the disk as there are two disks that have the same boundary C, the disk D and its complement D'. See Fig. 1.1. Using the Stokes's theorem in vector calculus, $\oint_C \vec{A} \cdot d\vec{l} = \int_S (\vec{\nabla} \times \vec{A}) \cdot d\vec{S}$ where $C = \partial S$ is the closed boundary of a finite area S, one can rewrite the expression (1.9) as

$$\exp\left(i\alpha \oint_C A_i dx^i\right) = \exp\left(i\alpha \int_D F_{ij} dS^{ij}\right)$$
$$= \exp\left(-i\alpha \int_{D'} F_{ij} dS^{ij}\right), \quad (1.10)$$

where $F_{ij} = \partial_i A_j - \partial_j A_i$ is the field strength tensor and $dS^{ij} = \epsilon^{ijk} dS_k$. The latter two expressions are manifestly well defined. We choose the right-hand sign convention of a disk, and positive for D and negative for D'. By combining the two expressions, we arrive

$$\exp\left(i\alpha \int_{D+D'} F_{ij} dS^{ij}\right) = 1, \quad (1.11)$$

which impose an interesting condition. As $D + D'$ is the whole sphere S^2 and $\int_{S^2} F_{ij} dS^{ij} = \int_{S^2} \hat{r} \cdot (\hat{r} \sin\theta d\theta d\phi) = 4\pi$, the solid angle for a sphere. The condition is $\alpha = N/2$, integer or half integer. This is the famous Dirac's quantization condition for the product of electric and magnetic charges.

While it is not required to have a specific form of the vector potential, we present an acceptable form

$$\vec{A}_v = g \frac{\cos\theta - 1}{r \sin\theta} \hat{\phi}, \quad (1.12)$$

in spherical polar coordinates (with the explicit factor r and a parameter g) as it is useful below. It is valid for the entire spherical surface except $\theta = \pi$, where $\vec{A}_v$ diverges due to the presence of the Dirac string. Then the action (1.9) has the form

$$S_B \propto \oint \frac{(\cos\theta - 1)}{r \sin\theta} \hat{\phi} \cdot (r \sin\theta d\phi \hat{\phi}) = \oint dt (\cos\theta - 1)\dot{\phi}, \quad (1.13)$$

where we use the infinitesimal displacement vector $d\vec{r} = dr\hat{r} + rd\theta\hat{\theta} + r\sin\theta d\phi\hat{\phi}$ for fixed r. This action is called Geometric Berry phase term, which is clear later on. As mentioned above, it breaks the parity symmetry while it is not transparent in the spherical polar coordinates. This broke symmetry is one of the underlying themes across the book.

1.2 Action formalisms for localized spins

In §1.1, we have introduced the action (1.9) from the symmetry and the low energy effective theory point of view. It turns out that this action is a specific example of the so-called geometric phase or Berry phase, which has far more general applicability across different disciplines. In this section, we introduce the celebrated Landau-Lifshitz equation for the magnetic moments in the context of ferromagnetic crystals and reformulate the equation for the localized spins using the action representing the geometric phase.

1.2.1 Landau-Lifshitz equation

Let us introduce the Landau-Lifshitz equation in the context of ferromagnetic materials following the original paper [2]. Throughout this book, we mostly focus on the ferromagnet except Chapter 7, the last chapter, when we discuss recent understanding on anti-ferromagnets and ferrimagnets. A ferromagnetic crystal has magnetized regions that are almost fully saturated to the saturation magnetization M_s. The boundaries between the oppositely magnetized layers are finite and dynamically movable regions where the direction of magnetic moments change gradually form one direction to the opposite.

Let us consider Fig. 1.2, especially the deep regions *I*, *II*, and *III* of the bulk where there exists magnetic easy axis along $\hat{z}$ direction. We shall find a stable configuration, the distribution of the direction of magnetic moment, by minimizing the magnetic energy. We consider $\vec{m} = \vec{M}/M_s$, where $\vec{m}^2 = 1$, as the magnitude of the magnetic moment $\vec{M}$ is constant ($|\vec{M}| = M_s$) through the whole ferromagnetic material.

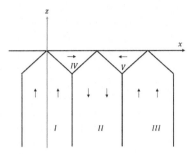

Figure 1.2 Ferromagnetic domains facing *x*-axis with the easy axis along *z*-axis.

The energy density has two contributions,

$$\mathcal{E} = \frac{a}{2}(\nabla \vec{m})^2 + \frac{b}{2}\left(m_x^2 + m_y^2\right). \tag{1.14}$$

The first term, with $(\nabla \vec{m})^2 = (\nabla m_x)^2 + (\nabla m_y)^2 + (\nabla m_z)^2$, is due to the inhomogeneity in the distribution of the direction of magnetic moments, meaning that spatial variation of magnetic moments costs energy. The second term is magnetic anisotropy energy term due to the presence of easy axis (*z* coordinate), which signifies that there are energy costs when the moments are not aligned with the easy axis.

Before proceeding further, we check that the symmetry of the energy density (Hamiltonian density) given in (1.14). The magnetic moment $\vec{m}(\vec{x}, t)$ is defined at every point of space-time. Without the constraint $\vec{m}^2 = m_x^2 + m_y^2 + m_z^2 = 1$, the Hamiltonian is symmetric under the shift $m_z \to m_z + c_z$ with a constant c_z and rotation along *z* direction, $m_\alpha \to O_{\alpha\beta} m_\beta$ for $\alpha, \beta = x, y$. The constraint further breaks the shift symmetry of m_z. The discrete symmetry, $m_\alpha \to -m_\alpha$ for $\alpha = x, y, z$, survives even after imposing the constraint. Yet, the discrete symmetry $m_z \to -m_z$ is further spontaneously broken in the presence of ferromagnetic layers in the regions *I*, *II*, and *III*.

To find the minimum, consider the left panel of Fig. 1.3, where the boundary surface (facing along *x*-axis in between the ferromagnetic regions *I* and *II*) between the regions with the opposite magnetic moments along the easy axis. Then, the distribution of the magnetic moment depends only on *x* coordinate, independent of *y* and *z*

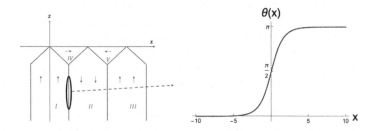

Figure 1.3 The boundary between ferromagnetic domains and the profile of the angle $\theta(x)$, that starts from $\theta = 0$ in the region I and ends at $\theta = \pi$ in the region II.

coordinates. The energy density (1.14) becomes

$$\mathcal{E} = \frac{a}{2}(\partial_x \vec{m})^2 + \frac{b}{2}(m_x^2 + m_y^2) \,. \tag{1.15}$$

Furthermore, deep in the bulk, the magnetic moments are distributed in yz-plane. Let us set the angle between z-axis and $\vec{m}$ as θ, which gives

$$\vec{m} = (0, m \sin\theta, m \cos\theta) \,, \tag{1.16}$$

with $m = 1$ and $\theta = \theta(x)$. Thus the problem reduces to minimize the energy density

$$\mathcal{E} = \frac{a}{2} m^2 (\partial_x \theta)^2 + \frac{b}{2} m^2 \sin^2\theta \,, \tag{1.17}$$

and the corresponding Euler equation is

$$a \partial_x (\partial_x \theta) - b \sin\theta \cos\theta = 0 \,. \tag{1.18}$$

This equation becomes $a(\partial_x \theta) \cdot \partial_x (\partial_x \theta) - b \sin\theta \cdot \partial_x (\sin\theta) = 0$ after multiplying $(\partial_x \theta)$ on both terms. It can be integrated to get $(\partial_x \theta)^2 - (b/a) \sin^2\theta = c_3$.

The size of the ferromagnetic layers I and II are large compared to the size of the boundary region in between them, and thus we have the boundary conditions:

$$\theta = 0 \text{ for } x = -\infty, \quad \text{and} \quad \theta = \pi \text{ for } x = \infty,$$
$$\partial_x\theta = 0 \text{ for } x = \pm\infty, \quad \text{or} \quad \partial_x\theta = 0 \text{ for } \theta = 0, \pi. \tag{1.19}$$

This boundary conditions at $x = \pm\infty$ force $c_3 = 0$, and the equation reduces to $\partial_x\theta = \pm\sqrt{b/a}\sin\theta$. The equation can be integrated to $\ln\left(\sqrt{(1-\cos\theta)/(1+\cos\theta)}\right) = \pm\sqrt{(b/a)}x + c_4$ by using $\int d\theta/\sin\theta = \int (\sin^2\theta/2 + \cos^2\theta/2)/(2\sin\theta/2\cos\theta/2)d\theta = \ln(\tan(\theta/2)) = \ln\left(\sqrt{(1-\cos\theta)/(1+\cos\theta)}\right)$. According to the boundary conditions, we choose the positive sign and $c_4 = 0$. Thus we get

$$\cos\theta = -\tanh\left(\sqrt{\frac{b}{a}}x\right). \tag{1.20}$$

This is depicted in the right panel of Fig. 1.3. The width of the boundary region can be estimated roughly as $\sqrt{b/a}$, which is taken as 1 in the figure. Further details can be found in [2].

Now we consider the ferromagnetic material in the presence of external magnetic field along the easy axis, z-axis. The boundaries between the ferromagnetic layers move, so that the layers with magnetic moments parallel to the field become wider. If the macroscopic field strength inside the material is H, the energy density of the material can be written as

$$\mathcal{E}_H = \frac{a}{2}(\partial_x\vec{m})^2 - \frac{b}{2}m_z^2 - \vec{H}\cdot\vec{m}. \tag{1.21}$$

The second term on the right-hand side is equivalent to that given in $(m_x^2 + m_y^2) = \vec{m}^2 - m_z^2$ for $\vec{m}^2 = 1$. Without the external magnetic field, $\vec{H} = 0$, the microscopic magnetic moments are random and cancel each other.

The system achieves equilibrium when the energy is minimized. After including the volume integral, we get

$$\delta E_H = -\int dV\left(\frac{a}{2}(\partial_x^2\vec{m}) + \frac{b}{2}m_z\hat{z} + \vec{H}\right)\cdot\delta\vec{m} = 0. \tag{1.22}$$

Here $\hat{z}$ is a unit vector along z coordinate. To achieve this, we need to require the quantity $\vec{H}_{\text{eff}} = (a/2)(\partial_x^2 \vec{m}) + (b/2)m_z\hat{z} + \vec{H}$ should be perpendicular to $\delta\vec{m}$. We also know that $\delta(\vec{m}^2) = \vec{m} \cdot \delta\vec{m} = 0$. As we are working on the boundary region that is effectively 2 dimensional, $\vec{H}_{\text{eff}}$ should be parallel to $\vec{m}$, and thus the role of the "effective field." In the presence of the effective field, the magnetic moment would act as a free moment and would rotate around $\vec{H}_{\text{eff}}$ as the change of the magnetic moment is perpendicular to the direction of $\vec{m}$. Thus

$$\dot{\vec{m}} = \mu_0 \vec{H}_{\text{eff}} \times \vec{m} = \mu_0 \vec{m} \times \frac{\delta \mathcal{E}_H}{\delta \vec{m}}, \tag{1.23}$$

where $\mu_0 = e/mc$ as they are related to spin moments and we use $\delta \mathcal{E}_H / \delta \vec{m} = -\vec{H}_{\text{eff}}$. This is the celebrated Landau-Lifshitz equation derived in [2].

How can we understand the physical origin of the Landau-Lifshitz equation and the associated Lagrangian (1.21)? There are two kinds of interaction between the magnetic moments in the crystal, exchange interaction and relativistic interaction. Contrast to the electronic crystal structure that is crucially related to the underlying lattice structure, the formation of magnetic structure is mainly due to the exchange interaction of the atom in the materials, which is quite independent of the total magnetic moments relative to the lattice. This exchange interaction cannot change the magnitude of the magnetic moment, and yields the equation (1.23).

The relativistic interaction can push the magnetic moment toward the effective field, yet is much weaker as the interactions are relativistic effects that are suppressed by $\sim \mathcal{O}(v^2/c^2)$, where v and c are the atomic velocity and the speed of light. As it is weaker, we can assume it does not change the coefficient of the contribution in (1.23) and simply add another contribution as

$$\dot{\vec{m}} = \mu_0 \vec{H}_{\text{eff}} \times \vec{m} + \mu_0 \lambda \left(\vec{H}_{\text{eff}} - \frac{(\vec{H}_{\text{eff}} \cdot \vec{m})\vec{m}}{\vec{m}^2} \right), \tag{1.24}$$

where $\lambda \ll |\vec{m}|$ as the relativistic interaction is much weaker than the exchange one. Here we can check that the equation (1.24) does not change the magnitude of $\vec{m}$ as we can check $\vec{m} \cdot \dot{\vec{m}} = 0$. We are not going to take into account of the relativistic interaction any further.

1.2.2 Landau-Lifshitz revisited

In §1.2.1, we considered the magnetic moments of the ferromagnetic material and its motion in the presence of external magnetic field. We are mostly interested in the dynamics of the magnetic moments that arise from electrons whose positions are localized in space and whose spins are free to rotate. Then,

$$\vec{M} = -\gamma_0 \vec{S}, \qquad (1.25)$$

where $\vec{S}$ is spin of the electrons. The gyromagnetic ratio $\gamma_0 = g|q_e|/2m > 0$ is given by the magnetic moment over the angular momentum, spin angular momentum for electron with g-factor $g = 2$ and $q_e = -e < 0$.

For completeness we provide an intuitive classical picture of the quantity in an elementary fashion. Consider a particle with a charge q orbiting with velocity v counterclockwise view from above with radius r in xy plane as in the left panel of Fig. 1.4. The magnetic dipole moment $\vec{\mu} = \mu_z \hat{z} = -(evr/2)\hat{z}$ is given by the area πr^2 times the current $I = q/T = -ev/2\pi r$, where the period is $T = 2\pi r/v$. The orbital angular momentum is $\vec{L} = \vec{r} \times \vec{p} = rmv\hat{z}$ for the electron orbit. $\gamma_0 = |\mu_z|/L_z = g|q_e|/2m = \mu_B/\hbar$, where μ_B is the Bohr magneton, $\hbar$ the Planck constant, e the absolute value of the electric charge of an electron, and m the mass of an electron. We added the g-factor for a quantum electron spin, which cannot be explained by this classical picture. In SI unit, $[\gamma_0] = C/Kg = [\vec{M}]/[\vec{L}] = (C \cdot m^2/s)/(Kg \cdot m^2/s)$, where C is Coulomb, the unit of electric charge. We consider various (spin) torques $\partial_t \vec{S} = -\partial_t \vec{M}/\gamma_0$, the time derivative of (spin) angular momentum $\vec{S}$, later in detail.

Here we consider the (localized and freely rotating) spin dynamics instead of the magnetic moment. Under the influence of an external magnetic field $\vec{B}$, the potential is given by $V = -\vec{M} \cdot \vec{B}$. For a localized and freely rotating spin, kinetic energy vanishes and this potential energy is the only contribution in this simple setup. The motion of the spin is described by the Zeeman Hamiltonian

$$H_Z = -\vec{M} \cdot \vec{B} = \gamma_0 \vec{S} \cdot \vec{B}, \qquad (1.26)$$

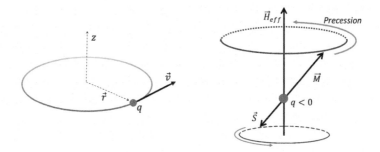

Figure 1.4 Left: A simple classical model of an electron orbiting in a circle provides an intuitive picture for gyromagnetic ratio. Right: Motion of the magnetic moment and spin due to precession in an effective field $\vec{H}_{\text{eff}}$, which also includes the external magnetic field $\vec{B}$.

where one can also use $H_Z = \mu_B \vec{\sigma} \cdot \vec{B}$ with $\mu_B = e\hbar/2m$ by expressing the electron spin in terms of the Pauli matrices as $\vec{S} = (\hbar/2)\vec{\sigma}$.

The spin operators satisfy the commutation relation $[S_\alpha, S_\beta] = i\hbar\epsilon_{\alpha\beta\gamma}S_\gamma$ with totally antisymmetric tensor $\epsilon_{\alpha\beta\gamma}$ defined as

$$\epsilon_{\alpha\beta\gamma} = \begin{cases} 1 & (\alpha\beta\gamma) = \text{cyclic permutation of } (xyz), \\ -1 & (\alpha\beta\gamma) = \text{anti-cyclic permutation of } (xyz), \\ 0 & \text{otherwise, two or more identical indices}, \end{cases} \quad (1.27)$$

where we frequently use $(xyz) = (123)$. The ϵ tensor is designed to describe the cross product $(\vec{A} \times \vec{B})_\alpha = \epsilon_{\alpha\beta\gamma}A_\beta B_\gamma$ in the index notation, while the delta symbol is for the dot product, $\vec{A} \cdot \vec{B} = \delta_{\alpha\beta}A_\alpha B_\beta = A_\alpha B_\alpha$, where $\delta_{\alpha\beta} = 1$ for $\alpha = \beta$. Thus

$$\delta_{\alpha\beta} = \begin{cases} 1 & \alpha = \beta, \\ 0 & \text{otherwise}. \end{cases} \quad (1.28)$$

The repeated indices are assumed to be summed over.

In the Heisenberg picture, time evolution of an operator $\vec{S}$ is given by the commutator of the Hamiltonian with the operator, $\partial_t \vec{S} = (i/\hbar)[H, \vec{S}]$. The computation for α-th component goes $[H, S_\alpha] = \gamma_0[S_\beta B_\beta, S_\alpha] = i\hbar\gamma_0 B_\beta \epsilon_{\beta\alpha\gamma}S_\gamma = i\hbar\gamma_0 \epsilon_{\alpha\gamma\beta}S_\gamma(B_\beta) = i\hbar\gamma_0(\vec{S} \times \vec{B})_\alpha$. We generalize the computation slightly to write the final result as

$$\dot{\vec{S}} = \vec{S} \times \left(-\frac{\delta H}{\delta \vec{S}}\right), \tag{1.29}$$

where we use $\vec{H}_{\text{eff}}$ instead of $\vec{B}$ as in (1.24). To be connected with the Landau-Lifshitz equation, we identify $\vec{M} = -\gamma_0 \vec{S} = M_s \vec{m}$. Then,

$$\dot{\vec{m}} = \frac{\gamma_0}{M_s} \vec{m} \times \left(\frac{\delta H}{\delta \vec{m}}\right) = \frac{\gamma_0}{M_s} \vec{H}_{\text{eff}} \times \vec{m}. \tag{1.30}$$

This is the Landau-Lifshitz equation (1.23), which has been derived for the case with the unit magnitude $M_s = 1$ (or the factor M_s can be absorbed into the definition of $\vec{H}_{\text{eff}}$ or H). It describes the precession motion of a localized spin along the direction of the effective field $\vec{H}_{\text{eff}}$ that includes the external magnetic field $\vec{B}$ as depicted in the right panel of Fig. 1.4.

Let us look into the equation more closely. The constraint $\vec{m}^2 = 1$, meaning that its magnitude is fixed, plays crucial roles. By taking a time derivative, we arrive an already familiar equation, $\vec{m} \cdot \dot{\vec{m}} = 0$, which means that $\vec{m}$ and $\dot{\vec{m}}$ are orthogonal to each other. One can see, in general, the unit magnetic moment vector can be decomposed as $\dot{\vec{m}} = \vec{m} \times \vec{A} + \vec{m} \times (\vec{m} \times \vec{B})$ for two vectors $\vec{A}$ and $\vec{B}$, that are parallel to each other $\vec{B} \propto \vec{A}$. One can check this by evaluating

$$\begin{aligned}(\vec{m} \times \vec{A}) &\cdot (\vec{m} \times (\vec{m} \times \vec{B})) \\ &= (\epsilon_{\alpha\beta\gamma} m_\beta A_\gamma)(\epsilon_{\alpha\delta\epsilon} m_\delta (\epsilon_{\epsilon\alpha'\beta'} m_{\alpha'} B_{\beta'})) \\ &= (\delta_{\beta\delta}\delta_{\gamma\epsilon} - \delta_{\beta\epsilon}\delta_{\gamma\delta}) m_\beta A_\gamma m_\delta (\epsilon_{\epsilon\alpha'\beta'} m_{\alpha'} B_{\beta'}) \\ &= A_\epsilon \epsilon_{\epsilon\alpha'\beta'} m_{\alpha'} B_{\beta'} - m_\epsilon m_\gamma A_\gamma \epsilon_{\epsilon\alpha'\beta'} m_{\alpha'} B_{\beta'} \\ &= \vec{m} \cdot (\vec{B} \times \vec{A}),\end{aligned} \tag{1.31}$$

where we use the frequently used identity

$$\epsilon_{ijk}\epsilon_{ilm} = \delta_{jl}\delta_{km} - \delta_{jm}\delta_{kl}, \tag{1.32}$$

$m_\alpha m_\alpha = 1$ and $\epsilon_{jpq} m_j m_p B_q = 0$ as ϵ_{jpq} is antisymmetric for the indices $\{jp\}$ and $m_j m_p$ is symmetric for (jp). The result vanishes when $\vec{A}$ and $\vec{B}$ are parallel to each other. It is clear that $\vec{m}$ and $(\vec{m} \times \vec{A})$ are orthogonal

14 | *Symmetries of Magnetic Skyrmions*

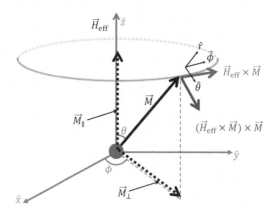

Figure 1.5 Various components of the magnetic moment and the effective field $\vec{H}_{\text{eff}}$.

to each other as $\vec{m} \cdot (\vec{m} \times \vec{A}) = \epsilon_{ijk} m_i m_j A_k = 0$, and the same is true for $\vec{m}$ and $\vec{m} \times (\vec{m} \times \vec{A})$. Thus the vectors $\vec{m}$, $(\vec{m} \times \vec{A})$, and $(\vec{m} \times (\vec{m} \times \vec{A}))$ are orthogonal to each other and span three dimensional space.

Back to the Landau-Lifshitz equation, we identify $\vec{A} \propto \vec{H}_{\text{eff}} = -\delta H/\delta \vec{m}$. Then we can see that the three vectors $\vec{m}$, $\vec{H}_{\text{eff}} \times \vec{m}$, and $(\vec{H}_{\text{eff}} \times \vec{m}) \times \vec{m}$ are orthogonal to each other. This is depicted in Fig. 1.5. In the spherical polar coordinates (r, θ, ϕ), these vectors have the following directions: $\vec{m} \propto \hat{r}$, $\vec{H}_{\text{eff}} \times \vec{m} \propto \hat{\phi}$, and $(\vec{H}_{\text{eff}} \times \vec{m}) \times \vec{m} \propto \hat{\theta}$. The vector $\vec{m}$ in the Landau-Lifshitz equation (1.30) can be conveniently decomposed into the components that are parallel $\vec{m}_\parallel$ and transverse $\vec{m}_\perp$ to the effective field direction $\vec{H}_{\text{eff}}$ as the right-hand side is only affected by the transverse component, $\vec{H}_{\text{eff}} \times (\vec{m}_\parallel + \vec{m}_\perp) = \vec{H}_{\text{eff}} \times \vec{m}_\perp$. Thus, the equation (1.30) turns into

$$\dot{\vec{m}}_\parallel + \dot{\vec{m}}_\perp = \frac{\gamma_0}{M_s} \vec{H}_{\text{eff}} \times \vec{m}_\perp . \tag{1.33}$$

As the two components, $\vec{m}_\parallel$ and $\vec{m}_\perp$, are perpendicular to each other (and they are also perpendicular to $\vec{H}_{\text{eff}} \times \vec{m}_\perp$), $\vec{m}_\parallel$ does not change as $\partial_t \vec{m}_\parallel = 0$. The equation turns into

$$\dot{\vec{m}}_\perp = \frac{\gamma_0}{M_s} \vec{H}_\text{eff} \times \vec{m}_\perp . \tag{1.34}$$

This motion with the moving transverse component and maintaining the parallel component to $\vec{H}_\text{eff}$ is the precession, depicted in the right panel of Fig. 1.4.

1.2.3 Spin Lagrangian

As we have discussed in the previous section §1.2.2, the dynamics of magnetic moments can be understood by the quantum spin dynamics. We present the Lagrangian in spin language that produces the Landau-Lifshitz equation. As a localized and freely rotating spin has a fixed magnitude, we set $\vec{S} = \hbar S \vec{n}$ and the dimensionless unit vector, $\vec{n} = (\sin\theta\cos\phi, \sin\theta\sin\phi, \cos\theta)$, is parallel to the radial unit vector. The action $\mathcal{S} = \int dt \mathcal{L}$ in an appropriate unit is given by

$$\begin{aligned}\mathcal{L} &= \hbar S(\cos\theta - 1)\dot{\phi} - \gamma_0 \hbar \mathcal{H}(\vec{n}) \\ &= \vec{A}_\text{v} \cdot \dot{\vec{n}} - \gamma_0 \hbar \mathcal{H}(\vec{n}) ,\end{aligned} \tag{1.35}$$

where the first term on the right-hand side is nothing but $\mathcal{L}_B$ introduced in (1.13) and the γ_0 is to produce the Landau-Lifshitz equation of motion. Another form in the second line with vector potential $\vec{A}_\text{v} = \hbar S(\cos\theta - 1)/(r\sin\theta)\hat{\phi}$ introduced in (1.13) is useful for variation of the action in terms of $\vec{n}$. We already considered a few examples of energy function such as (1.14) and (1.21) along with their symmetries, which can be served as $\mathcal{H}(\vec{n})$ after taking into account of the difference between magnetization and spin. We consider a specific Hamiltonian that accommodates skyrmions later in this chapter.

We note that the orthonormal set of unit vectors of the spherical polar coordinates is relevant for the spin dynamics. The unit vectors are $\{\vec{n} = \hat{r}, \hat{\theta}, \hat{\phi}\}$. In terms of the coordinates,

$$\begin{aligned}\vec{n} = \hat{r} &= (\sin\theta\cos\phi,\ \sin\theta\sin\phi,\ \cos\theta)\,,\\ \hat{\theta} &= (\cos\theta\cos\phi,\ \cos\theta\sin\phi,\ -\sin\theta)\,,\\ \hat{\phi} &= (-\sin\phi,\ \cos\phi,\ 0)\,,\end{aligned} \quad (1.36)$$

where $\hat{\theta}$ and $\hat{\phi}$ can be obtained by setting $\hat{\theta} = \hat{r}(\theta \to \theta + \pi/2)$ and $\hat{\phi} = \hat{r}(\theta = \pi/2, \phi \to \phi + \pi/2)$. These basis vectors satisfy the cyclic cross products, $\vec{n} \times \hat{\theta} = \hat{\phi}$, $\hat{\theta} \times \hat{\phi} = \vec{n}$, and $\hat{\phi} \times \vec{n} = \hat{\theta}$. Thus,

$$\begin{aligned}\frac{\partial \vec{n}}{\partial \theta} &= \hat{\theta}\,,\\ \frac{\partial \vec{n}}{\partial \phi} &= \sin\theta\,\hat{\phi}\,,\\ \dot{\vec{n}} &= \dot{\theta}\hat{\theta} + \sin\theta\,\dot{\phi}\,\hat{\phi}\,.\end{aligned} \quad (1.37)$$

The last equation gives $\dot{\vec{n}} \cdot \hat{\theta} = \dot{\theta}$ and $\dot{\vec{n}} \cdot \hat{\phi} = \sin\theta\,\dot{\phi}$.

The variation of this action (1.35) gives

$$\delta S = \int dt \left(-\hbar S\left[\sin\theta\,\dot{\phi} + \gamma_0 \frac{\delta\mathcal{H}}{\delta\theta}\right]\delta\theta \\ + \hbar S\left[\sin\theta\,\dot{\theta} - \gamma_0 \frac{\delta\mathcal{H}}{\delta\phi}\right]\delta\phi \right), \quad (1.38)$$

where we use (1.37) to evaluate the variations. The equation associated with $\delta\theta$ gives

$$0 = \sin\theta\,\dot{\phi} + \gamma_0 \frac{\delta\mathcal{H}}{\delta\theta} = \left(\dot{\vec{n}} + \gamma_0 \vec{n} \times \frac{\delta\mathcal{H}}{\delta\vec{n}}\right) \cdot \hat{\phi}\,, \quad (1.39)$$

where we use $\delta\mathcal{H}/\delta\theta = (\delta\vec{n}/\delta\theta) \cdot (\delta\mathcal{H}/\delta\vec{n}) = (\delta\mathcal{H}/\delta\vec{n}) \cdot \hat{\theta} = (\delta\mathcal{H}/\delta\vec{n}) \cdot (\hat{\phi} \times \vec{n})$. The equation of motion associated with $\delta\phi$ gives the same equation after a similar computation. Thus we obtain the equation of motion from the spin action (1.35). The equation associated with $\delta\theta$ gives

$$\dot{\vec{n}} - \gamma_0 \left(\frac{\delta\mathcal{H}}{\delta\vec{n}}\right) \times \vec{n} = 0\,. \quad (1.40)$$

This is the same as (1.29) and equivalent to (1.30) as $\vec{n} = -\vec{m}$. Thus we have demonstrated that the action (1.35) provides the Landau-Lifshitz equation. Here we focus on deriving the equations of motion. Interested readers can consult an excellent and comprehensive derivation of the Lagrangian using Coherent-State Path Integral for Spins along with various alternative actions for spin including CP^1 action and Wess-Zumino action (Chapter 1 of [3]).

1.2.3.1 CP^1 action

Before moving on, we introduce the so-called CP^1 action that is useful for describing the emergent electrodynamics in the context of skyrmions. We introduce

$$\mathbf{z} = \begin{pmatrix} \cos\theta/2 \\ e^{i\phi}\sin\theta/2 \end{pmatrix}, \tag{1.41}$$

which is a spin-1/2 coherent state. We define $\vec{n} = \mathbf{z}^\dagger \vec{\sigma} \mathbf{z}$, where $\vec{\sigma}$ is the Pauli matrices,

$$\sigma_1 = \begin{pmatrix} 0 & 1 \\ 1 & 0 \end{pmatrix}, \quad \sigma_2 = \begin{pmatrix} 0 & -i \\ i & 0 \end{pmatrix}, \quad \sigma_3 = \begin{pmatrix} 1 & 0 \\ 0 & -1 \end{pmatrix}. \tag{1.42}$$

One can check that

$$\begin{aligned} n_1 &= \mathbf{z}^\dagger \sigma_1 \mathbf{z} = \sin\theta\cos\phi, \\ n_2 &= \mathbf{z}^\dagger \sigma_2 \mathbf{z} = \sin\theta\sin\phi, \\ n_3 &= \mathbf{z}^\dagger \sigma_3 \mathbf{z} = \cos\theta, \end{aligned} \tag{1.43}$$

satisfies the usual representation of $\vec{n}$. There are several useful quantities that come into play in various parts of this book.

Now, one can define a gauge field,

$$\mathbf{a}_\mu = -i\mathbf{z}^\dagger \partial_\mu \mathbf{z} = \frac{1}{2}(1 - \cos\theta)\partial_\mu \phi, \tag{1.44}$$

where μ, ν are space-time indices that include time. This tells that we can rewrite the Lagrangian in terms of CP^1 representation as

$$\mathcal{L} = 2i\hbar S(\mathbf{z}^\dagger \partial_t \mathbf{z}) - \gamma_0 \hbar \mathcal{H}(\mathbf{z}), \tag{1.45}$$

where $\mathcal{H}(\mathbf{z})$ is rewritten in terms of $\mathbf{z}$ as well. Using the gauge field, one can check the following identity

$$\frac{1}{2}\vec{n}\cdot(\partial_\mu\vec{n}\times\partial_\nu\vec{n}) = \frac{1}{2}\sin\theta\Big((\partial_\mu\theta)(\partial_\nu\phi) - (\partial_\nu\theta)(\partial_\mu\phi)\Big)$$
$$= \partial_\mu\mathbf{a}_\nu - \partial_\nu\mathbf{a}_\mu \,. \qquad(1.46)$$

The last quantity is the field strength tensor for the electromagnetic gauge field $\mathbf{a}_\mu$. It is now clear that the emergent electric and magnetic fields can be written as

$$\frac{1}{2}\vec{n}\cdot(\partial_i\vec{n}\times\partial_j\vec{n}) = \partial_i\mathbf{a}_j - \partial_j\mathbf{a}_i = \epsilon_{ijk}\mathbf{b}_k \,,$$
$$\frac{1}{2}\vec{n}\cdot(\partial_t\vec{n}\times\partial_i\vec{n}) = \partial_t\mathbf{a}_i - \partial_i\mathbf{a}_t = \mathbf{e}_i \,. \qquad(1.47)$$

Here we have the emergent electric field $\vec{e}$ and magnetic field $\vec{b}$.

Later we will see that quantity $(1/2)\vec{n}\cdot(\partial_\mu\vec{n}\times\partial_\nu\vec{n})$ in the left-hand side of (1.46) turns out to be the topological skyrmion charge density. We define 2+1 dimensional topological current.

$$J_\mu = \frac{1}{8\pi}\epsilon_{\mu\nu\rho}\vec{n}\cdot(\partial_\nu\vec{n}\times\partial_\rho\vec{n}) = \frac{1}{2\pi}\epsilon_{\mu\nu\rho}\partial_\nu\mathbf{a}_\rho \,. \qquad(1.48)$$

In component form, the topological current has the form

$$\vec{J} = \frac{1}{2\pi}(\partial_x\mathbf{a}_y - \partial_y\mathbf{a}_x, \partial_y\mathbf{a}_t - \partial_t\mathbf{a}_y, \partial_t\mathbf{a}_x - \partial_x\mathbf{a}_t)$$
$$= \frac{1}{2\pi}(\mathbf{b}_z, -\mathbf{e}_y, \mathbf{e}_x) \,. \qquad(1.49)$$

Here the time component of the topological current is the emergent magnetic field $\mathbf{b}_z$, while the spatial components are the electric fields $(-\mathbf{e}_y, \mathbf{e}_x)$. These identifications have an interesting implication in the conservation law of the topological current.

$$\partial_\mu J_\mu \propto \partial_t \mathbf{b}_z - \partial_x \mathbf{e}_y + \partial_y \mathbf{e}_x = 0 \,. \qquad(1.50)$$

This is nothing but the Faraday's law, $\vec{\nabla}\times\vec{E} = -\partial_t\vec{B}$.

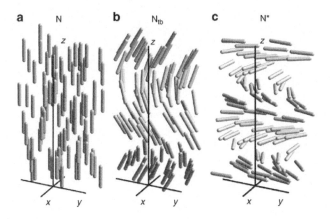

Figure 1.6 Phases of liquid crystal. Left: Nematic N phase, uniaxial alignment, $\theta_0 = 0$, Middle: N$_{tb}$, with oblique helicoid $0 < \theta_0 < \pi/2$ and Right: chiral nematic N* phase, right helicoid, $\theta_0 = \pi/2$ (the twist is right-handed or left-handed, depending on molecular chirality. Credit: Nature Communications [4].

1.3 Stability of magnetic skyrmions in chiral magnet

The fact that the spin or magnetic moment is a vector has important consequences. In a ferromagnetic ground state, all the spins are aligned in one direction. So they describe the homogeneous spins $\vec{n}$ = const. or magnetic moments $\vec{m}$ = const. Inhomogeneous distributions $\vec{n}(\vec{r})$ describes various deformed states of the spin. We describe various new ground states in a progressive way. We first consider the helicoidal states by allowing the parity symmetry breaking and then the skyrmion states with an additional ingredient on top of the broken parity. Chiral magnetic skyrmions had been predicted to exist much before they were demonstrated in experiments [5][6][7].

1.3.1 Helicoidal state in Nematic liquid crystals

We start with a nematic liquid crystal as its physical description is essentially the same as that of the spin spiral state. The orientational symmetry of the nematic liquid crystal is uniaxial with a preferred orientation of the molecules at every point. A molecule's macroscopic state can be described by specifying a preferred direction with a unit vector $\vec{n}(\vec{r})$ at each point. We use the same vector $\vec{n}$ as the underlying physical descriptions are essentially the same. In a complete equilibrium, the state is homogeneous and one can use a constant vector to describe the state. Various deformed states can be described by the inhomogeneous vector $\vec{n}(\vec{r})$ that depends on the position vector $\vec{r}$.

$$\vec{n} = (\sin\theta_0 \cos\phi, \sin\theta_0 \sin\phi, \cos\theta_0), \tag{1.51}$$

See Fig. 1.6. Here we describe the N* phase systematically using the symmetry following Chapter XIII of [8].

1.3.1.1 Free energy: symmetry considerations

We consider a macroscopic deformation $\vec{n}(\vec{r})$ that varies slowly meaning that the characteristic dimensions of the deformation are large compared to the typical atomic spacing, the distance the molecules are separated. Thus the derivatives of $\vec{n}(\vec{r})$ with respect to the coordinates are small. This also justifies us to use the continuous function $\vec{n}(\vec{r})$. We can consider the total free energy $F = \int dV \mathcal{F}$ of the deformed configurations by including small numbers of the derivatives of the vector $\vec{n}$. Here we include up to forth order of the vector, $\vec{n}(\vec{r})$, and to the second order of its derivatives.

The free energy can be constructed with only scalar quantities, combinations of the vector $\vec{n}$ and its derivatives, as they are invariant. There are only two possible terms that have one derivative, $\nabla \cdot \vec{n}$ and $\vec{n} \cdot (\nabla \times \vec{n})$. The scalar $\nabla \cdot \vec{n}$ is total derivative and becomes a surface term that is not important for understanding the bulk properties. The second one, $\vec{n} \cdot (\nabla \times \vec{n})$, is a pseudo scalar and break the parity symmetry, carrying explicitly the ϵ tensor as in (1.6). To keep the parity invariance, this term should be discarded.

To construct the scalars with two derivatives, we consider the fourth rank tensor

$$\frac{\partial n_k}{\partial x_i} \frac{\partial n_l}{\partial x_m}, \tag{1.52}$$

which can form scalar by contracting pairs of indices or be multiplied by appropriate component(s) of the vector $\vec{n}$. Note that the deformation vector still satisfies the constraint $\vec{n}^2 = 1$ and thus $n_i \partial_j n_i = 0$. First we list the terms with indices contracted to have scalars with two vectors.

$$\frac{\partial n_k}{\partial x_i}\frac{\partial n_k}{\partial x_i}, \quad \frac{\partial n_k}{\partial x_i}\frac{\partial n_i}{\partial x_k}, \quad \frac{\partial n_i}{\partial x_i}\frac{\partial n_k}{\partial x_k}. \tag{1.53}$$

The last one is nothing but $(\partial n_i/\partial x_i)(\partial n_k/\partial x_k) = (\partial_i n_i)(\partial_k n_k) = (\nabla \cdot \vec{n})^2$. The middle one is essentially the same as the last one as their difference is a total derivative: $(\partial_i n_i)(\partial_k n_k) - (\partial_k n_i)(\partial_i n_k) = \partial_i(n_i \partial_k n_k - n_k \partial_k n_i)$, where we use $n_i \partial_i \partial_k n_k - n_k \partial_k \partial_i n_i = (\vec{n}\cdot\nabla)(\nabla\cdot\vec{n}) - (\vec{n}\cdot\nabla)(\nabla\cdot\vec{n}) = 0$ as repeated indices are dummy indices.

To understand the first term, we evaluate a new term $(\vec{n}\cdot(\nabla\times\vec{n}))^2$, which is square of one of the terms we consider above.

$$\begin{aligned}(\vec{n}\cdot(\nabla\times\vec{n}))^2 &= (\epsilon_{ijk}n_i\partial_j n_k)(\epsilon_{lmn}n_l\partial_m n_n)\\ &= (\partial_j n_k)(\partial_j n_k) - (\partial_j n_k)(\partial_k n_j) - (n_i\partial_i n_k)(n_j\partial_j n_k),\end{aligned} \tag{1.54}$$

where we use $n_i n_i = 1$, $n_i \partial_k n_i = \partial_k(\vec{n})^2/2 = 0$, and the tensor identity

$$\begin{aligned}\epsilon_{ijk}\epsilon_{lmn} &= \delta_{il}(\delta_{jm}\delta_{kn} - \delta_{jn}\delta_{km})\\ &\quad + \delta_{im}(\delta_{jn}\delta_{kl} - \delta_{jl}\delta_{kn}) + \delta_{in}(\delta_{jl}\delta_{km} - \delta_{jm}\delta_{kl}).\end{aligned} \tag{1.55}$$

Thus we have in a vector notation

$$(\nabla\vec{n})^2 = (\vec{n}\cdot(\nabla\times\vec{n}))^2 + \frac{\partial n_k}{\partial x_i}\frac{\partial n_i}{\partial x_k} + ((\vec{n}\cdot\nabla)\vec{n})^2, \tag{1.56}$$

where we use the notation $(\nabla\vec{n})^2 = (\partial n_k/\partial x_i)(\partial n_k/\partial x_i)$, $((\vec{n}\cdot\nabla)\vec{n})^2 = ((\vec{n}\cdot\nabla)\vec{n})\cdot((\vec{n}\cdot\nabla)\vec{n})$ as we do $\vec{n}^2 = \vec{n}\cdot\vec{n}$. The middle term in the right-hand side is equivalent to $(\nabla\cdot\vec{n})^2$ up to a surface term as mentioned. This exercise show us the advantage of the index notation, which is popular for dealing with complex calculations.

The result (1.56) shows that two derivative terms with contracted indices produce three different invariant terms

$$(\nabla \cdot \vec{n})^2, \quad (\vec{n} \cdot (\nabla \times \vec{n}))^2, \quad ((\vec{n} \cdot \nabla)\vec{n})^2. \tag{1.57}$$

It turns out that there is one more scalar that we can construct from two first derivate terms, $(\nabla \cdot \vec{n})(\vec{n} \cdot (\nabla \times \vec{n}))$. This term is odd function in the deformation vector. We remove this term by imposing the symmetry that the free energy is invariant under $\vec{n} \to -\vec{n}$. (This invariance can be also imposed for the spin state below in the presence of the external magnetic field $\vec{B}$ by imposing the transformation $\vec{B} \to -\vec{B}$ simultaneously.)

We are ready to put all these scalar terms to construct the free energy. We first consider the case with parity invariance. If we would include a term $b(\vec{n} \cdot (\nabla \times \vec{n}))$ with a coefficient b. Parity invariance requires to have $b = -b$ as the rest of the term pick up $-$ sign under the parity transform. So $b = 0$. Then,

$$\begin{aligned}\mathcal{F} = \mathcal{F}_0 + a_1 (\nabla \cdot \vec{n})^2 \\ + a_2 (\vec{n} \cdot (\nabla \times \vec{n}))^2 + a_3 \big((\vec{n} \cdot \nabla)\vec{n}\big)^2,\end{aligned} \tag{1.58}$$

for the positive coefficients, a_1, a_2, a_3. We include a constant term $\mathcal{F}_0$ that is present in general. To minimize the free energy, we need to impose the conditions, $\nabla \cdot \vec{n} = 0, \vec{n} \cdot (\nabla \times \vec{n}) = 0$ and $(\vec{n} \cdot \nabla)\vec{n} = 0$. From vector calculus, we know that the only solution is $\vec{n} = $ const. as a vector is completely determined by its divergence and curl according to Helmholtz theorem.

1.3.1.2 Spiral state with broken parity

When the parity symmetry is broken, a new term $b\vec{n} \cdot (\nabla \times \vec{n})$ is allowed. As there is also the corresponding quadratic term, we can rewrite them into a complete square. Then,

$$\begin{aligned}\tilde{\mathcal{F}} = \tilde{\mathcal{F}}_0 + a_1 (\nabla \cdot \vec{n})^2 \\ + a_2 (\vec{n} \cdot (\nabla \times \vec{n}) + q_0)^2 + a_3 \big((\vec{n} \cdot \nabla)\vec{n}\big)^2,\end{aligned} \tag{1.59}$$

where $q_0 = b/(2a_2)$, and the constant term is modified. The ground state can be described by

$$\nabla \cdot \vec{n} = 0, \qquad (\vec{n} \cdot \nabla)\vec{n} = 0, \qquad \vec{n} \cdot (\nabla \times \vec{n}) = -q_0, \qquad (1.60)$$

along with the constraint $(\vec{n})^2 = 1$, $\nabla(\vec{n})^2 = 0$. To solve the equations, we consider a simple case where the vector depends on one coordinate, say z, as $\vec{n}(z) = n_x(z)\hat{x} + n_y(z)\hat{y} + n_z(z)\hat{z}$. Then, the first equation in (1.60) gives $\nabla \cdot \vec{n} = \partial_z n_z(z) = 0$. We simply choose $n_z = 0$. The second equation in (1.60) is trivially satisfied as $(\vec{n} \cdot \nabla)\vec{n} = (n_x \partial_x + n_y \partial_y)(n_x(z)\hat{x} + n_y(z)\hat{y}) = 0$.

The last equation gives $\vec{n} \cdot (\nabla \times \vec{n}) = \epsilon_{ijk} n_i \partial_j n_k = n_y(\partial_z n_x) - n_x(\partial_z n_y) = -q_0$. Thus we have a set of coupled equations including the constraint.

$$\begin{aligned} n_y(\partial_z n_x) - n_x(\partial_z n_y) &= -q_0, \\ n_x(\partial_z n_x) + n_y(\partial_z n_y) &= 0. \end{aligned} \qquad (1.61)$$

Using $n_x^2 + n_y^2 = 1$ and simple algebra, one can rewrite them as $\partial_z n_x = -q_0 n_y$ and $\partial_z n_y = q_0 n_x$, which can be converted into second order equations as $\partial_z^2 n_x = -q_0 \partial_z n_y = -q_0^2 n_x$ and $\partial_z^2 n_y = -q_0^2 n_y$. The solutions are $\sin(q_0 z)$ and $\cos(q_0 z)$. A specific solution depends on the boundary condition. We simply choose

$$n_x = \cos(q_0 z), \qquad n_y = \sin(q_0 z). \qquad (1.62)$$

This solution satisfied $\vec{n}^2 = 1$. The vectors are periodically arranged along the z coordinate with the periodicity $2\pi/q_0$. This is helical motion representing the spiral state.

1.3.2 Stability of spin spiral state

We are interested in establishing the stability of the helical spin state of the material MnSi that have been performed in [10][11]. The particular crystal structure is the so-called B20 and the corresponding space group is $P2_13$. They are related to a *cubic unit cell* with dimensions with one side 4.558 Angstrom. It is also important to note that the relative locations of the atoms in the unit cell are given by (p, p, p), $(1/2 + p, 1/2 - p, -p)$, $(-p, 1/2 + p, 1/2 - p)$ and $(1/2 - p, -p, 1/2 + p)$, where $p_{Mn} = 0.137$ for Mn atoms and

24 | *Symmetries of Magnetic Skyrmions*

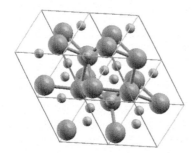 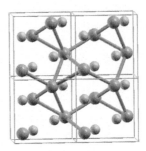

Figure 1.7 Crystal structure of MnSi as viewed from [111] (Left) and [100] (Right) crystal directions. Larger ones are Mn atoms according to the Miller index. Reproduced with permission from [9].

$p_{Si} = 0.845$ for Si atoms. Thus the cubic lattice of MnSi does not have a parity (an inversion) symmetry, which attributes the name 'chiral'.

These atoms are depicted in Fig. 1.7 in two different directions, one along the body diagonal direction [111] and and another along [100] direction. It turns out that MnSi crystal reveals a helical magnetic structure, whose wave vector is aligned in the [111] direction. The wave vector is rather small, $k \sim 0.036$ inverse Angstrom, indicating a long period, $\lambda \sim 175$ Angstrom, about the length of 22 cubic unit cells along the diagonal direction.

1.3.2.1 Free energy

With the knowledge of §1.3.1, we consider the cubic crystal respecting the cubic discrete rotation $(S_x, S_y, S_z) \to (S_y, S_z, S_x)$ acting on the spin vector $\vec{S}$ (or the magnetic moment $\vec{n}$) with a fixed length, $\vec{S}^2 = S_0^2$. Due to this constraint the quadratic potential term $\vec{S}^2$ is irrelevant when constructing the free energy or the relevant Lagrangian. We can keep track of the constraint by introducing a Lagrange multiplier $(u/4)(\vec{S}^2 - S_0^2)^2$. Then the discrete rotation symmetry allows only one inequivalent fourth order invariant $S_x^4 + S_y^4 + S_z^4$, or equivalently $S_x^2 S_y^2 + S_y^2 S_z^2 + S_z^2 S_x^2$, to the free energy. This is from the anisotropic energy. We note that a ferromagnetic cubic crystal ceases to have this symmetry when it develops a spontaneous magnetization along a particular direction.

There are also energies associated with the non-uniform configurations of the spins. Macroscopically, they contribute to the kinetic energies that are expressed in terms of derivatives of the spin vector $\vec{S}$ with respect to the coordinates. We already considered them above in (1.52). The largest contribution comes from the exchange interaction, which is further explained in §4.1.1. In the presence of the cubic symmetry, the most general form is $(J/2)(\partial S_i/\partial x_j)(\partial S_i/\partial x_j)$. For the ferromagnetic order to be stable, the coefficient J should be positive definite. The cubic symmetry allows another term $(A_1/2)((\partial S_x/\partial x)^2 + (\partial S_y/\partial y)^2 + (\partial S_z/\partial z)^2)$. This term is invariant under the simultaneous rotations of both the spins and the coordinates. See a general discussion of ferromagnetism and cubic lattice in [12].

Repackaging them together, we have a general free energy for the spin $\vec{S}$ that respects the cubic crystal symmetry as

$$\tilde{\mathcal{F}}_S = \mathcal{F}_{\text{FM}} + \mathcal{F}_A + \mathcal{F}_{\text{DM}}, \tag{1.63}$$

where $\mathcal{F}_{\text{DM}}$ also contributes when the parity symmetry is broken. Let us consider these terms one by one.

$$\mathcal{F}_{\text{FM}} = \frac{J}{2}(\nabla \vec{S})^2 + \frac{u}{4}((\vec{S})^2 - S_0^2)^2. \tag{1.64}$$

This is the free energy for the Landau-Ginsberg theory of the ferromagnetic moments. For the infinitely strong $u \to \infty$ and $S_0 = 1$, this reduces to the classical non-linear sigma model described in terms of $\vec{n}$. Here we use the notation: $(\nabla \vec{S})^2 = (\nabla_i \vec{S}) \cdot (\nabla_i \vec{S}) = (\nabla_i S_\alpha)(\nabla_i S_\alpha)$ with $i = x, y, \cdots$ referring the spatial coordinates and $\alpha = x, y, \cdots$ referring to the components of the spin in this section. Later we mix these indices and use i, j, k for coordinates and spin components.

The spins (or magnetic moments) are anchored to the atomic sites that form the cubic crystal lattice, for example MnSi, with discrete cubic symmetry. Due to the spin-orbit coupling, there are anisotropic spin terms in the free energy.

$$\mathcal{F}_A = \frac{A_1}{2}\left((\partial_x S_x)^2 + (\partial_y S_y)^2 + (\partial_z S_z)^2\right) \\ + A_2(S_x^4 + S_y^4 + S_z^4). \tag{1.65}$$

These are the lowest anisotropic terms up to the quadratic power of derivatives and quartic power of the spin field $\vec{S}$.

As a result of the locations of the Mn and Si atoms in the cubic MnSi crystal, there is no inversion center and the parity symmetry is broken. The spin-orbit coupling generates a term, known as the Dzyaloshinskii–Moriya (DM) interaction [13][14], and given by

$$\mathcal{F}_{\text{DM}} = b\vec{S} \cdot (\nabla \times \vec{S}). \tag{1.66}$$

The sign of b depends on the materials. We consider the parity transformation given by $\vec{r} \to -\vec{r}$ in all the three dimensions and thus $\nabla \to -\nabla$. Thus $\vec{S} \cdot (\nabla \times \vec{S})$ breaks the parity symmetry.

1.3.2.2 Periodic solution

Motivated by the periodic solution (1.62), we consider a periodic spin configuration

$$\vec{S}(\vec{r}) = \frac{1}{\sqrt{2}}\left(\vec{S}_k e^{i\vec{k}\cdot\vec{r}} + \vec{S}_k^* e^{-i\vec{k}\cdot\vec{r}}\right), \tag{1.67}$$

which is a Fourier transform for a vector. One can check the vector $\vec{S}(\vec{r})$ is real. The materials such as MiSi have physical parameters so that the anisotropic terms (1.65) provide small contributions compared to the other terms. Putting (1.67) into (1.63) (without $\mathcal{F}_A$ given in (1.65)), and integrating over the volume for $F = \int dV \mathcal{F}$, we obtain the following expression as only the terms with vanishing exponents contribute.

$$F_{\text{FDM}}(k) = \frac{J}{2}\vec{k}^2|\vec{S}_k|^2 + ib\vec{S}_k^* \cdot (\vec{k} \times \vec{S}_k) \\ + \frac{u}{4}(|\vec{S}_k|^2 - S_0^2)^2 + \frac{u}{8}(\vec{S}_k \cdot \vec{S}_k)(\vec{S}_{-k} \cdot \vec{S}_{-k}). \tag{1.68}$$

Note that the last term comes from the quartic term $(u/4)(\vec{S} \cdot \vec{S})^2$ whose scalar product has the same exponent $i\vec{k} \cdot \vec{x}$. The last two terms are independent of the wave vector $\vec{k}$ and can be used to impose $|\vec{S}_k|^2 = \vec{S}_k^* \cdot \vec{S}_k = S_0^2$ and $\vec{S}_k \cdot \vec{S}_k = 0$.

To minimize the free energy, we explicitly write $\vec{S}_k = \vec{\alpha}_k + i\vec{\beta}_k$ with two real vectors $\vec{\alpha}_k$ and $\vec{\beta}_k$, which are also constrained by $|\vec{S}_k|^2 = |\vec{\alpha}_k|^2 + |\vec{\beta}_k|^2 = S_0^2$. The DM contribution becomes $i\vec{S}_k^* \cdot (\vec{k} \times$

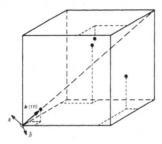

Figure 1.8 Crystal structure of MnSi with wave vectors along with the locations of atoms in the unit cell given by $(p,p,p), (1/2+p, 1/2-p, -p), (-p, 1/2+p, 1/2-p)$, and $(1/2-p, -p, 1/2+p)$ with $p_{Mn} = 0.137$ for Mn atoms and $p_{Si} = 0.845$ for Si atoms. Reproduced with permission from [10].

$\vec{S}_k) = 2b\vec{k}\cdot(\vec{\alpha}_k \times \vec{\beta}_k)$, which can be checked using the index notation. Thus,

$$F_{\text{FDM}}(k) = \frac{J}{2}\vec{k}^2(|\vec{\alpha}_k|^2 + |\vec{\beta}_k|^2) + 2b\vec{k}\cdot(\vec{\alpha}_k \times \vec{\beta}_k). \quad (1.69)$$

The first term is positive definite for the ferromagnetic materials. The contributions from the second term depend on the sign of b. When $b > 0$, we want $\vec{k}\cdot(\vec{\alpha}_k \times \vec{\beta}_k) < 0$. This means the $\vec{k}$ is anti-parallel to $\vec{\alpha}_k \times \vec{\beta}_k$, which describes the 'right-handed' spiral state. When $b < 0$, the free energy is minimized by $\vec{k}\cdot(\vec{\alpha}_k \times \vec{\beta}_k) > 0$. This 'left-handed' spiral state satisfies that the wave vector $\vec{k}$ is parallel to $(\vec{\alpha}_k \times \vec{\beta}_k)$. See Fig. 1.8 for the realization of the left-handed case. Thus the left-handed and right-handed chiral symmetry is explicitly broken by the DM interaction term (1.66).

To minimize the free energy explicitly, we choose $\vec{\alpha}_k \perp \vec{\beta}_k$ and $|\vec{\alpha}_k| = |\vec{\beta}_k|$ as their squared sum is fixed. Then (1.69) for both right-handed and left-handed states become

$$\begin{aligned}F_{\text{FDM}}(k) &= \frac{J}{2}S_0^2\left(|\vec{k}|^2 - 2\frac{|b|}{J}|\vec{k}|\right)\\ &= \frac{J}{2}S_0^2\left(|\vec{k}| - \frac{|b|}{J}\right)^2 - \frac{S_0^2}{2J}|b|^2.\end{aligned} \quad (1.70)$$

where $b = -|b|$ for the left-handed spiral state. Thus we can fix the length of the wave vector and the corresponding energy of the spiral configuration as

$$k = \frac{2\pi}{\lambda} = \frac{|b|}{J}, \qquad E_{\text{Spiral}} = -\frac{S_0^2}{2J}|b|^2, \qquad (1.71)$$

where $k = |\vec{k}|$. Thus the wave length of the magnetic spiral state is fixed by the strength of DM interaction b over the strength of the exchange interaction J.

1.3.2.3 Anisotropic term to fix the direction $\vec{k}$

Our analysis until now is independent of the direction of the wave vector $\vec{k}$ as we do not include the anisotropy term (1.65). The terms (1.64) and (1.66) are also invariant under the simultaneous rotation of spin and space. To figure out the orientation of $\vec{k}$ with respect to $\vec{S}$, we can consider the first term in (1.65). Putting (1.67) into the term, we obtain

$$F_A(k) = \frac{A_1}{2}\left(k_x^2|S_{kx}|^2 + k_y^2|S_{ky}|^2 + k_z^2|S_{kz}|^2\right). \qquad (1.72)$$

From the analysis above, we know that the magnitudes of the vectors, $\vec{k}$ and $\vec{S}_k$, are fixed.

Minimizing the term $F_A(k)$ crucially depends on the sign of A_1. For $A_1 > 0$, we prefer this term to vanish. This can be achieved if these two vectors are orthogonal to each other. For example, (1.72) vanishes for $\vec{k} = (k, 0, 0)$ and $S_{kx} = 0$. Then the wave vector is aligned with one of the three crystallographic directions, [100] with $S_{kx} = 0$, [010] with $S_{ky} = 0$, or [001] with $S_{kz} = 0$. On the other hand, for $A_1 < 0$, we want to maximize the magnitude of the term $k_x^2|S_{kx}|^2 + k_y^2|S_{ky}|^2 + k_z^2|S_{kz}|^2$. This can be achieved by equating all the components of the wave and the spin vectors, respectively. Yet, due to the symmetry, we can achieve the goal by setting $k_x \approx k_y \approx k_z$ and thus $\vec{k} \parallel [111]$. This state with $A_1 < 0$ turns out to be realized in the real material MnSi. This is depicted in Fig. 1.8. The anisotropy terms are weak, for example $k_x^2 \approx k_y^2 \approx k_z^2$, and do not change the qualitative analysis performed above, such as (1.71), without the anisotropy term.

While we use the free energy (1.63) fully to analyze the helical spin state of MnSi, it is clear from our analysis that not all the ingredients are necessary to establish that the spin spiral state is stable. In the language of the non-linear sigma model $\vec{n}$ by imposing $S_0^2 = 1$ at $u \to \infty$ in (1.64), the following two terms

$$\frac{J}{2}(\nabla \vec{n})^2 + b\vec{n} \cdot (\nabla \times \vec{n}) \tag{1.73}$$

are essential to establish that the ground state is the helical spin state as these lead to (1.69)–(1.71). In the following section, we introduce another term to establish the existence of a length scale that allows us to have stable skyrmions.

1.3.3 Stability of skyrmion crystals

In the previous section, §1.3.2, we consider the spin state in 3 dimensional materials. Here we change our attention to a 2 dimensional film geometry, where skyrmions seem to be stable in a larger region of the temperature (including the zero temperature) and the megnetic field phase diagram. According to Derrick's theorem, the free energy given in (1.63) is not enough to establish a stable, time-independent, and localized solution, such as a skyrmion. We follow the original argument of Derrick and apply that in the context of the spin Hamiltonian that contains (1.63).

1.3.3.1 Derrick's theorem

The original account of Derrick [15] considers a nonlinear sigma model for a field $\Theta(t, \vec{x})$ with the following (d+1)-dimensional action.

$$S = \int d^d x dt \left(\frac{1}{c^2} (\partial_t \Theta)^2 - (\nabla \Theta)^2 - f(\Theta) \right), \tag{1.74}$$

whose field equation is $\nabla^2 \Theta - (1/c^2)\partial_t^2 \Theta = f'(\Theta)/2$. Time independent equation is $\nabla^2 \Theta = \sin(2\Theta)/2$ for $f(\Theta) = \sin^2 \Theta$. The localized solution means that the total energy for the entire space is finite.

For a time-independent solution we consider the energy that is

$$E = \int d^d x \left((\nabla \Theta)^2 + f(\Theta) \right). \tag{1.75}$$

To have a stable solution, we need $\delta E = 0$ and $\delta^2 E > 0$, meaning that there exists a local minimum under a small change of the length scale. Now we suppose $\Theta(\vec{x})$ is a localized solution. For the finite energy condition, we examine the scaling property whether the solution exist at a finite length scale by considering $\Theta_\lambda(\vec{x}) = \Theta(\lambda\vec{x})$. Then, for the two terms $I_1 = \int d^d x (\nabla\Theta)^2$ and $I_2 = \int d^d x f(\Theta)$, the energy becomes $E_\lambda = \int d^d x \left((\nabla\Theta_\lambda)^2 + f(\Theta_\lambda)\right) = \lambda^{2-d} I_1 + \lambda^{-d} I_2$ after rescaling the variable $\lambda\vec{x} \to \vec{x}'$. Then,

$$(\partial_\lambda E_\lambda)|_{\lambda \to 1} = (2-d)I_1 - dI_2 ,$$
$$(\partial_\lambda^2 E_\lambda)|_{\lambda \to 1} = (2-d)(1-d)I_1 + d(d+1)I_2 . \quad (1.76)$$

The first equation, which vanishes as $\Theta_{\lambda \to 1}(\vec{x})$ is a solution, provides $I_2 = [(2-d)/d] I_1$. The second equation gives us

$$(\partial_\lambda^2 E_\lambda)|_{\lambda \to 1} = 2(2-d)I_1 . \quad (1.77)$$

We know $I_1 > 0$ as it is an integral of a positive quantity. In one spatial dimension, the simple non-linear sigma model (1.74) can provide a localized, time-independent, finite energy solution. This is not the case for two or higher spatial dimensions. The Derrick's argument has been generalized for various different systems with different fields, such as electromagnetic gauge fields.

1.3.3.2 Derrick's theorem for spiral states

Let us consider the free energy (1.73) that admits helical spin states in 2 spatial dimensions. The first term with two derivatives becomes a constant and does not contribute to the Derrick's scaling argument as can be seen in (1.76) for $d = 2$. The second (DM interaction) term alone cannot support a finite energy solution.

Skyrmion crystals are typically stabilized in the presence of the external magnetic field $\vec{B}$. Thus we add the Zeeman term $\vec{B} \cdot \vec{n}$ to the free energy. Here we use the language of magnetization $\vec{m} = -\vec{n}$.

$$\mathcal{F}^{2D} = \frac{J}{2}(\nabla\vec{m})^2 + b\vec{m} \cdot (\nabla \times \vec{m}) - \vec{B} \cdot \vec{m} . \quad (1.78)$$

Note that the differential operator here is 2 dimensional one, $\nabla = \hat{x}\partial_x + \hat{y}\partial_y = (\partial_x, \partial_y, 0)$. The spin field $\vec{m}$ has three components with a fixed length, while it only depends on two spatial dimensions $\vec{m}(x,y)$ or $\vec{m}(\rho,\varphi)$ in the polar coordinates. Let us consider R_s as a typical size of a stabilized skyrmion crystal. Then the free energy, the two dimensional integral of $\mathcal{F}^{2D}$, is supported by the crystal and gives

$$F^{2D} = E_0 - c_b R_s + c_Z R_s^2 \,. \tag{1.79}$$

The first term is constant as it is from the interchange interaction and independent of scale. The second term is due to the DM interaction that can provide positive or negative contributions depending on the sign of b. The last Zeeman term typically push the magnetization to be parallel (spins to be anti-parallel) to the external magnetic field B. At the skyrmion core, the magnetization (spin) directions are opposite and anti-parallel (parallel) to the magnetic field, and thus the coefficient c_Z is positive for the skyrmions. For positive constants c_b and c_Z, a localized finite energy solution is stabilized. This argument is essentially that of Derrick in a slightly different language.

To be quantitative, we consider the spin vector in a more concrete way,

$$\vec{m} = \left(\sin[\theta]\cos[\phi],\ \sin[\theta]\sin[\phi],\ \cos[\theta]\right) \tag{1.80}$$

$$= \left(\sin[f(u)]\cos[N(\varphi+\varphi_0)],\ \sin[f(u)]\sin[N(\varphi+\varphi_0)],\ \cos[f(u)]\right),$$

where we rewrite the non-linear fields, $\theta(\vec{x}) = f(\rho)$ in terms of the reduced radial coordinate $u = \rho/R_s$ that is dimensionless and $\phi(\vec{x}) = N(\varphi+\varphi_0)$ that signifies the winding number N. The extra phase factor φ_0 plays an important role that distinguishes different types of skyrmions as illustrated in Fig. 1.9.

Using the explicit form of the spin function (1.80), the free energy (1.78) can be evaluated using the 2 dimensional integral $\int dxdy = \int ud u d\varphi$. The first term gives

$$E_0 = \pi J \int_0^\infty du \left(u(\partial_u f(u))^2 + \frac{N^2}{u}(\sin f(u))^2 \right). \tag{1.81}$$

We note that the expression is independent of R_s as the integral is constant, which is consistent with the Derrick's argument. The

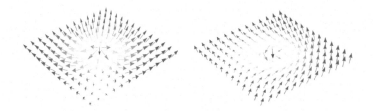

Figure 1.9 Two dimensional magnetic moment vectors (with the parallel magnetic field at the boundary) configurations for Néel-type skyrmions (left with $\varphi_0 = 0$) and Bloch-type skyrmions (right with $\varphi_0 = \pi/2$).

second term can be computed as

$$E_{\text{DM}} = bR_s \int_0^{2\pi} d\varphi \, \sin\left[(N-1)\varphi + N\varphi_0\right] \\ \times \int_0^\infty du \left(u(\partial_u f(u)) + \frac{N}{2} \sin[2f(u)] \right). \quad (1.82)$$

One can also verify this using Mathematica program by defining $u[x,y]$ and $\varphi[x,y]$ and the vectors $m = \{m_x, m_y, m_z\}$ as in (1.80) and $x = \{x,y,z\}$ (okay as u and φ are independent of z)

Sum[Sum[Sum[
   LeviCivitaTensor[3][[i,j,k]] × n[[i]] × D[n[[k]], x[[j]]],
   {i, 1, 3}], {j, 1, 3}], {k, 1, 3}] .

Of course, one needs to take care of changing the coordinates as $u = \sqrt{x^2 + y^2}$ and $\varphi = \text{ArcTan}[y/x]$ and the inverse relations $x = u \cos[\varphi]$ and $y = u \sin[\varphi]$.

The first line in (1.82) shows that the term is non-zero only for $N = 1$ and in the presence of φ_0.

$$E_{\text{DM}}^{N=1} = 2\pi b R_s \, \sin[\varphi_0] \int_0^\infty du \left(u(\partial_u f(u)) + \frac{1}{2} \sin[2f(u)] \right). \quad (1.83)$$

As this term vanishes for $\varphi_0 = 0$, only the Bloch-type skyrmions have the non-zero contribution from the DM interaction given in

(1.78) in 2 spatial dimensions. The term maximizes for $\varphi_0 = \pm\pi/2$. For $\varphi = \pm\pi/2$, $\sin(\varphi) = \pm 1$. We use $\varphi = \pi/2$ for the Bloch-type skyrmion illustrated in the right side of Fig. 1.9. For this, the term (1.83) becomes negative for $b < 0$ that we already discussed above. Néel-type skyrmions also have contributions from different DM interaction in 2 spatial dimensions that is given by

$$\mathcal{F}_{DM}^{2D,\text{Néel}} = b_N\left[(\vec{m}\cdot\nabla)m_z - m_z(\nabla\cdot\vec{m})\right]. \tag{1.84}$$

This term is part of the 3 dimensional DM interaction (1.66) decomposed into 2 dimensional ones along with the term $b\vec{m}\cdot(\nabla\times\vec{m})$.

The last term in (1.78) can be computed for $\vec{B} = B\hat{z}$ as

$$E_Z = 2\pi B R_s^2 \int_0^\infty du\, u\left(1 - \cos f(u)\right), \tag{1.85}$$

where we add a constant so that $\cos f(u) = 1$ provides zero Zeeman energy.

Combining all the contributions, we have

$$\begin{aligned}F^{2D} = 2\pi J \int_0^\infty du\, u &\left(\frac{1}{2}(\partial_u f(u))^2 + \frac{1}{2u^2}(\sin f(u))^2\right.\\ &\left.+ \frac{bR_s}{J}\left((\partial_u f(u)) + \frac{1}{2u}\sin[2f(u)]\right) + \frac{BR_s^2}{J}(1-\cos f(u))\right).\end{aligned} \tag{1.86}$$

We can rescale u without affecting the first line as these two terms are independent of the scale. By rescaling $u \to \tilde{u} = (bR_s/J)u = (b/J)\rho$, we rewrite the expression to

$$\begin{aligned}F^{2D} = 2\pi J \int_0^\infty d\tilde{u}\, \tilde{u} &\left(\frac{1}{2}(\partial_{\tilde{u}} f(\tilde{u}))^2 + \frac{1}{2\tilde{u}^2}(\sin f(\tilde{u}))^2\right.\\ &\left.+ (\partial_{\tilde{u}} f(\tilde{u})) + \frac{1}{2\tilde{u}}\sin[2f(\tilde{u})] + \frac{BJ}{b^2}(1-\cos f(\tilde{u}))\right).\end{aligned} \tag{1.87}$$

Thus there is only one free parameter $BJ/(b^2)$, and the strength of the DM interaction is related to the external magnetic field B for the overall fixed scale J. More appropriate dimensionless radial coordinate is $\tilde{u} = (bR_s/J)u = (b/J)\rho$. Thus the length scale of the skyrmion is given by J/b, which is also the length scale of the helical spin state we discussed earlier.

1.3.3.3 Analysis of differential equation

We are ready to study the free energy (1.87) by taking its variation with respect to the radial function f, $\delta F^{2D}/\delta f = 0$. The terms with derivatives require the integration by parts, and we have

$$\tilde{u}^2 f'' + \tilde{u} f' - \frac{1}{2}\sin 2f + 2\tilde{u}\sin^2 f - \frac{BJ}{b^2}\tilde{u}^2 \sin f = 0, \qquad (1.88)$$

where $'$ is the differentiation with respect to $\tilde{u}$. This equation is complicated and can be solved numerically. Instead, we try to get analytic understanding of the differential equation for $0 \leq \tilde{u} < \infty$.

For the magnetic field $\vec{B} = B\hat{z}$, the magnetization vectors are parallel away from the skyrmion ($\tilde{u} \to \infty$) and anti-parallel at the skyrmion core ($\tilde{u} = 0$) to the magnetic field $\vec{B}$. They translate into $f(\tilde{u} = \infty) = 0$ and $f(\tilde{u} = 0) = \pi$. With the normalization chosen in (1.85), only the anti-parallel magnetization (skyrmion) costs energy.

We look into the degree (effective power) of $\tilde{u}$ of the terms in (1.88). The first three terms have $\tilde{u}^0$, while the last two terms are linear and quadratic in $\tilde{u}$. For we try to solve the differential equation, we can not ignore the first two terms, while the last three terms can be discarded with physical justifications. For $\tilde{u} \to \infty$, we keep the last term and discard the third and fourth terms. Furthermore, $\sin f \sim f$ as $f \sim 0$. Then the differential equation is

$$\tilde{u}^2 f'' + \tilde{u} f' - \frac{BJ}{b^2}\tilde{u}^2 f = 0. \qquad (1.89)$$

This is nothing but the modified Bessel equation (due to the $-$ sign in the last term).

The Bessel equation has the typical form $\rho^2 \partial_\rho^2 R_\nu(k\rho) + \rho \partial_\rho R_\nu(k\rho) - (k^2 \rho^2 + \nu^2) R_\nu(k\rho) = 0$ and the corresponding solutions $R_\nu(k\rho) = c_1 I_\nu(k\rho) + c_2 K_\nu(k\rho)$. The second kind of modified Bessel function approaches 0 for $\rho \to \infty$. Thus the solution for our equation goes as

$$f \sim K_0(\sqrt{BJ/b^2}\,\tilde{u}) = K_0(\sqrt{B/J}\,\rho), \qquad (1.90)$$

which is monotonically decreasing to vanish for $\rho \to \infty$.

We return for $\tilde{u} \to 0$. We keep the middle term and discard the last two terms. Moreover, $\sin 2f \sim 2(f - \pi)$ as $f \sim \pi$. Then the differential equation is

$$\tilde{u}^2(f - \pi)'' + \tilde{u}(f - \pi)' - (f - \pi) = 0, \qquad (1.91)$$

which is homogeneous 2nd order differential equation. We try $f - \pi = \tilde{u}^a$. Then we have $a(a - 1) + a - 1 = 0$, whose solutions are $a = \pm 1$. Thus, $f - \pi = c_1 \tilde{u} + c_2/\tilde{u}$. We choose $c_2 = 0$ as the solution diverges as $\tilde{u} \to 0$. Thus the solutions for our differential equation is $f \sim \pi - c_1(b/J)\rho$ for $\rho \to 0$. These analytic approach only can provide qualitative features near the core and far infinity, but not for the interesting intermediate region.

1.3.3.4 Phase diagram

Let us zoom out to see a bigger picture related to the stability of the skyrmion crystals. As we have already encountered, there is also the helical magnet (spin) state when $\vec{B} = 0$. We also know that the magnetic moments (spins) will align (anti-align) when the external magnetic field is strong enough as the Zeeman term dominates. Thus it is expected for the skyrmion crystals to be stable in a range of magnetic field $B_{c1} < B < B_{c2}$. For comparing the free energies appropriately, we need to properly normalize the free energies for their relevant sizes for a given sample size, setting here to be a unit area.

For the helical state, we set $\theta = f(y)$ and $\phi = \phi_0 = 0$ by choosing the periodic boundary condition along y-coordinate. The magnetic moment (1.80) simply becomes

$$\vec{m} = \left(\sin[f(y)], 0, \cos[f(y)] \right). \qquad (1.92)$$

It is straightforward to compute the free energy, the integral of $\mathcal{F}^{2D}$ given in (1.78), for the unit cell with the periodicity $0 \leq y < \ell$ with the boundary condition $f(0) = \pi$ and $f(\ell) = 0$ and for the unit length sample size along x-direction.

$$F_H = \frac{J}{\ell} \int_0^\ell dy \left(\frac{1}{2}(\partial_y f)^2 - \frac{b}{J}\partial_y f + \frac{B}{J}(1 - \cos f) \right). \qquad (1.93)$$

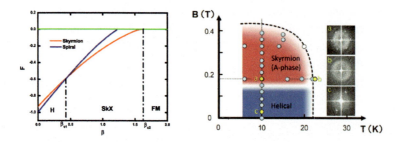

Figure 1.10 Left: Free energies of three different states, i) helical state (H, blue), ii) skyrmion crystal (SkX, red), and iii) ferromagnetic state (FM, green) that is normalized to be zero. In the middle region, $\beta_{c1} < \beta = B/(2J) < \beta_{c2}$, skyrmion crystals have the lowest energy and are stable. Reproduced with permission from [16]. Right: Phase diagram of the MnSi thin film as functions of the magnetic field B and temperature (T). Right insets represent the diffractograms of different states. Credit: American Chemical Society [17].

Finally the ferromagnetic state is described by a constant field $\vec{m} = (0, 0, 1)$ and thus we choose $f(x,y) = 0$ as it gives vanishing energy $F_{FM} = 0$ for magnetic moment up state $\vec{m} \parallel \vec{B}$.

Careful investigation requires numerical studies [16], which is expected to be valid for the wide variety of materials. There are two critical magnetic fields, B_{c1} and B_{c2}.

$$B_{c1} = 0.2(b^2/J), \qquad B_{c2} = 0.8(b^2/J). \tag{1.94}$$

For $0 \leq B < B_{c1}$, the helical spiral state is stable, while the skyrmion crystal state is stable for the region $B_{c1} < B < B_{c2}$. For the large magnetic field $B > B_{c2}$, the ferromagnetic state is the ground state. At the phase boundaries, both relevant states coexist, and the phase transitions are first order. The result is depicted in the left panel of Fig. 1.10.

Our discussion until now is done for a fixed temperature. We know that all the spins are randomly distributed at a very high temperature. The phase diagram of the MnSi thin film in a 2 dimensional phase space, magnetic field (in Tesla) and temperature (in Kelvin), shows the phase transitions among the three states (helical, skyrmion crystal, and ferromagnetic states) as the magnetic field is increased for a fixed temperature at $T = 10K$. This is illustrated in the right panel of Fig. 1.10. Diffractograms representing the skyrmion (6-fold) and helical (2-fold) phases are displayed at the points indicated by the point a and c (yellow circles). These two different patterns of the diffractograms indicate that the skyrmion crystal is a different state of matter compared to the helical state.

For a fixed magnetic field at $B = 0.18T$, increasing temperature changes the ground state from the skyrmion crystals to a disordered state that is captured by the broad halo ring in the middle diffractogram (at the point b, yellow circle). This 2 dimensional thin film MnSi phase diagram is considerably different from that of the 3 dimensional bulk MnSi, where the middle region is replaced by a conical phase with a small portion of skyrmion crystals near the phase boundary between the conical and disordered phases. It has been argued that the skyrmion crystal state is a true ground state as it extends toward zero temperature that is clear in the phase diagram depicted in the right panel of Fig. 1.10.

1.3.4 Topological nature and skyrmion charge

After establishing the stability of a skyrmion crystal along with the other stable phases, such as helical and ferromagnetic phases, depending on tunable parameters such as the external magnetic field and temperature, we look into the topological nature of skyrmion.

We have already introduced the skyrmion topological current in (1.48).

$$J_\mu = \frac{1}{8\pi} \epsilon_{\mu\nu\rho} \epsilon_{\alpha\beta\gamma} (m_\alpha \partial_\nu m_\beta \partial_\rho m_\gamma) \,, \tag{1.95}$$

where $\mu, \nu, \rho = t, x, y$ are the 2+1 dimensional space-time indices and $\alpha, \beta, \gamma = 1, 2, 3$ (or x, y, z) are indices for the magnetic moment components. We check the conservation equation

$$\partial_\mu J_\mu = \frac{1}{8\pi} \epsilon_{\mu\nu\rho} \epsilon_{\alpha\beta\gamma} (\partial_\mu m_\alpha)(\partial_\nu m_\beta)(\partial_\rho m_\gamma) \equiv 0 \,. \tag{1.96}$$

The equation identically vanishes due to the constraint $\vec{m}^2 = 1$. To see it, we rewrite the constraint as $\partial_\rho m_\gamma = \partial_\rho \sqrt{1 - m_\alpha^2 - m_\beta^2} = \partial_\rho m_\alpha (\cdots) + \partial_\rho m_\beta (\cdots)$ (as $\alpha \neq \beta \neq \gamma$) and observe $\epsilon_{\mu\nu\rho}(\partial_\mu m_a)(\partial_\rho m_a)(\cdots) = 0$. The latter can be checked as $(\partial_1 m_a)(\partial_3 m_a) - (\partial_3 m_a)(\partial_1 m_a) = 0$ for $\nu = 2$. The same is true for the other term $\partial_\rho m_\beta(\cdots)$.

The topological charge is an integral over the 'volume' (area) of the time component $J_t = (1/8\pi)\epsilon_{\alpha\beta\gamma} m_\alpha(\partial_x m_\beta \partial_y m_\gamma - \partial_y m_\beta \partial_x m_\gamma) = (1/4\pi)\epsilon_{\alpha\beta\gamma} m_\alpha \partial_x m_\beta \partial_y m_\gamma = (1/4\pi)\vec{m}\cdot(\partial_x\vec{m}\times\partial_y\vec{m})$.

$$Q = \int d^2x J_t = \int dxdy \frac{1}{4\pi}\vec{m}\cdot\left(\frac{\partial\vec{m}}{\partial x}\times\frac{\partial\vec{m}}{\partial y}\right)$$
$$= \int dxdy \frac{\sin\theta}{4\pi}\left(\frac{\partial\theta}{\partial x}\frac{\partial\phi}{\partial y} - \frac{\partial\theta}{\partial y}\frac{\partial\phi}{\partial x}\right). \quad (1.97)$$

The expression is nothing but the Jacobian of the coordinate transformation from 2 dimensional Euclidean space (x,y) to the spherical coordinate (θ,ϕ) given in the spin vector (1.80). To find this map's connection to the topological description, we need to impose that the moment field $\vec{m}$ is single-valued at the 2 dimensional infinity, $\rho = \sqrt{x^2+y^2}\to\infty$. (A two dimensional space with its infinity identified is topologically equivalent to a spherical surface.) Then, the identity map $\vec{m}(\hat{r}) = \hat{r}$ is a topological configuration with a unit winding number, $\vec{m}(x,y) = \hat{r}(x,y)$. This is known as the second homotopy group, $\pi_2(S^2) = \mathbb{Z}$, and the topological number is an integer. The particular map for $\pi_2(S^2) = 1$ is achieved by the so-called stereographic projection depicted in Fig. 1.11.

More concretely, we consider the coordinates of the Euclidean space as $(x,y) = \rho(\cos\varphi,\sin\varphi)$ and those of the spherical surface as $R(\sin\theta\cos\phi, \sin\theta\sin\phi, \cos\theta)$. Under the stereographic projection, the angle φ is identified with ϕ. From the triangle connecting three points S, N, and the point (x,y) and the fact that the angle associated with the south and north poles are the right angle and $(180-\theta)/2$, we get $\cot(\theta/2) = \rho/(2R)$. Thus, $\cos(\theta/2) = \rho/\sqrt{\rho^2+4R^2}$ and $\sin(\theta/2) = 2R/\sqrt{\rho^2+4R^2}$. Using the identities $\cos\theta = \cos^2(\theta/2) - \sin^2(\theta/2) = (\rho^2-4R^2)/(\rho^2+4R^2)$ and $\sin\theta = 2\cos(\theta/2)\sin(\theta/2) = 4R\rho/(\rho^2+4R^2)$, we arrive

$$\vec{m}(\rho,\varphi) = \hat{r}(\rho,\varphi) = \left(\frac{4R\rho\cos\varphi}{\rho^2+4R^2},\frac{4R\rho\sin\varphi}{\rho^2+4R^2},\frac{\rho^2-4R^2}{\rho^2+4R^2}\right). \quad (1.98)$$

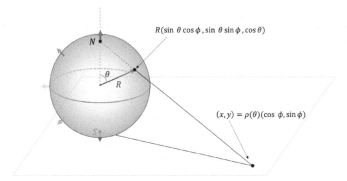

Figure 1.11 Stereographic projection of a spherical surface to 2 dimensional Euclidean space (x,y). The south pole becomes the origin, while the north pole is mapped to the spatial infinity of the Euclidean space that is identified as a single point. Five magnetic moment vectors are depicted along a great circle that can be connected to the Néel-type skyrmion depicted in Fig. 1.9.

This mapping corresponds to the unit winding Néel-type skyrmion with 5 magnetic moment vector of equal size. One can match these vectors in the left side of Fig. 1.9, where the vectors point up and down at the origin $\rho = 0$ and the infinity $\rho = \infty$, respectively. The extension with higher winding numbers and that of the Bloch-type skyrmion can be achieved with modifications in the mapping of the angle $\phi(\varphi)$ as in (1.80).

In the polar coordinates, the topological charge can be computed (after changing the coordinates) as

$$\begin{aligned} Q &= \int_0^\infty d\rho \int_0^{2\pi} d\varphi \, \frac{\sin\theta(\rho)}{4\pi} \frac{\partial\theta(\rho)}{\partial\rho} \frac{\partial\phi(\varphi)}{\partial\varphi} \\ &= \frac{(-1)}{4\pi} \int_{\rho=0}^{\rho=\infty} d(\cos\theta(\rho)) \int_{\varphi=0}^{\varphi=2\pi} d\phi(\varphi) \\ &= \frac{(-1)}{4\pi} \left[\cos[\theta(\infty)] - \cos[\theta(0)]\right] \left[\phi(2\pi) - \phi(0)\right] \\ &= -1 \,. \end{aligned} \qquad (1.99)$$

where we use $\theta(\infty) = 0$, $\theta(0) = \pi$, $\phi(2\pi) = 2\pi$ and $\phi(0) = 0$. Thus the topological number associated with the magnetic moment (1.98) is -1. If one uses the spin field $\vec{n} = -\vec{m}$, the topological number is $+1$ as it is odd power of the field.

1.4 Symmetries of skyrmion action

From the previous sections, we introduce the essential and relevant ingredients for describing the physics of skyrmions. Here we put them together to consider the full action and the corresponding symmetries and conservation laws following [18]

$$\mathcal{L} = \mathcal{L}_B - \mathcal{L}_H = \vec{A}_v \cdot \dot{\vec{m}} - \gamma_0 \hbar \mathcal{H}(\vec{m}) \,, \tag{1.100}$$

where we express the first term in the Lagrangian for the magnetic moment in terms of the vector potential as in (1.35) that is given by $\mathcal{L}_B = \vec{A}_v \cdot \dot{\vec{m}} = \hbar S(\cos\theta - 1)\dot{\phi}$ with $\vec{A}_v = \hbar S(\cos\theta - 1)/(r\sin\theta)\hat{\phi}$. We note that there is only one time derivative as mentioned above. The Hamiltonian is given by

$$\mathcal{H}(\vec{m}) = \frac{J}{2}(\nabla \vec{m})^2 + b\vec{m} \cdot (\nabla \times \vec{m}) - \vec{B} \cdot \vec{m} \,, \tag{1.101}$$

whose terms have been introduced to analyze the stability of the skyrmion crystals. In component forms, the Hamiltonian density reads

$$\mathcal{H}(\vec{m}) = \frac{J}{2}(\partial_\mu m_\alpha)^2 + b\epsilon_{\alpha j\gamma} m_\alpha (\partial_j m_\gamma) - B m_3 \,, \tag{1.102}$$

where $\mu, \nu, \cdots = t, x, y$ refer to the space-time indices, $i, j, k, \cdots = x, y$ space indices, and $\alpha, \beta, \gamma = 1, 2, 3$ the indices of the magnetic vector $\vec{m}$.

Before working on the Lagrangian (1.100), we digress on the stress energy tensor or energy momentum tensor as it is one of the important physical quantities we consider throughout the book. This section contains detailed exposition of somewhat complicated yet important matter, Noether theorem, stress energy tensor and conservation equations. While sufficient technical details are provided, it is more important to grasp the main ideas underlying all those details.

1.4.1 Stress energy tensor

In 2+1 dimensions, the stress energy tensor, $T^{\mu\nu}$, has 9 components as the indices $\mu, \nu = t, i$ run over the space-time coordinates, $t = 0$ and $i, j = 1, 2$. It can be conveniently written in the matrix form.

$$T^{\mu\nu}(x^0, x^i) = \begin{pmatrix} T^{00} & T^{01} & T^{02} \\ T^{10} & T^{11} & T^{12} \\ T^{20} & T^{21} & T^{22} \end{pmatrix}, \quad \eta^{\mu\nu} = \begin{pmatrix} -1 & 0 & 0 \\ 0 & 1 & 0 \\ 0 & 0 & 1 \end{pmatrix}. \quad (1.103)$$

Here T^{00} is the energy density, T^{i0} the energy flux across a surface perpendicular to i direction, T^{0j} the momentum density along the direction j, T^{ij} the flux of j directional momentum across a surface perpendicular to i direction. It is straightforward to generalize the tensor to the general $d+1$ dimensions. The space-time indices can be raised or lowered by the flat metric $\eta_{\mu\nu} = \text{diag}\{-1, 1, 1\}$. The mixed metric is nothing but the identity matrix, $\eta^{\mu}{}_{\nu} = \delta^{\mu}{}_{\nu} = \text{diag}\{1, 1, 1\}$.

There are several different ways to construct the stress energy tensor. We use the Noether's theorem that establishes the relation between the symmetries and conservation laws. Consider a function $f(x^\mu)$ located in a space-time point P in an inertial frame. The function can be written as $f'(x'^\mu)$ in another inertial frame because the functional relationship is frame dependent in general. Under the small deformation $x^\mu \to x'^\mu = x^\mu + \delta x^\mu$, the function changes as

$$\begin{aligned} \delta f &= f'(x') - f(x) = f'(x + \delta x) - f(x) \\ &= f'(x) - f(x) + \delta x^\mu \partial_\mu f'(x) + \mathcal{O}(\delta x^2) \\ &= \delta_0 f(x) + \delta x^\mu \partial_\mu f(x) \,. \end{aligned} \quad (1.104)$$

The last line is the result of keeping up to $\mathcal{O}(\delta x)$ for an infinitesimal δx. As a result, $\delta x^\mu \partial_\mu f'(x) = \delta x^\mu \partial_\mu f(x))$, which is the transport term. The other term $\delta_0 f(x) = f'(x) - f(x)$ describes the functional change. Under the uniform translation, $x^\mu \to x^\mu + a^\mu$, there is no change in a local field. Thus $\delta f = 0$. This gives $\delta_0 f(x) = -a^\mu \partial_\mu f(x) = -ia^\mu P_\mu f$, where we use $\delta x^\mu = a^\mu$ and define $P_\mu = -i\partial_\mu$ that is the translation generator.

Consider a scalar field $\phi'(x') = \phi(x)$ and another infinitesimal transformation $\delta x_i = \epsilon_{ij} x_j$ with $\epsilon_{ij} = -\epsilon_{ji}$. Then $\delta_0 \phi(x) = -\delta x^i \partial_i \phi(x) = -\epsilon_{ij} x_j \partial_i \phi(x) = (1/2)(\epsilon_{ij} x_i \partial_j - \epsilon_{ij} x_j \partial_i) \phi(x) \equiv (i/2) \epsilon_{ij} L_{ij} \phi(x)$. $L_{ij} = i(x_i \partial_j - x_j \partial_i)$ is the infinitesimal rotation generator, which can be generalized to the Lorentz transformation by including time coordinate $L_{\mu\nu} = i(x_\mu \partial_\nu - x_\nu \partial_\mu)$. The transformation is local if it depends on the values of the fields and their derivatives only at a given point x^μ.

We look into a general action for a local field $\Phi(x^\mu)$ in $d+1$ dimensional space-time.

$$S = \int d^{d+1}x \mathcal{L}(\Phi, \partial_\mu \Phi) \,. \tag{1.105}$$

Under an arbitrary change $\delta \Phi$, the action changes as $\delta \mathcal{L} = (\cdots) \delta \Phi + (\cdots) \delta(\partial_\mu \Phi)$.

$$\delta S = \int d^{d+1}x \left[\frac{\partial \mathcal{L}}{\partial \Phi} \delta \Phi + \frac{\partial \mathcal{L}}{\partial [\partial_\mu \Phi]} \delta(\partial_\mu \Phi) \right] \,. \tag{1.106}$$

Here we only examine the change of the field Φ. The variation of the field and its derivatives commute, $\delta(\partial_\mu \Phi) = \partial_\mu (\delta \Phi)$. Thus,

$$\begin{aligned}\delta S &= \int d^{d+1}x \left[\frac{\partial \mathcal{L}}{\partial \Phi} - \partial_\mu \frac{\partial \mathcal{L}}{\partial [\partial_\mu \Phi]} \right] \delta \Phi \\ &+ \int d^{d+1}x \, \partial_\mu \left[\frac{\partial \mathcal{L}}{\partial [\partial_\mu \Phi]} \delta \Phi \right] \,.\end{aligned} \tag{1.107}$$

The last term is a total derivative and can be changed into a surface integral as $\int d^d S_\mu (\partial \mathcal{L}/\partial [\partial_\mu \Phi]) \delta \Phi$ where a surface element $d^d S_\mu$ has a vector index μ as it has a preferred direction. Usually, the variational principle demands the surface term to vanish. Then the action is stationary under an arbitrary change $\delta \Phi$ if the following Euler-Lagrange equation is satisfied:

$$\partial_\mu \left(\frac{\partial \mathcal{L}}{\partial [\partial_\mu \Phi]} \right) - \frac{\partial \mathcal{L}}{\partial \Phi} = 0 \,. \tag{1.108}$$

This equation is nothing but the condition that the functional derivative of the action with respect to Φ vanishes. Note that this is true only because we discarded the surface term. As far as we discard surface terms, we can add a arbitrary total derivative term, such as $\mathcal{L}' = \mathcal{L} + \partial_\mu \Lambda^\mu$ with arbitrary Λ^μ. In classical mechanics, the transformation between $\mathcal{L}'$ and $\mathcal{L}$ is called a canonical transformation.

Now we perform a combined variation of the action with the changes in the coordinates δx^μ and the field $\delta \Phi$. Then the integration measure changes as $\delta(d^{d+1}x) = d^{d+1}x \partial_\mu(\delta x^\mu)$, while the function $\mathcal{L}$ has the variation $\delta \mathcal{L} = \delta_0 \mathcal{L} + \delta x^\mu \partial_\mu \mathcal{L}$. Then, $\delta_0 \mathcal{L} = \frac{\partial \mathcal{L}}{\partial \Phi}\delta_0\Phi + \frac{\partial \mathcal{L}}{\partial[\partial_\mu\Phi]}\delta_0(\partial_\mu\Phi) = [\frac{\partial \mathcal{L}}{\partial \Phi} - \partial_\mu(\frac{\partial \mathcal{L}}{\partial[\partial_\mu\Phi]})]\delta_0\Phi + \partial_\mu(\frac{\partial \mathcal{L}}{\partial[\partial_\mu\Phi]}\delta_0\Phi) = \partial_\mu(\frac{\partial \mathcal{L}}{\partial[\partial_\mu\Phi]}\delta_0\Phi)$, where we impose the Euler-Lagrange equation and use the fact δ_0 is just a functional change. Combining the results of the coordinate and field variations, we get (after imposing the equation of motion)

$$\delta S = \int d^{d+1}x \left[\partial_\mu(\delta x^\mu)\mathcal{L} + \delta x^\mu \partial_\mu \mathcal{L} + \partial_\mu \left(\frac{\partial \mathcal{L}}{\partial[\partial_\mu\Phi]}\delta_0\Phi \right) \right]$$
$$= \int d^{d+1}x \partial_\mu \left[(\mathcal{L}\eta^\mu_\nu - \frac{\partial \mathcal{L}}{\partial[\partial_\mu\Phi]}\partial_\nu\Phi)\delta x^\nu + \frac{\partial \mathcal{L}}{\partial[\partial_\mu\Phi]}\delta\Phi \right]. \quad (1.109)$$

where we use $\delta_0 \Phi = \delta\Phi - \delta x^\nu \partial_\nu \Phi$ and shuffle the indices. If the action is invariant under the combined variations of the coordinate and field that can be written in terms of constant parameter w^a,

$$\delta x^\nu = \frac{\delta x^\nu}{\delta w^a}\delta w^a, \qquad \delta\Phi = \frac{\delta\Phi}{\delta w^a}\delta w^a, \quad (1.110)$$

then $\delta S = \int d^{d+1}x\,(\partial_\mu j^\mu_a)w^a = 0$. The conserved current j^μ_a is given by

$$j^\mu_a = \left(\frac{\partial \mathcal{L}}{\partial[\partial_\mu\Phi]}\partial_\nu\Phi - \mathcal{L}\eta^\mu_\nu \right)\frac{\delta x^\nu}{\delta w^a} - \frac{\partial \mathcal{L}}{\partial[\partial_\mu\Phi]}\frac{\delta\Phi}{\delta w^a}. \quad (1.111)$$

For $\delta x^\nu = a^\nu$ and $\delta w^a = a^a$, we have $\delta x^\nu/\delta w^a = \eta^\nu_a$ and $\delta\Phi/\delta w^a = 0$. After renaming a to ν and j^μ_ν to T^μ_ν, we get

$$T^\mu_\nu = \frac{\partial \mathcal{L}}{\partial[\partial_\mu\Phi]}\partial_\nu\Phi - \mathcal{L}\eta^\mu_\nu. \quad (1.112)$$

This is a general formula as Φ can be any field, such as a scalar, a vector, a spinor, or a collection of fields. For the rotation, one can use $\delta x^i = \epsilon^{ij} x_j$ and $\delta w^a = \epsilon^{kl}$, we have $\delta x^i/\delta w^a = \delta(\epsilon^{ij}x_j)/\delta(\epsilon^{kl}) = (1/2)(\eta^i{}_k \eta^j{}_l - \eta^i{}_l \eta^j{}_k)x_j = (1/2)(\eta^i{}_k x_l - \eta^i{}_l x_k)$ and $\delta \Phi/\delta w^a = 0$. This can be extended to the Lorentz symmetry by including the boost, which is 'rotation' including the time component when there is the relativistic symmetry.

If the action is not invariant under the transformation, the conservation equation is not valid. For example, when $\delta x^\rho = 0$ in (1.111), we get

$$\partial_\mu j_a^\mu = -\partial_\mu \left(\frac{\partial \mathcal{L}}{\partial[\partial_\mu \Phi]} \frac{\delta \Phi}{\delta w^a} \right)$$
$$= -\left(\frac{\partial \mathcal{L}}{\partial \Phi} \right) \frac{\delta \Phi}{\delta w^a} - \frac{\partial \mathcal{L}}{\partial[\partial_\mu \Phi]} \partial_\mu \left(\frac{\delta \Phi}{\delta w^a} \right) = -\frac{\delta \mathcal{L}}{\delta w^a}, \quad (1.113)$$

where we use the Euler-Lagrange equation (1.108) and $\partial_\mu(\delta \Phi) = \delta(\partial_\mu \Phi)$. We encounter an important example of this type below.

Looking into the conservation equation, we explicitly write it as

$$0 = \int_{T_i}^{T_f} dt \frac{\partial}{\partial t} \left(\int d^d x j_a^0 \right) + \int_{T_i}^{T_f} dt \int d^d x (\partial_i j_a^i)$$
$$= \int d^d x j_a^0(T_f, \vec{x}) - \int d^d x j_a^0(T_i, \vec{x}) \equiv \tilde{Q}_a(T_f) - \tilde{Q}_a(T_i). \quad (1.114)$$

The last term in the first line vanishes for a suitably chosen boundary, such as infinity. Then the remaining term with the surface integral is nothing but the conserved charge, $\tilde{Q}_a = \int d^d x j_a^0(\vec{x})$. This charge is different from the topological charge Q we have considered above. The existence of the latter is independent of the action.

1.4.2 Translation symmetry

The action (1.100) is invariant under the uniform translation of space and time coordinates, $x^\mu \to x^\mu + a^\mu$ when the magnetic field $\vec{B}$ is uniform. The stress energy tensor, using (1.112), is given by

$$T^\mu{}_\nu = \frac{\partial \mathcal{L}}{\partial[\partial_\mu \vec{m}]} \cdot \partial_\nu \vec{m} - \mathcal{L} \eta^\mu{}_\nu. \quad (1.115)$$

The dot product in the first term means that there are three independent terms for each component of $\vec{m}$. The conservation equations are given by $\partial_\mu T^\mu_\nu = 0$.

For the time translation invariance and the corresponding quantity, we consider the time component, $\nu = 0$. Then, $\partial_0 T^0_0 + \partial_i T^i_0 = 0$. The Lagrangian has two separated parts, $\mathcal{L}_B$ with one time derivative and $\mathcal{L}_H$ without time derivatives. Then, the energy density and energy flux (current) are

$$T^0_0 = \frac{\partial \mathcal{L}_B}{\partial \dot{\vec{m}}} \cdot \dot{\vec{m}} - \mathcal{L} = \mathcal{L}_B - \mathcal{L} = \mathcal{H},$$
$$T^i_0 = \frac{\partial \mathcal{L}_H}{\partial [\partial_i \vec{m}]} \cdot \dot{\vec{m}} = -\frac{\partial \mathcal{H}}{\partial [\partial_i \vec{m}]} \cdot \dot{\vec{m}}.$$
(1.116)

Here and for the rest of the chapter, we simplifies our expressions by taking $\gamma_0 \hbar = 1$. The energy conservation equation gives

$$\dot{\mathcal{H}} - \partial_i \left(\frac{\partial \mathcal{H}}{\partial [\partial_i \vec{m}]} \cdot \dot{\vec{m}} \right) = 0.$$
(1.117)

Following (1.114), we can integrate this equation over the space-time. With a suitably large space, the second term can be shown to vanish as we consider the physical configuration that does not require infinitely large energy. Then,

$$E = \int d^2x \int dt \dot{\mathcal{H}} = \int d^2x \mathcal{H}.$$
(1.118)

The integral is independent of time, and thus the energy is conserved.

For the spatial translation invariance, we consider the space components, $\nu = i$. Then, $\partial_0 T^0_i + \partial_k T^k_i = 0$. Then, the momentum density is

$$T^0_i = \frac{\partial \mathcal{L}_B}{\partial [\partial_0 \vec{m}]} \cdot \partial_i \vec{m} = \vec{A}_v \cdot \partial_i \vec{m} = -2S a_i,$$
(1.119)

where we use $\partial_i \vec{m} = (\partial_i \theta) \partial_\theta \vec{m} + (\partial_i \phi) \partial_\phi \vec{m} = (\partial_i \theta) \hat{\theta} + (\partial_i \phi) \sin \theta \hat{\phi}$ and $\vec{A}_v \cdot \partial_i \vec{m} = (\cos \theta - 1) \partial_i \phi = -2S a_i$, which is defined in (1.44). In this language, $\mathcal{L}_B = \vec{A}_v \cdot \dot{\vec{m}} = -2S a_0$. We also note that T^0_i entirely come from $\mathcal{L}_B$. Thus, we are not able to define a conserved total momentum as the term $\mathcal{L}_H$ in (1.100) does not contribute to the stress energy tensor, $T^0_{Hi} = 0$. We try to understand whether it is possible to define the total momentum.

We look into the momentum flux that goes as

$$T^k_i = \frac{\partial \mathcal{L}_H}{\partial[\partial_k \vec{m}]} \cdot \partial_i \vec{m} - \mathcal{L}\eta^k_i = -T^k_{Hi} - \mathcal{L}_B \eta^k_i. \tag{1.120}$$

where $T^k_{Hi} = \frac{\partial \mathcal{H}}{\partial[\partial_k \vec{m}]} \cdot \partial_i \vec{m} + \mathcal{H}\eta^k_i$ and $\eta^k_i = \delta^k_i$. Then the momentum conservation equation gives $\partial_0(T^0_{Bi} + T^0_{Hi}) + \partial_k(T^k_{Bi} + T^k_{Hi}) = -2\dot{\mathbf{a}}_i + 0 - \partial_i(\mathcal{L}_B) - \partial_k T^k_{Hi} = 0$. Using this, the parts related to $\mathcal{L}_H$ can be rearranged as

$$\begin{aligned}\partial_\mu T^\mu_{Hi} = \partial_k T^k_{Hi} &= \partial_t(\vec{A}_v \cdot \partial_i \vec{m}) - \partial_i(\vec{A}_v \cdot \partial_t \vec{m}) \\ &= 2S\partial_i \mathbf{a}_0 - 2S\partial_0 \mathbf{a}_i \\ &= S\vec{m} \cdot (\partial_i \vec{m} \times \partial_0 \vec{m}),\end{aligned} \tag{1.121}$$

where we use (1.47) and (1.48). The left-hand side is the spatial divergence of momentum flux (of the static Lagrangian) as the time derivative of momentum density (of the static Lagrangian) vanishes. This means that we can integrate the equation over a spatial volume so that the left side vanishes. Then, $\int d^2x \vec{m} \cdot (\partial_i \vec{m} \times \partial_0 \vec{m}) \propto \int d^2x \epsilon_{i0j} J_j = 0$, where $\vec{J}$ is the topological current.

Searching for the charge, conserved momentum, associated with the spatial translation, the right-hand side (the topological current J_j) can be thought to provide a time derivative of conserved momentum. With a reasonable assumption that the product $x_i J_j$ also vanishes at the boundary and the fact that the topological current satisfies $\partial_\mu J^\mu = \partial_0 J^0 + \partial_j J^j = 0$, we can write

$$\begin{aligned}\int d^2x J_i = \int d^2x \big(-\partial_j(x_i J^j) + J_i\big) &= -\int d^2x\, x_i \partial_j J^j \\ &= \partial_0 \int d^2x\, x_i J^0.\end{aligned} \tag{1.122}$$

We identify the total conserved momentum [20] as

$$P_i = 4\pi \epsilon_{i0j} \int d^2x\, x_j J^0. \tag{1.123}$$

Thus, we define the conserved momentum in terms of the topological charge J^0. This is the first moment of the topological charge distribution.

1.4.3 Rotation symmetry

Here we consider the invariance of the action (1.100) under the rotation of the spin along the direction of the magnetic field $\vec{B} = B\hat{z}$. The action has the DM interaction (1.66) that is invariant under the simultaneous rotations of the spin and the space. The rotation of spin vector can be written as $m_i \to R_{ij}m_j$. For example, $\vec{m}^2$ is invariant under the rotation as $\vec{m}^2 = m_i m_i \to (R_{ik}m_k)(R_{il}m_l) = m_k(R_{ik}R_{il})m_l = m_k(R_{ki}^T R_{il})m_l = m_k \delta_{kl} m_l = m_l m_l = \vec{m}^2$. Here we use the fact that rotational matrices are orthogonal, $R_{ki}^T = R_{ki}^{-1}$.

The DM interaction (1.66) can be written as $b\vec{m} \cdot (\nabla \times \vec{m}) = b\epsilon_{ijk} m_i \partial_j m_k$. If we rotate only the magnetic moment vector as $m_i \to R_{il}m_l$, the DM term is not invariant. With the simultaneous rotation $\partial_j \to R_{jm}\partial_m$, we can check $\epsilon_{ijk} m_i \partial_j m_k \to \epsilon_{ijk} R_{il} R_{jm} R_{kn}(m_l \partial_m m_n) = \epsilon_{lmn} m_l \partial_m m_n$ because the ϵ symbol transforms as a tensor under the rotation $\epsilon_{lmn} = \epsilon_{ijk} R_{il} R_{jm} R_{kn}$. Thus we consider both the rotation of the vector $\vec{m}$ and spatial rotation of the coordinate $\vec{x}$ in the presence of the DM interaction.

The rotation of the vector $\vec{m}$ with an infinitesimal parameter ω (related to the spin angular momentum) is described by

$$\delta m_i = \omega \epsilon_{izk} m_k . \tag{1.124}$$

In the vector notation, $\delta \vec{m} = \omega \hat{z} \times \vec{m}$. To obtain the Euler-Lagrange equation, we perform the functional variation on two parts $\mathcal{L}_B$ and $\mathcal{L}_H$ separately of the action (1.100) similar to (1.106). The variation for $\mathcal{L}_B$ is simple as we can identify $\delta \phi = \omega$ for the rotation taking place in xy plane.

$$\delta S_B = \int d^2x dt \delta(S(\cos\theta - 1)\dot\phi) = -\int d^2x dt\, \omega S \partial_t m_z , \tag{1.125}$$

where we identify $m_z = \cos\theta$ after performing integration by parts.

The variation for $\mathcal{S}_H$ is given by (1.107),

$$\delta \mathcal{S}_H = \int d^2x dt \omega \hat{z} \cdot \left(\vec{m} \times \left[\frac{\partial \mathcal{L}_H}{\partial \vec{m}} - \partial_\mu \left(\frac{\partial \mathcal{L}_H}{\partial [\partial_\mu \vec{m}]}\right)\right]\right), \quad (1.126)$$

where we use the Euler-Lagrange form to do the job on the action (1.100). Using $\delta \Phi = \delta \vec{m} = \omega \hat{z} \times \vec{m}$ in (1.107) and $(\hat{z} \times \vec{m}) \cdot \vec{A} = \hat{z} \cdot (\vec{m} \times \vec{A})$ for any vector $\vec{A}$, which can be checked using the index notation. Explicitly, the expression in the square bracket is given by

$$\frac{\partial \mathcal{L}_H}{\partial \vec{m}} - \partial_i \left(\frac{\partial \mathcal{L}_H}{\partial [\partial_i \vec{m}]}\right) = -J\partial_i^2 \vec{m} + 2b(\nabla \times \vec{m}) - \vec{B}, \quad (1.127)$$

where $\partial_\mu (\partial \mathcal{L}_H / \partial [\partial_\mu \vec{m}]) = \partial_i (\partial \mathcal{L}_H / \partial [\partial_i \vec{m}])$ as there is no time derivatives in $\mathcal{L}_H$. The related computations are straightforward. For example, $\partial (b\vec{m} \cdot (\nabla \times \vec{m}))/\partial \vec{m} = b(\nabla \times \vec{m})$ and

$$\partial_i \frac{\partial (b\epsilon_{\beta j\gamma} m_\beta (\partial_j m_\gamma))}{\partial [\partial_i m_\alpha]} = \partial_i (b\epsilon_{\beta j\gamma} m_\beta \delta_{ij} \delta_{\alpha\gamma}) \quad (1.128)$$

$$= b\epsilon_{\beta i\alpha} \partial_i m_\beta = -b\epsilon_{\alpha i\beta} \partial_i m_\beta,$$

which is $-b(\nabla \times \vec{m})$, where the index α is the vector index in the computation. We also note that the direct functional variation gives the same result, as $\delta \mathcal{L}_H / \delta \vec{m} = -J\partial_i^2 \vec{m} + 2b(\nabla \times \vec{m}) - \vec{B}$.

By combining the variations $\delta \mathcal{S}_B$ from (1.125) and $\delta \mathcal{S}_H$ from (1.126), we obtain the Euler-Lagrange equation.

$$0 = -\omega S \partial_t m_z - \omega \hat{z} \cdot \left(\vec{m} \times \left[-J\partial_i^2 \vec{m} + 2b(\nabla \times \vec{m}) - \vec{B}\right]\right) \quad (1.129)$$

$$= -\omega \left[S \partial_t m_z - J\hat{z} \cdot (\vec{m} \times \partial_i^2 \vec{m}) - 2b(\vec{m} \cdot \nabla) m_z\right],$$

where we use $\hat{z} \cdot (\vec{m} \times \vec{B}) = \vec{m} \cdot (\vec{B} \times \hat{z}) = 0$ and $\hat{z} \cdot (\vec{m} \times (\nabla \times \vec{m})) = \epsilon_{zjk} m_j \epsilon_{klm} (\partial_l m_m) = m_j \partial_z m_j - m_j \partial_j m_z = -(\vec{m} \cdot \nabla) m_z$. One can also check the Euler-Lagrange equation can be obtained by performing functional derivative $\delta \vec{m} \cdot [\delta(\mathcal{S}_B - \mathcal{S}_H)/\delta \vec{m}] = 0$ and using $\delta \vec{m} = \omega \hat{z} \times \vec{m}$.

The current is defined by (1.111) (for $\delta x = 0$).

$$j^\mu_{\delta \vec{m}} = -\frac{\partial \mathcal{L}}{\partial [\partial_\mu \vec{m}]} \frac{\delta \vec{m}}{\delta \omega} = -\hat{z} \cdot \left(\vec{m} \times \frac{\partial \mathcal{L}}{\partial [\partial_\mu \vec{m}]}\right), \quad (1.130)$$

which can be expressed in components as

$$j^0_{\delta \vec{m}} = -Sm_z ,$$
$$j^i_{\delta \vec{m}} = J\hat{z} \cdot (\vec{m} \times \partial_i \vec{m}) + bm_i m_z .$$
(1.131)

Then,

$$\begin{aligned}\partial_\mu j^\mu_{\delta \vec{m}} &= -S\dot{m}_z + J\hat{z} \cdot (\vec{m} \times \partial^2_i \vec{m}) + b\partial_i(m_i m_z) \\ &= bm_z(\partial_i m_i) - b(m_j \partial_j)m_z ,\end{aligned}$$
(1.132)

where we use the Euler-Lagrange equation. Thus the current $j^\mu_{\delta \vec{m}}$ is not conserved as the DM interaction term with the parameter b is not invariant under the change of $\vec{m}$ alone.

We change our attention to the orbital angular momentum. The vector $\vec{m}$ changes with the same infinitesimal parameter ω through the spatial rotation of the coordinate $\vec{r} \to \vec{r} + \delta\vec{r}$ with $\delta\vec{r} = \omega\hat{z} \times \vec{r}$. Under this change, the vector $\vec{m}$ change as $\delta\vec{m} = \vec{m}(\vec{r} + \delta\vec{r}) - \vec{m}(\vec{r}) = (\delta\vec{r} \cdot \nabla)\vec{m} = (\omega[\hat{z} \times \vec{r}] \cdot \nabla)\vec{m} = \omega[\epsilon_{jzk}x_k]\partial_j\vec{m}$. Thus in component notation, we have

$$\delta m_\alpha = \omega \epsilon_{zjk} x_j \partial_k m_\alpha .$$
(1.133)

Note that $j, k = x, y$ as the rotation takes place only in xy plane.

Using $\delta\Phi = \delta\vec{m} = \omega(x\partial_y - y\partial_x)\vec{m}$ in (1.107) (without the total derivative term),

$$\begin{aligned}\delta S &= \int d^2x dt\, \omega(x\partial_y - y\partial_x)\vec{m} \cdot \left(\frac{\partial \mathcal{L}}{\partial \vec{m}} - \partial_\mu\left(\frac{\partial \mathcal{L}}{\partial[\partial_\mu \vec{m}]}\right)\right) \\ &= -\int \omega(x\partial_\mu T^\mu_y - y\partial_\mu T^\mu_x) ,\end{aligned}$$
(1.134)

where we use $\partial_x \mathcal{L} = (\partial_x \vec{m}) \cdot (\partial \mathcal{L}/\partial \vec{m}) + (\partial_x[\partial_\nu \vec{m}]) \cdot (\partial \mathcal{L}/\partial[\partial_\nu \vec{m}])$ and $(\partial_x \vec{m}) \cdot (\partial \mathcal{L}/\partial \vec{m}) - (\partial_x \vec{m}) \cdot (\partial_\mu[\partial \mathcal{L}/\partial[\partial_\mu \vec{m}]]) = \partial_x \mathcal{L} - \partial_\mu((\partial_x \vec{m}) \cdot [\partial \mathcal{L}/\partial[\partial_\mu \vec{m}]]) = -\partial_\mu T^\mu_x$. The last equality is valid by the definition (1.115). Thus the Euler-Lagrange equation is

$$x\partial_\mu T^\mu_y - y\partial_\mu T^\mu_x = 0 ,$$
(1.135)

which is expressed in terms of the stress energy tensor.

The corresponding current is defined by the first part of (1.111) (for $\delta\Phi = 0$).

$$j^\mu_{\delta\vec{r}} = \left(\frac{\partial \mathcal{L}}{\partial[\partial_\mu \vec{m}]}\partial_\nu \vec{m} - \mathcal{L}\eta^\mu_\nu\right)\frac{\delta x^\nu}{\delta w} = T^\mu_i \epsilon^{izj}x_j = xT^\mu_y - yT^\mu_x, \quad (1.136)$$

where we use $\delta\vec{r} = \omega\hat{z} \times \vec{r}$ and thus $\delta x^i = \omega\epsilon^{izj}x_j$. Then,

$$\begin{aligned}\partial_\mu j^\mu_{\delta\vec{r}} &= \partial_\mu(xT^\mu_y - yT^\mu_x) = T_{xy} - T_{yx} \\ &= bm_\alpha(\epsilon_{x\alpha\gamma}\partial_y m_\gamma - \epsilon_{y\alpha\gamma}\partial_x m_\gamma) \\ &= b(m_j\partial_j)m_z - bm_z(\partial_j m_j),\end{aligned} \quad (1.137)$$

where we use the Euler-Lagrange equation. Thus the current $j^\mu_{\delta\vec{r}}$ is not conserved either. The stress energy is not symmetric due to the DM interaction, which is clear from this direct computation.

Now we consider the orbital angular momentum, $\ell_{\text{orb}} = j^0_{\delta\vec{r}}$, time component of the conserved current associated with the spatial rotation in (1.136).

$$\ell_{\text{orb}} = xT^0_y - yT^0_x = -2S(x\mathbf{a}_y - y\mathbf{a}_x) = -2S\vec{r}\times\vec{a}, \quad (1.138)$$

where we use $T^0_i = -2S\mathbf{a}_i$. By integrating the angular momentum density over the volume, we get

$$\begin{aligned}L_{\text{orb}} &= \int d^2x\, \ell_{\text{orb}} \\ &= -S\int d^2x\left(\partial_x(x^2+y^2)\mathbf{a}_y - \partial_y(x^2+y^2)\mathbf{a}_x\right) \\ &= S\int d^2x(x^2+y^2)(\partial_x\mathbf{a}_y - \partial_y\mathbf{a}_x) \\ &= 2\pi S\int d^2x\, \rho^2 j^0,\end{aligned} \quad (1.139)$$

where we assume that the product $\rho^2 j^0$ vanishes sufficiently fast at the boundary so that the surface integral vanishes. Here we see that the second moment of the topological charge distribution is the total orbital angular momentum, which is similar to the linear momentum that is the first moment given in (1.123).

As we already have alluded, we can consider the full expression of (1.111). Under the direct variations,

$$\delta \vec{r} = \omega \hat{z} \times \vec{r}, \qquad \delta \vec{m} = \omega \hat{z} \times \vec{m}, \qquad (1.140)$$

the following combined currents can be verified to be conserved $\partial_\mu j_\omega^\mu = 0$ from (1.132) and (1.137).

$$\begin{aligned} j_\omega^\mu &= \left(\frac{\partial \mathcal{L}}{\partial[\partial_\mu \vec{m}]} \partial_\nu \vec{m} - \mathcal{L}\eta_\nu^\mu\right)\frac{\delta x^\nu}{\delta \omega} - \frac{\partial \mathcal{L}}{\partial[\partial_\mu \vec{m}]}\frac{\delta \vec{m}}{\delta \omega} \\ &= xT_y^\mu - yT_x^\mu - Sm_z \delta^{\mu 0} + \left(J\hat{z} \cdot (\vec{m} \times \partial_i \vec{m}) + bm_i m_z\right)\delta^{\mu i}, \end{aligned} \qquad (1.141)$$

where the last line is given in (1.130) with the explicit expressions listed in (1.131). We note that the variation (1.133) is indirect change of the vector $\vec{m}$ due to the space variation $\delta \vec{r}$. The conservation equation in terms of the components is given by

$$\partial_\mu j_\omega^\mu = \partial_0(\ell_{\text{orb}} - Sm_z) + \partial_i(j_{\delta \vec{r}}^i + j_{\delta \vec{m}}^i) = 0. \qquad (1.142)$$

Thus the total angular momentum density $\ell_{\text{tot}} = \ell_{\text{orb}} - Sm_z$ is conserved and is the combination of the orbital and spin angular momenta.

1.4.4 (Angular) Momentum of the skyrmion

We compute the momentum (1.123) and angular momentum ((1.131) and (1.139)) that are related to the topological properties of the skyrmion. Due to the translation invariance, the energy of the skyrmion is independent of its position. Thus there are two zero modes in the xy plane. We denote them as $\vec{R}(t) = (R_x(t), R_y(t))^T$. While the momenta are independent of a specific form of the skyrmion solution, one can use the explicit form given in (1.98) with $\vec{x} \to \vec{x} - \vec{R}(t)$. This form has the skyrmion charge $Q = -1$.

The linear momentum is

$$\begin{aligned} P_i &= 4\pi \epsilon_{i0j} \int d^2x \left[(x_j - R_j) + R_j\right] J^0(\vec{x} - \vec{R}) \\ &= 4\pi \epsilon_{i0j} R_j Q = -4\pi \epsilon_{i0j} R_j. \end{aligned} \qquad (1.143)$$

Here we use $\int d^2x \ (x - R_x) J^0(\vec{x} - \vec{R}) = \int d^2x' \ x' J^0(\vec{x}') \propto \int_0^{2\pi} d\varphi' \cos\varphi' = 0$ for x component. For y component, we similarly have $\int_0^{2\pi} d\varphi' \sin\varphi' = 0$. The corresponding conservation equation is

$$\partial_0 P_i = -4\pi\epsilon_{i0j}\partial_0 R_j = -4\pi\epsilon_{izj}\hat{z}\partial_0 R_j$$
$$\propto (\hat{z} \times \partial_0 \vec{R})_i = 0 \ . \tag{1.144}$$

This equation can be identified as the Thiele equation without Gilbert damping term that we develop later in §4.4.

We examine the total angular momentum of the skyrmion ℓ_{tot} from (1.131) and (1.139). As the skyrmion we consider forms in the ferromagnetic material, there is already non-zero spin angular momentum. We subtract this contribution by $m_z \to m_z - 1$. We also use $\ell_{orb} = 2\pi S r^2 J^0$.

$$\begin{aligned} L_{tot} &= \int d^2x \, S\left(2\pi\rho^2 J^0(\vec{x} - \vec{R}) - (m_z(\vec{x} - \vec{R}) - 1)\right) \\ &= \int d^2x' \, S\left(2\pi\rho'^2 J^0(\vec{x}') - (\cos\theta - 1)\right) \\ &\quad + 2\pi S \int d^2x' \, (2\vec{x}' \cdot \vec{R} + \vec{R}^2) J^0(\vec{x}') \\ &= 2\pi S \int d^2x \, \vec{R}^2 J^0(\vec{x}) = 2\pi S \vec{R}^2 Q = -2\pi S \vec{R}^2 \ , \end{aligned} \tag{1.145}$$

where $\vec{x}' = \vec{x} - \vec{R}$, $\int d^2x' \, \vec{x}'' J^0(\vec{x}') = 0$ and

$$\begin{aligned} &\int d^2x' \left(2\pi\rho'^2 J^0(\vec{x}') - (\cos\theta - 1)\right) \\ &= \int d\varphi \int d\rho \rho \left(\frac{2\pi\rho^2}{4\pi\rho} \sin\theta[\rho](\partial_\rho\theta[\rho]) - (\cos\theta[\rho] - 1)\right) \quad (1.146) \\ &= \frac{\rho^2}{2}(1 - \cos\theta[\rho])\bigg|_{\rho=0}^{\rho=\infty} = \rho^2 \sin^2\left(\frac{\theta[\rho]}{2}\right)\bigg|_{\rho=0}^{\rho=\infty} = 0, \end{aligned}$$

where we use $\theta = 0$ as $\rho \to \infty$. Thus the total angular momentum is proportional to $\vec{R}^2$. The angular momentum conservation can be verified as

$$\partial_t L_{\text{tot}} = -4\pi S R_i \partial_t R_i = S R_i \epsilon_{zik} \partial_t P_k = S(\vec{R} \times \partial_t \vec{P})_z, \quad (1.147)$$

where we use $\partial_t R_i = \epsilon_{i0k}\partial_t P_k/(4\pi)$ that can be derived by multiplying ϵ_{0ik} to the (1.144) followed by manipulation of two epsilon tensors. Here we also change the index 0 to z. The result vanishes as in (1.144).

1.5 Mutually compatible observables

When the parity symmetry is broken, some physical systems can generate an angular momentum spontaneously [21]. This spontaneously generated angular momentum contains the expectation value of the linear momentum operator. Due to this, one can not keep the angular momentum and translation invariance at the same time, which puts specific constraints on maximum compatible quantities in Ward identities considered in §5.3 of this book. See also [22]. Here we provide some basic idea related to this.

For simplicity, we consider a 2 dimensional system without a time dependence so that a conservation equation is reduced to $\partial_i T^{0i}(\vec{x}) = 0$. Reminded with the antisymmetric property of the epsilon tensor ϵ^{ij}, one can check the conservation equation has an obvious solution.

$$\langle T^{0i}(\vec{x}) \rangle = \epsilon^{ij}\partial_j l(\vec{x}). \quad (1.148)$$

The solution is independent of the choice of $l(\vec{x})$. Because the physical properties are independent of the details, we can choose $l(\vec{x})$ such that the computation becomes simple.

$$l(\vec{x}) = \begin{cases} \ell/2 & (|x|, |y| \leq b) \\ 0 & (\text{otherwise}) \end{cases}, \quad (1.149)$$

where ℓ is a constant. This is illustrated in Fig. 1.12. Thus $T^{0i}(\vec{x})$ vanishes both inside and outside of the square region with its sides

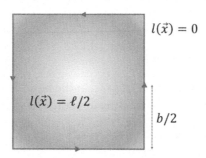

Figure 1.12 Spontaneously generated angular momentum $l(\vec{x}) = \ell/2$ inside a square with length b. $l(\vec{x}) = 0$ outside. There is a momentum current going around along the boundary. Can we detect the angular momentum as the boundary is pushed to the infinity?

with a length b. Nevertheless, it does not vanish along the boundary.

$$\langle T^{0i}(\vec{x})\rangle = (\ell/2)\epsilon^{ij}\left[-\delta(x^j - b/2) + \delta(x^j + b/2)\right] \\ \times \theta(b/2 - |x|)\theta(b/2 - |y|) \,. \qquad (1.150)$$

This corresponds to an edge current, a momentum flow around the boundary of the square. The magnitude and direction of the edge current are ℓ and along the boundary, either clockwise or counterclockwise, depending on the sign of ℓ. This boundary contribution is related to the topological nature of the underlying field theory.

Now we consider the infinite volume limit, $b \to \infty$. Normally we discard the effects of T^{0i} because it is a boundary contribution. However, a careful treatment shows that it indeed makes a contribution to the total angular momentum [21].

$$L = \int d^2\vec{x}\,\epsilon_{ij}x^i\langle T^{0j}\rangle = \int d^2\vec{x}\,\ell(\vec{x})\partial_i x^i = \ell \int d^2\vec{x} = \ell V_2 \,, \quad (1.151)$$

where V_2 is the volume enclosed by the boundary. The total angular momentum is independent of the shape of the boundary. This example illustrates that identifying spontaneously generated angular momentum is subtle in the system with a broken parity symmetry and, in particular, can exists even in a system without boundary. (For a torus, we further identify the left and right boundaries and top and bottom boundaries. There seems to be a tension for the directions of the momentum at the boundaries. While we can transfer this tension into other quantities, careful investigation is required to clarify the compatibility.)

Note that there is even more subtle point when we consider maximum compatible symmetries in the presence of the rotational symmetry, the translation symmetry, and the angular momentum. The translation symmetry is not compatible with the presence of the spontaneously generated angular momentum, which is manifested with the expectation value of the momentum operator $\langle T^{0i}(\vec{x}) \rangle \neq 0$. (This is also true in the presence of a boundary. We do not explicitly consider the boundary effect in this book.) Thus we have two independent options that are incompatible with each other.

- Option I: keeping the rotation symmetry and the angular momentum without the translation symmetry.
- Option II: keeping both the translation and rotation symmetries without angular momentum.

These two different options lead to two highly non-trivial and mutually exclusive sets of Ward identities as reviewed in quantum Hall systems in Chapter 3 and also in the skyrmion physics considered in §5.3 [22].

Chapter 2

Hydrodynamics

The magnetic skyrmions are particle-like objects that can move around under certain circumstances and can be controlled by electric current or temperature gradient. One major theme of the book is the transport properties of the skyrmions such as conductivities and viscosities, which are governed by hydrodynamics. This chapter provides an introduction for this subject that is crucial for understanding the later chapters.

2.1 Introduction: Relativistic hydrodynamics

Hydrodynamics is an effective theory that is useful as it captures the universal features of the interesting systems depending on the underlying symmetries. It has been proved to be powerful to describe the corresponding dynamics at a long distance and a large time scale and to incorporate the dissipative effects, for which the Lagrangian description is not useful. Hydrodynamics is described, at best, by conservation equations as they incorporate the dissipative effects.

The central object of hydrodynamics is the stress energy tensor $T^{\mu\nu}$ that is introduced in §1.4.1. In the relativistic hydrodynamics with Lorentz invariance, the hydrodynamic equations are given

---

Skyrmions and Hall Transport
Bom Soo Kim
Copyright © 2023 Jenny Stanford Publishing Pte. Ltd.
ISBN 978-981-4968-34-8 (Hardcover), 978-1-003-37253-0 (eBook)
www.jennystanford.com

by the conservation equation $\partial_\mu T^{\mu\nu} = 0$ in $d+1$ dimensions. The parameters of the stress energy tensor, $T^{\mu\nu}(T, u^\mu)$, are the temperature T and the fluid velocity u_μ with the time and space components $u^\mu = (u^0, u^i)$, $i = 1, \cdots, d$. We normalize the velocity as $\eta_{\mu\nu} u^\mu u^\nu = -1$. For the Lorentz invariant case, the number of the conservation equations, $n = d + 1$, is the same as the number of unknowns T and u^μ. Thus system of equations is closed, and one can solve the equations. In equilibrium, the parameters T and u^μ are constants and thus the conservation equations are trivially satisfied.

There exist ambiguities for choosing the velocity vector with given condition. One frequent choice is the Eckart frame, where the flow of fluid is defined by the particle flow. We explain this further in the section of charged hydrodynamics. Here we choose to impose a different condition, the Landau frame condition,

$$T^{\mu\nu} u_\nu = -\varepsilon u^\mu, \tag{2.1}$$

which signifies that the flow of the fluid is the energy flow. We further impose the local second law of thermodynamics, which connects the hydrodynamics to the thermodynamics. The corresponding equation is given by

$$(\partial_\mu T^{\mu\nu}) u_\nu = 0. \tag{2.2}$$

This equation, while seemingly empty due to the conservation equation, leads to the so-called entropy current. The entropy current is useful because it is required to be positive definite and constrains the values of various transport coefficients.

We solve the hydrodynamic conservation equations order by order in a derivative expansion for the systems with a local thermal equilibrium. The local equilibrium enforces that the details of the thermal fluctuations can be smoothened out, which is important for revealing universal physical properties. For the relativistic hydrodynamics with the Lorentz invariance, the possible ingredients are $T, u^\mu, \partial_\mu, \eta^{\mu\nu}$.

2.1.1 Ideal hydrodynamics

At the leading (ideal) order without any derivatives, one can construct only two symmetric second rank tensors $u^\mu u^\nu$ and the projection tensor $P^{\mu\nu} = u^\mu u^\nu + \eta^{\mu\nu}$. They are orthogonal to each other, $P^{\mu\nu}u_\mu = P^{\mu\nu}u_\nu = 0$. We require the tensor to be symmetric because rotation and boost symmetries are parts of the Lorentz symmetry. Then the stress energy tensor can be written as

$$T^{\mu\nu}_{N0}(T,u) = \varepsilon(T)\, u^\mu u^\nu + p(T)\, P^{\mu\nu}, \tag{2.3}$$

where the coefficients of the two tensors are called the thermodynamic functions, the energy density $\varepsilon(T)$ and the pressure $p(T)$. The expression can be written as a slightly different form $T^{\mu\nu} = (\varepsilon + p)\, u^\mu u^\nu + p\, \eta^{\mu\nu}$. The combination $w = \varepsilon + p$ is the heat function per unit volume.

The energy density is a scalar (independent of the choice of the velocity vector) that can be obtained by contracting the equation with u_μ, $\varepsilon = u_\mu T^{\mu\nu} u_\nu$. Pressure, another scalar, is also given by contracting the stress energy tensor with the projection operator $p = (1/d)T^{\mu\nu}P_{\mu\nu}$, where the number of spatial dimensions, d, comes from $P^{\mu\nu}P_{\mu\nu} = (\eta^{\mu\nu} + u^\mu u^\nu)(\eta_{\mu\nu} + u_\mu u_\nu) = \delta^\mu_\mu + u^\mu u_\mu + u^\nu u_\nu + (u^\mu u_\mu)^2 = (d+1) - 1 - 1 + 1 = d$. In the local rest frame $u^0 = 1, u^i = 0$, the stress energy tensor in $2+1$ dimensions has a diagonal form

$$T^{\mu\nu} = \begin{pmatrix} \varepsilon & 0 & 0 \\ 0 & p & 0 \\ 0 & 0 & p \end{pmatrix}. \tag{2.4}$$

One can check them, for example, $T^{00} = \varepsilon u^0 u^0 + p P^{00} = \varepsilon$ as $P^{00} = u^0 u^0 + \eta^{00} = 0$.

Typically, the full hydrodynamic equations involve the conservation of particle number. Here we focus on the simplest hydrodynamics where the notion of the particle number and its conservation are irrelevant, such as the ultra relativistic case or the hydrodynamics of the quantum critical point. For the latter, the strong coupling dynamics prevent particle-like descriptions, yet its hydrodynamics can be formulated with appropriate symmetries [26]. It is straightforward to check that (2.3) satisfies the Landau frame condition (2.1) by using $u_\mu u^\mu = -1$ after contracting velocity to the stress energy tensor. We also evaluate the second law of thermodynamics equation (2.2) for (2.3) explicitly.

$$\begin{aligned}(\partial_\mu T^{\mu\nu}_{N0})u_\nu &= -(\partial_\mu \varepsilon)u^\mu - (\varepsilon + p)(\partial_\mu u^\mu) \\ &= -T\partial_\mu(su^\mu) + \big(-\partial_\mu \varepsilon + T(\partial_\mu s)\big)u^\mu = 0,\end{aligned}$$ (2.5)

where we use $u^\mu(\partial_\nu u_\mu) = -u_\mu(\partial_\nu u^\mu) = (1/2)\partial_\nu(u^\mu u_\mu) = 0$, $P^{\mu\nu}u_\nu = 0$, and $u^\mu u_\mu = -1$ to derive the first line, and the heat function $w = Ts = \varepsilon + p$ and the entropy current $s^\mu = su^\mu$ with the entropy density s to derive the second line. Thus, we have the following equation for the entropy current,

$$T(\partial_\mu s^\mu) = \big(-\partial_\mu \varepsilon + T(\partial_\mu s)\big)u^\mu \equiv 0.$$ (2.6)

At the ideal order, the entropy current is required to vanish. This gives the usual thermodynamic relation $d\varepsilon = Tds$. Then the total differential of the heat function, $dw = Tds + sdT$ gives $dp = sdT$.

Let us summarize the ideal hydrodynamics. Originally we state that there are 4 equations $\partial_\mu T^{\mu\nu} = 0$ and 4 unknowns that are T and u^μ (with $u_\mu u^\mu = -1$). In the form (2.3), there are 5 unknowns that are ε, p, and u^μ. While solving, we supplied the equation of state, $\varepsilon + p = Ts$. This introduce the entropy density s, which is also constrained by the entropy current equation (2.6). Thus we can also say that our hydrodynamic system has 6 equations and 6 unknowns, that can be uniquely solved.

2.1.2 First derivative order

At the next, first derivative, order, we can construct two symmetric Lorentz covariant second rank tensors using the velocity and a derivative, the divergence of the velocity $P^{\mu\nu}(\partial_\alpha u^\alpha)$ and the shear tensor

$$\sigma^{\mu\nu} = \partial^\mu u^\nu + \partial^\nu u^\mu - P^{\mu\nu}(\partial_\delta u^\delta).$$ (2.7)

The shear tensor is traceless as $\eta^{\rho\sigma}[\partial_\rho u_\sigma + \partial_\sigma u_\rho - (2/d)P_{\rho\sigma}(\partial_\delta u^\delta)] = (2 - (2/d) \cdot d)(\partial_\delta u^\delta) = 0$, where we use $\eta^{\rho\sigma}P_{\rho\sigma} = \eta^{\rho\sigma}(\eta_{\rho\sigma} + u_\rho u_\sigma) = \delta^\rho_\rho + u^\rho u_\rho = (d+1) - 1 = d$. We choose $d = 2$ in $2+1$ dimensions. They are multiplied by the projection operator $P^{\mu\nu}$ so that the definition of the energy ε and pressure p would not change after adding the first order derivative terms. By construction the Landau frame condition (2.1) is satisfied up to the first order. By including these two tensors, we have two more terms in the stress energy tensor. The corresponding coefficients are the shear and bulk viscosities, respectively. Combining them together,

$$\begin{aligned} T^{\mu\nu}_{N1} &= \varepsilon u^\mu u^\nu + p P^{\mu\nu} + \pi^{\mu\nu}_S, \\ \pi^{\mu\nu}_S &= -\eta(T) \, P^{\mu\rho} P^{\nu\sigma} \sigma_{\rho\sigma} - \zeta(T) \, P^{\mu\nu} (\partial_\delta u^\delta), \end{aligned} \tag{2.8}$$

where η is the shear viscosity and ζ is the bulk viscosity. This is the well known relativistic hydrodynamics of Landau and Lifshitz [23].

We check the condition for the local second law of thermodynamics (2.2).

$$\begin{aligned} (\partial_\mu T^{\mu\nu}_{N1}) u_\nu &= -(\partial_\mu \varepsilon) u^\mu - (\varepsilon + p)(\partial_\mu u^\mu) + (\partial_\mu \pi^{\mu\nu}_S) u_\nu \\ &= -T \partial_\mu (s u^\mu) - \pi^{\mu\nu}_S (\partial_\mu u_\nu) = 0, \end{aligned} \tag{2.9}$$

where we use the ideal order relations including $d\varepsilon = Tds$ and the fact that the first order contribution $\pi^{\mu\nu}_S$ is transverse to the velocity vector $\pi^{\mu\nu}_S u_\nu = 0$. In the first derivative order, the entropy current is positive semidefinite.

$$\begin{aligned} \partial_\mu (s u^\mu) &= -\frac{1}{T} \pi^{\mu\nu}_S (\partial_\mu u_\nu) \\ &= \frac{\eta}{2T} P^{\mu\rho} P^{\nu\sigma} \sigma_{\mu\nu} \sigma_{\rho\sigma} + \frac{\zeta}{T} (\partial_\mu u^\mu)(\partial_\nu u^\nu), \end{aligned} \tag{2.10}$$

where we rewrite $\partial_\mu u_\nu \to (1/2)(\partial_\mu u_\nu + \partial_\nu u_\mu - P_{\mu\nu}(\partial_\delta u^\delta))$ as $\pi^{\mu\nu}_S$ is symmetric and $\pi^{\mu\nu}_S P_{\mu\nu}(\partial_\delta u^\delta) = P^{\mu\rho} P^{\nu\sigma} \sigma_{\rho\sigma} P_{\mu\nu}(\partial_\delta u^\delta) = P^{\rho\sigma} \sigma_{\rho\sigma}(\partial_\delta u^\delta) = 0$ to derive the first term with η. The two terms in the last line are expressed as complete squares. They are positive semidefinite with the conditions

$$\eta \geq 0 \quad \text{and} \quad \zeta \geq 0. \tag{2.11}$$

The transport coefficients are constrained by the entropy current equation.

There are many possible extensions by considering different physical context and different set of symmetries, especially less symmetric cases. We consider some of them in this chapter and the rest of the book. Before moving on, we comment on the non-relativistic limit, where the relativistic velocity, $u^\mu = \gamma_v(1, v^i/c)^T = (1/\sqrt{1 - v_i v^i/c^2})(1, v^i/c)^T$ with T as the transpose and $\gamma_v = 1/\sqrt{1 - v_i v^i/c^2}$, can be expanded for small non-relativistic ones for $v^i/c \ll 1$. One can also introduce the conserved current associated with a particle number, whose relativistic hydrodynamics is slightly more involved, yet straightforward. Interested readers can work out the details and also consult [23]. The particle number current can be readily generalized to the charged hydrodynamics under the influence of electromagnetic fields that we will consider here and later in this book. We visit the hydrodynamics with a smaller number of symmetries.

2.2 Hydrodynamics: New developments

The relativistic hydrodynamics of Landau-Lifshitz is simple and elegant, allowing only two transport coefficients, shear and bulk viscosities, in the first derivative order due to the restrictive Lorentz symmetry. During the period 2011–2014, the relativistic hydrodynamics with some broken symmetries, such as parity [24][25] and boost [26][27], was formulated in the same field theoretic formulations as in §2.1. On one hand, it is surprising to find that these have not been done long time ago. On the other hand, it reveals that physical motivation is crucial to think in certain directions. The formulation of the relativistic hydrodynamics with broken parity and/or boost symmetries have been considered by the recent developments that involve with physical systems with those broken symmetries.

2.2.1 Hydrodynamics with broken parity

Parity transformation is the flip of a sign of one spatial dimension. Thus we can call it a mirror transformation as well. Usually, the presence of ϵ tensor signifies the parity symmetry breaking

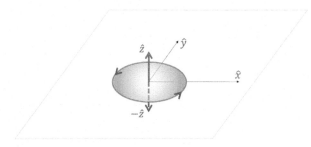

Figure 2.1 A quasi-two dimensional disk in *xy* plane can be attached with an orientation, which can be understood as the arrow attached at its boundary going around counterclockwise (right-handed) viewing from the *z* axis. This orientation is built in the right-handed cross product rule, $\hat{x} \times \hat{y} = \hat{z}$. Thus we see that the area element $d\vec{S} = \hat{z}dxdy$ in *xy* plane is actually a vector and has a direction according to right-handed rule. For a closed surface such as a spherical surface, we assign the positive direction as an outward normal one.

as the fully antisymmetric tensor, for example $\epsilon^{\mu\nu\rho}$ for $2+1$ dimensions, carries all the space-time indices. In hydrodynamics, we usually use the velocities and their derivatives to construct the tensors. For example, we consider $\epsilon^{\mu\nu\rho}u_\mu\partial_\nu u_\rho = u_t(\partial_x u_y - \partial_y u_x) + u_x(\partial_y u_t - \partial_t u_y) + u_y(\partial_t u_x - \partial_x u_t)$, where we use $\epsilon^{txy} = -\epsilon^{yxt} = 1$ and their cyclic permutations. Under the parity transformation $x \to -x$, the quantity changes into $-u_t(\partial_x u_y - \partial_y u_x) - u_x(\partial_y u_t - \partial_t u_y) - u_y(\partial_t u_x - \partial_x u_t) = -\epsilon^{\mu\nu\rho}u_\mu\partial_\nu u_\rho$ with $\partial_x \to -\partial_x$ and $u_x \to -u_x$. Thus, $\epsilon^{\mu\nu\rho}u_\mu\partial_\nu u_\rho \to -\epsilon^{\mu\nu\rho}u_\mu\partial_\nu u_\rho$ under the parity transformation and signifies the parity symmetry breaking as the transformed quantity is not the same as before.

In the ideal order, we cannot construct the totally antisymmetric tensor using the velocity vector. Thus we have the same stress energy tensor (2.3) at this order. In the first derivative order, one can construct another tensors using the epsilon tensor from the already available tensors, the shear tensor $\sigma^{\mu\nu}$ and a derivative of a velocity $\partial_\mu u_\nu$. Thus two independent tensors are

$$\tilde{\sigma}^{\mu\nu} = \epsilon^{\rho\sigma(\mu}u_\rho\sigma_\sigma^{\nu)}, \qquad P^{\mu\nu}\tilde{\Omega} = -P^{\mu\nu}(\epsilon^{\rho\sigma\delta}u_\rho\nabla_\sigma u_\delta). \qquad (2.12)$$

The new tensor $\tilde{\sigma}^{\mu\nu}$ is transverse to the velocity, $\tilde{\sigma}^{\mu\nu}u_\mu = \tilde{\sigma}^{\mu\nu}u_\nu = 0$. The other one is nothing but the vorticity. Thus the first order parity breaking hydrodynamics is described by the stress energy tensor

$$T_{P1}^{\mu\nu} = T_{N1}^{\mu\nu}$$
$$- \eta_H \tilde{\sigma}^{\mu\nu} - \zeta_H P^{\mu\nu} \tilde{\Omega} \,, \tag{2.13}$$

where the first line $T_{N1}^{\mu\nu}$ is given in (2.8). The first transport coefficient in the second line is called Hall viscosity η_H. This Hall viscosity is an interesting quantity that is one of the major subjects of this book. We review this in §2.3 and explicitly compute the Hall viscosity for some simple systems in Chapter 3. The last term is called the vorticity in the so-called vortical frame, and the corresponding coefficient is the Hall bulk viscosity, ζ_H, the Hall analogue of the bulk viscosity. These exciting recent developments can be found in [24][25]. This program has been extended to the second order in the derivative expansion and also to the charged case (with an additional conserved current) that is slightly more complicated.

2.2.2 Hydrodynamics with broken boost

Lorentz symmetry is composed of the rotation symmetry among the spatial coordinates and the boost symmetry that mixes the time and space directions. For a system with rotational symmetry, the stress energy tensor is symmetric $T^{ij} = T^{ji}$ and can be put into a diagonal form by orthogonal transformation, for example, $T^{ij} = \text{diag}\{T^{xx}, T^{yy}\}$ in 2 dimensional space. The same is true for the boost invariance. The tensor is symmetric $T^{i0} = T^{0i}$. The full stress energy tensor can be put into a diagonal form (2.4) using the Lorentz transformations.

If the boost symmetry is broken, the stress energy tensor has antisymmetric components in addition to the symmetric ones, and is no longer symmetric, $T^{0i} \neq T^{i0}$ along the space direction $i = x, y$. The stress energy tensor $T_{N1}^{\mu\nu}$ in (2.8) is symmetric. We can add an antisymmetric part. In the ideal order, there is nothing we can add as mentioned before. In the first derivative order, we have [26] [27]

$$T_{B1}^{\mu\nu} = T_{N1}^{\mu\nu}$$
$$+ \pi_A^{[\mu\nu]} + (u^\mu \pi_A^{[\nu\sigma]} + u^\nu \pi_A^{[\mu\sigma]}) u_\sigma \,, \tag{2.14}$$

where $T_{N1}^{\mu\nu} = \varepsilon u^\mu u^\nu + pP^{\mu\nu} + \pi_S^{(\mu\nu)}$ and $(\mu\nu)$ and $[\mu\nu]$ are symmetric and antisymmetric combinations. Similar to the symmetric part that is transverse to velocity, $\pi_S^{(\mu\nu)} u_\nu = 0$, the set of three terms in the second line of (2.14) is also transverse to u_μ as $\left(\pi_A^{[\mu\nu]} + (u^\mu \pi_A^{[\nu\sigma]} + u^\nu \pi_A^{[\mu\sigma]}) u_\sigma\right) u_\nu = \pi_A^{[\mu\nu]} u_\nu - \pi_A^{[\mu\sigma]} u_\sigma = 0$, where we use $\pi_A^{[\nu\sigma]} u_\sigma u_\nu = 0$. Thus the Landau frame condition, $T^{\mu\nu} u_\nu = -\varepsilon u^\mu$, is satisfied.

We consider the divergence of the entropy current that comes from (2.2). The Landau condition suggests that differentiation by part helps the expression become simple.

$$\begin{aligned}(\partial_\mu T_{B1}^{\mu\nu}) u_\nu &= -T\partial_\mu(su^\mu) + \cdots - \left(\pi_A^{[\mu\nu]} + (u^\mu \pi_A^{[\nu\sigma]} + u^\nu \pi_A^{[\mu\sigma]}) u_\sigma\right)(\partial_\mu u_\nu) \\ &= -T\partial_\mu(su^\mu) + \cdots - \pi_A^{[\mu\nu]}\left(\partial_{[\mu} u_{\nu]} - u_{[\mu} u^\sigma \partial_\sigma u_{\nu]}\right) \\ &= 0,\end{aligned} \quad (2.15)$$

where $\cdots$ contains the symmetric contributions including the shear and bulk viscosities that are independently positive semidefinite. The last term in the first line vanishes as $u^\nu \partial_\mu u_\nu = 0$. The antisymmetric notation means $\partial_{[\mu} u_{\nu]} = (1/2)(\partial_\mu u_\nu - \partial_\nu u_\mu)$. Thus the equation for the entropy current is

$$\partial_\mu(su^\mu) = -\frac{1}{T}\pi_A^{[\mu\nu]}\left(\partial_{[\mu} u_{\nu]} - u_{[\mu} u^\sigma \partial_\sigma u_{\nu]}\right) + \cdots. \quad (2.16)$$

We demand the right-hand side to be positive semidefinite. This can be achieved by the following unique choice.

$$\pi_A^{[\mu\nu]} = -\alpha^{\mu\nu\rho\sigma}\left(\partial_{[\rho} u_{\sigma]} - u_{[\rho} u^\delta \partial_\delta u_{\sigma]}\right) \equiv \alpha^{\mu\nu\rho\sigma} D_{[\rho}^a u_{\sigma]}, \quad (2.17)$$

where $D_\mu^a = \partial_\mu - u_\mu u^\delta \partial_\delta$. Here we note the first two indices $[\mu\nu]$ and last two indices $[\rho\sigma]$ are antisymmetric, while it is symmetric for exchanging those two sets $[\mu\nu] \leftrightarrow [\rho\sigma]$. As we see below, the Hall viscosity is part of the viscosity tensor $\eta^{\mu\nu\rho\sigma}$ that is antisymmetric for exchanging the two sets $(\mu\nu) \leftrightarrow (\rho\sigma)$, while the first two and last two indices are symmetric.

The term $\alpha^{\mu\nu\rho\sigma}$ in (2.17) contains all possible transport coefficients to the first dissipative order. If all the indices are along the spatial directions, this transport coefficient indicates the broken rotation symmetry. Here we focus on the boost symmetry. Time-like direction is captured by the fluid velocity u^μ and space-like one by the projection operator. Thus we can organize the tensor as $\alpha^{\mu\nu\rho\sigma} = \alpha u^{[\mu} P^{\nu]}{}^{[\rho} u^{\sigma]}$. Then, explicit evaluation gives

$$\begin{aligned}\pi_A^{[\mu\nu]} &= -\alpha u^{[\mu} P^{\nu]}{}^{[\rho} u^{\sigma]} \left(\partial_{[\rho} u_{\sigma]} - u_{[\rho} u^\delta \partial_\delta u_{\sigma]} \right) \\ &= -\alpha u^{[\mu} u^\delta \partial_\delta u^{\nu]} = -\alpha u^{[\mu} a^{\nu]} ,\end{aligned} \qquad (2.18)$$

where $a^\nu = u^\delta \partial_\delta u^\nu$ is an acceleration. The entropy current demands $\alpha \geq 0$. Thus α is the *universal* transport coefficient due to the boost symmetry breaking in the neutral system. There might exist other system specific transport coefficients depending on particular systems. There exist other *universal* transport coefficients if the rotation symmetry is broken. As the rotation involves with two spatial direction, one can compute the first order contribution with $\alpha^{\mu\nu\rho\sigma} = \tilde{\alpha} P^{[\mu|[\rho} P^{\sigma]|\nu]}$, where $\tilde{\alpha} \neq \alpha$ and the antisymmetric combinations are only for the pairs $[\mu\nu]$ and $[\rho\sigma]$.

2.3 Hall viscosity: Introduction

General considerations of 2+1 dimensional hydrodynamics with the broken parity in §2.2.1 suggest the presence of the Hall viscosity η_H. We review the Hall viscosity in a slightly different setting so that its physical meaning is more clearly displayed. The reference [28] provides the theory of elasticity that describes the deformation of a solid body under an external force, which can be used to describe fluid as well. The Hall viscosity was first introduced in [29]. These two references are useful for this section.

2.3.1 Another view of stress tensor

Under an external force, a solid or a fluid is deformed, meaning that their shape and volume are changed. Under an infinitesimal deformation, a point of a fluid x^i changed to x'^i. The difference is the displacement vector $\xi^i = x'^i - x^i$. We consider two adjacent points of a fluid, whose distance vector is dx^i. Then the vector joining these two points after the deformation can be written as $dx'^i = dx^i + d\xi^i$. The distances between the two points before and after the deformation are related as

$$dl'^2 = dx'^i dx'^i = (dx^i + d\xi^i)^2$$

$$= dl^2 + \left(\frac{\partial \xi^i}{\partial x^j} + \frac{\partial \xi^j}{\partial x^i} + \frac{\partial \xi^k}{\partial x^i}\frac{\partial \xi^k}{\partial x^j}\right) dx^i dx^j \quad (2.19)$$

$$= dl^2 + 2\xi_{ij} dx^i dx^j,$$

where we use $d\xi^i = (\partial \xi^i/\partial x^j) dx^j$. The collection of the three terms is the so-called strain tensor. For an infinitesimal deformation, we drop the last term as it is the square of a small quantity. Thus the strain tensor

$$\xi_{ij} = \frac{1}{2}\left(\frac{\partial \xi_i}{\partial x^j} + \frac{\partial \xi_j}{\partial x^i}\right) \quad (2.20)$$

plays the role of describing the deformation of the fluid.

When a body (solid or fluid) is not deformed, it is in the state of equilibrium, meaning that the net force on each small portion of the body is zero. When small deformation happens, the fluid tends to return to the original state of the equilibrium. This internal forces that derive the body into equilibrium are called internal stress. Quantitatively, the total force on a small volume dV is given by $\int \vec{f} dV$ with the force per unit volume f_i. The length scale of the molecular forces of the body is small compared to that of the macroscopic elastic theory. Thus the internal stresses responsible for the macroscopic theory are due to "near-action" forces that act from any point to its neighboring points. Thus, the force exerts on any parts of the body by the surrounding parts act only on the surface of that part. Thus the volume integral can be transformed into a surface integral. Reminded that there is equivalent description using the divergence theorem of vector calculus, we rewrite the force into a divergence of a second rank tensor as $f_i = \partial T_{ji}/\partial x^j$.

$$\int f_i dV = \int \frac{\partial T_{ji}}{\partial x^j} dV = \oint T_{ji} dS^j, \quad (2.21)$$

where dS^j is the component of the surface element vector $d\vec{S}$ pointing outward normal of the surface. As the notation indicates T_{ij} is the stress tensor with only spatial components. From the derivation, $T_{ji}dS^j$ is the i-th component of the force acting on the surface element $d\vec{S}$. This means that T_{ji} is the i-th component of the force acting on the unit area perpendicular to x^j axis.

Let us consider the stress tensor of a body undergoing uniform compression from all sides, a pressure p (force per unit area) on its surface along the inward direction. The force $-pdS_i$ acts on the surface dS_i. Then $T_{ji}dS_j = -pdS_i = -p\delta_{ji}dS_j$. Thus the stress tensor for the uniform expansion is

$$T_{ij} = -p\delta_{ij}. \tag{2.22}$$

At this point we consider the work done by the internal stress and make a contact with the thermodynamics. Multiplying the force by an infinitesimal displacement, we consider the work $\delta w = f_i \delta \xi^i$.

$$\begin{aligned} \int f_i \delta \xi^i dV &= \int \frac{\partial T_{ji}}{\partial x^j} \delta \xi^i dV = \oint T_{ji} \delta \xi^i dS^j - \int T^{ji} \frac{\partial (\delta \xi_i)}{\partial x^j} dV \\ &= -\frac{1}{2} \int T^{ji} \delta \left(\frac{\partial \xi_i}{\partial x^j} + \frac{\partial \xi_j}{\partial x^i} \right) dV = -\int T^{ji} \delta \xi_{ji} dV, \end{aligned} \tag{2.23}$$

where we perform the surface integral at the infinity where deformation vanishes and use the symmetric properties of stress tensor with the rotation symmetry. Thus the work is given by the change due to the strain tensor as

$$\delta w = -T^{ij} \delta \xi_{ij}. \tag{2.24}$$

If the deformation is small, the body returns to the original undeformed state. Such deformations are called elastic, which is assumed to be the case here. (If there remains residual deformation after a large deformation, it is called plastic.) Furthermore, if the process of the deformation is slow enough the thermodynamic equilibrium is established at every moment. Then the thermodynamic process is also called reversible. An infinitesimal change $d\varepsilon$ of the internal energy is the difference between the heat Tds acquired by the reversible process and work done dw.

$$d\varepsilon = Tds - dw = Tds + T^{ij}d\xi_{ij}. \tag{2.25}$$

This is the thermodynamic relation of deformed bodies. For hydrostatic compression, the stress is given in (2.22). Thus

$$d\varepsilon = Tds - p\delta^{ij}d\xi_{ij} = Tds - pdv, \tag{2.26}$$

where the diagonal contribution of the strain tensor the relative volume change, $\delta^{ji}d\xi_{ij} = d\xi_{ii} = dv$. Here we consider the unit volume and thus, ε and s are energy and entropy densities.

2.3.1.1 Free energies

By taking the Legendre transformation from the internal energy to generalized free energy $F = \varepsilon - Ts$, we find $dF = -sdT + T^{ji}d\xi_{ij}$, where the independent variables are T and ξ_{ij}. Thus we can get the stress tensor by differentiating the internal energy or the free energy with respect to the strain tensor for constant entropy or temperature respectively.

$$T^{ij} = \left(\frac{\partial \varepsilon}{\partial \xi_{ij}}\right)_s = \left(\frac{\partial F}{\partial \xi_{ij}}\right)_T. \tag{2.27}$$

The thermodynamic potential $G = \varepsilon - Ts - T^{ji}\xi_{ij}$ is the generalization of the Gibbs free energy $G = \varepsilon - Ts + pv$. Its independent variables are T, T^{ij}, and the strain tensor can be obtained as $\xi_{ij} = -(\partial G/\partial T^{ji})_T$.

The change in free energy in isothermal compression of a crystal is a quadratic function of the strain tensor. The general form of the free energy of the deformed crystal is

$$F = \frac{1}{2}\lambda_{ijkl}\xi^{ij}\xi^{kl}, \tag{2.28}$$

where the coefficient λ_{ijkl} is the elastic modulus tensor of rank four. Due to the symmetric combination of two symmetric strain tensor, $\lambda_{ijkl} = \lambda_{jikl} = \lambda_{ijlk} = \lambda_{klij}$. There are only 6 independent components with the symmetry for $d = 2$, as two symmetric indices ij have $d \times (d+1)/2 = 3$ independent components, whose symmetric combination with kl is also $3 \times 4/2 = 6$. The stress tensor can be computed using (2.27).

$$T^{ij} = \frac{\partial F}{\partial \xi_{ij}} = \lambda_{ijkl}\xi^{kl} . \tag{2.29}$$

For rotational invariant systems, one can constrain these four indices tensors efficiently. $\lambda_{ijkl} = \lambda \delta_{ij}\delta_{kl}$, where $\lambda = v(\partial p/\partial v)$ is the elastic modulus that describes the change of pressure as the volume v varies.

Until now we assume that the deformation is reversible with the thermodynamic process with infinitesimal speed. Actual motion has finite velocities and the body is not in equilibrium at every moment of the process. Thus actual process is irreversible and mechanical energy is dissipated into heat eventually through the thermal conduction and the internal friction. We consider the latter, the viscosity, here.

The process of internal friction occur in a body when different parts of the body (fluid or solid) move with different velocities, so that there is a relative motion between various parts of the fluid with dissipative forces. This means that resulting stress involves with the space derivative of relative velocity $\partial \dot{\xi}_i/\partial x^j$, which is assumed to be small. We note that there should be no friction and the stress must vanish when the body undergoes uniform translation or rotation motions. These can be described as the constant velocity $\partial_t \xi^i = $ const. and the uniform rotation $\partial_t \vec{\xi} = \omega \times \vec{r}$ or $\partial_t \xi^i = \epsilon^{ijk}\omega_j x_k$ with the angular velocity $\vec{\omega}$. The latter condition puts constraint on the form of stress as

$$\dot{\xi}_{ij} = \frac{1}{2}\left(\frac{\partial \dot{\xi}_i}{\partial x^j} + \frac{\partial \dot{\xi}_j}{\partial x^i}\right), \tag{2.30}$$

which satisfies $(\partial \dot{\xi}_i/\partial x^j) + (\partial \dot{\xi}_j/\partial x^i) = \epsilon^{ikl}\omega_k(\partial x_l/\partial x^j) + \epsilon^{jkl}\omega_k(\partial x_l/\partial x^i) = (\epsilon^{ikj} + \epsilon^{jki})w_k = 0$ for the uniform rotation.

Anticipating the dissipative function similar to the free energy is quadratic function of $\dot{\xi}_{ij}$, the general form is given by

$$\tilde{F} = \frac{1}{2}\eta_{ijkl}\dot{\xi}^{ij}\dot{\xi}^{kl} . \tag{2.31}$$

This symmetric form constrains the shear tensor η_{ijkl} as $\eta_{ijkl} = \eta_{jikl} = \eta_{ijlk} = \eta_{klij}$. This condition is valid for the systems with rotational symmetry. Then the viscosity tensor can be put into $\eta_{ijkl} = -\eta(\delta_{ik}\delta_{jl} + \delta_{il}\delta_{jk}) - (\zeta - \eta)\delta_{ij}\delta_{kl}$ with shear η and bulk ζ viscosities. If the parity symmetry is broken, there would be more transport coefficients that exist in the presence of rotational symmetry, yet are invisible in this symmetric and quadratic form as we see below. Similar to (2.28), the free energy gives the stress tensor in terms of shear tensor and the time derivative of strain tensor $\dot{\xi}_{ij}$.

$$T^{ij} = \frac{\partial \tilde{F}}{\partial \dot{\xi}_{ij}} = \eta_{ijkl}\dot{\xi}^{kl}. \tag{2.32}$$

Combining all together

$$T_{ij} = p\delta_{ij} + \lambda_{ijkl}\xi_{kl} + \eta_{ijkl}\dot{\xi}_{kl}, \tag{2.33}$$

where p is pressure, λ_{ijkl} the elastic modulus tensor, and η_{ijkl} the viscosity tensor.

2.3.2 Hall viscosity: A geometric picture

For systems with broken parity, for example in the presence of a background magnetic field, the odd part of the shear tensor is allowed as we have introduced in §2.2.1. Hall viscosity was introduced in [29] and its explicit form in 2 spatial dimensions was explicitly written down in [30].

In 2 spatial dimensions there is a natural basis for 4-th rank tensor, the Pauli matrices, that is already introduced in (1.42) in the previous chapter.

$$\sigma^0 = \begin{pmatrix} 1 & 0 \\ 0 & 1 \end{pmatrix}, \quad \sigma^1 = \begin{pmatrix} 0 & 1 \\ 1 & 0 \end{pmatrix}, \quad \sigma^2 = \begin{pmatrix} 0 & -i \\ i & 0 \end{pmatrix}, \quad \sigma^3 = \begin{pmatrix} 1 & 0 \\ 0 & -1 \end{pmatrix}.$$

As the σ^2 is imaginary and $i\sigma^2$ plays the role of rotation generator, we use the other three Pauli matrices to construct the real fourth rank tensor as $\boldsymbol{\eta} = \eta_{ab}\sigma^a \otimes \sigma^b$ for $a, b \in 0, 1, 3$ with the tensor product $\otimes$.

In a component form,

$$\eta_{ijkl} = \eta_{ab}\sigma_{ij}^a \sigma_{kl}^b, \quad \text{where } a,b \in 0,1,3. \tag{2.34}$$

The space of the tensor is 9 dimensional, which is decomposed into 6 dimensional symmetric space and 3 dimensional antisymmetric space. There are three dimensional subspace that are isotropic, which is invariant under the rotation operator $\sigma^2 \otimes \sigma^2$. This can be checked using the properties of Pauli matrices, $\sigma^a \sigma^b = \delta^{ab} I + i\epsilon^{abc}\sigma^c$. For example, $(\sigma^2 \otimes \sigma^2)(\sigma^3 \otimes \sigma^3) = -(\sigma^1 \otimes \sigma^1)$. Thus the tensor $(\sigma^3 \otimes \sigma^3)$ is not invariant under the rotation. One can construct the following symmetric isotropic space.

$$\boldsymbol{\eta} = \eta(\sigma^1 \otimes \sigma^3 + \sigma^3 \otimes \sigma^1) + \zeta \sigma^0 \otimes \sigma^0, \tag{2.35}$$

which is invariant under the the action of $\sigma^2 \otimes \sigma^2$. The η and ζ are the viscosities. This observation is assuring for the relevance of the Pauli matrices in the context of hydrodynamics and shear tensor.

It turns out that there is the odd isotropic part that picks − sign under the rotation. There are only three possible non-zero components in 2 spatial dimensions. The first two and last two indices of are symmetric, $\eta_{ijkl} = \eta_{jikl} = \eta_{ijlk}$. Thus there are $2 \times 3/2 = 3$ possible combinations. Those three combinations are antisymmetric with each other $\eta_{ijkl} = -\eta_{klij}$, meaning that there are $3 \times 2/2 = 3$ nontrivial components. Explicitly,

$$\boldsymbol{\eta}^A = \eta_H(\sigma^1 \otimes \sigma^3 - \sigma^3 \otimes \sigma^1)$$
$$= \eta_H \left[\begin{pmatrix} 0 & 1 \\ 1 & 0 \end{pmatrix} \otimes \begin{pmatrix} 1 & 0 \\ 0 & -1 \end{pmatrix} - \begin{pmatrix} 1 & 0 \\ 0 & -1 \end{pmatrix} \otimes \begin{pmatrix} 0 & 1 \\ 1 & 0 \end{pmatrix} \right]. \tag{2.36}$$

The three non-trivial combinations are

$$\begin{aligned}\eta^A_{1112} &= \eta^A_{1121} = -\eta^A_{1211} = -\eta^A_{2111} = -\eta_H, \\ \eta^A_{1222} &= -\eta_H, \qquad \eta^A_{1122} = 0.\end{aligned} \tag{2.37}$$

These can be put into a general expression for two dimensional indices $i,j = 1,2$,

$$\eta^A_{ijkl} = -\eta^A_{klij} = -\frac{\eta_H}{2}(\epsilon_{ik}\delta_{jl} + \epsilon_{jl}\delta_{ik} + \epsilon_{il}\delta_{jk} + \epsilon_{jk}\delta_{il}). \tag{2.38}$$

 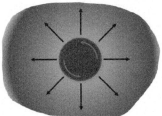

Figure 2.2 A solid cylinder rotating counterclockwise inside a fluid. Left: The direction of the shear viscosity, the black arrows, acts against that of the cylinder. Right: The Hall viscosity acts perpendicular to the motion of the middle cylinder. Despite its name, it does not produce any dissipation.

One can check that this quantity is symmetric under the exchange of the indices i and j as well as k and l. On the other hand, it is antisymmetric under the exchange of (ij) and (kl). Thus the Hall viscosity η_H reveals itself in the antisymmetric part of the shear tensor in the context of fluid dynamics (2.38).

2.3.2.1 Geometric picture

We offer a geometric description of the Hall viscosity in a realistic setting. Let us imagine a cylinder surrounded by a fluid. The cylinder is rotating counterclockwise with a constant frequency in the middle of the fluid as in Fig. 2.2. The well known shear viscosity acts as a clockwise force along the surface of the cylinder, anti-parallel to the direction of the motion, trying to slow down the cylinder as the work done by the fluid is negative. The situation is described in the left figure. On the other hand, Hall viscosity produces a force perpendicular to the rotating direction of the cylinder, outward or inward depending on the direction of the cylinder rotation. See the right side of Fig. 2.2. In particular, the Hall viscosity does not provide a dissipation to the motion of the cylinder.

One can compute the change of energy due to the viscosity tensor under a small deformation of the fluid. This is described by $\delta \varepsilon = -T_{ij} \delta \xi_{ij}$ given in (2.24). After using the constitute relations, one obtains

$$T\dot{s} = \eta_{ijkl}\dot{\xi}_{ij}\dot{\xi}_{kl} + \eta^A_{ijkl}\dot{\xi}_{ij}\dot{\xi}_{kl}, \qquad (2.39)$$

where s is the entropy density. From the fact that the entropy is positive semidefinite, one can constrain the value of the shear and bulk viscosities in η_{ijkl}, especially in the presence of the rotational invariance. Note that the second term $\eta_{ijkl}^A \dot{\xi}_{ij} \dot{\xi}_{kl}$ in (2.39) actually vanishes because the strain rates are symmetric under the exchange of the indices ij and kl, while η_{ijkl}^A is antisymmetric as in (2.38). This demonstrates that η_H is dissipationless as mentioned above. Thus, the Hall viscosity can exist even at zero temperature. It has a better chance to be observed in low temperature experiments because other dissipative effects are suppressed.

We compute the Hall viscosity explicitly for a quantum Hall systems in Chapter 3 and draw some exciting physical insights there. In later chapters, we also introduce this Hall viscosity in the chiral magnetic systems with skyrmions in Chapter 7.

2.4 Charged hydrodynamics

In this section we generalize the neutral hydrodynamics considered in §2.1 to include an additional conserved current $\partial_\mu J^\mu = 0$, where J^μ is $U(1)$ symmetry current, such as electric charge current or particle number current. Then the conservation equations are

$$\partial_\mu T^{\mu\nu} = F^{\nu\sigma} J_\sigma, \qquad \partial_\mu J^\mu = 0. \tag{2.40}$$

Here we also allow the fluid to couple to an external non-dynamical gauge field A_μ by introducing the coupling $F^{\nu\sigma} J_\sigma$ to the equation of the stress energy tensor. The gauge invariant field strength is given by $F_{\mu\nu} = \partial_\mu A_\nu - \partial_\nu A_\mu$. The gauge fields, electric and magnetic fields, are among the main tools to manipulate the physical systems of interest. One can also couple the external background metric $g_{\mu\nu}$ by generalizing the partial derivative ∂_μ to the covariant derivative ∇_μ.

By including the $U(1)$ current J^μ, we also introduce an additional variable μ, the chemical potential, in addition to the usual macroscopic parameters, temperature T and the velocity u^μ with $u^\mu u_\mu = -1$. For given $F^{\mu\nu}$, there are same number of conservation equations and the unknown variables, and the system can be solved. The Landau frame condition (2.1) is generalized to

$$T^{\mu\nu}u_\nu = -\varepsilon u^\mu, \qquad J^\mu u_\mu = -\rho, \qquad (2.41)$$

where ρ is the number (charge) density of $U(1)$ symmetry. The local second law of thermodynamics (2.2) becomes

$$(\partial_\mu T^{\mu\nu})u_\nu + \mu \partial_\mu J^\mu = u_\nu F^{\nu\sigma} J_\sigma. \qquad (2.42)$$

This is a scalar combination of the conservation equations given in (2.40).

As the discussion becomes quickly complicated, we consider the case without the external gauge field $F^{\mu\nu} = 0$ for the rest of this subsection. We list the results in a full detail including the gauge fields in §2.4.1 and §2.4.2.

The stress energy tensor of the Lorentz invariant hydrodynamics does not seem to change. Nevertheless (2.2) is modified due to the different constitute relation $w = \varepsilon + p = Ts + \mu\rho$. At the ideal order, the Landau frame condition is trivially satisfied for $T_{J0}^{\mu\nu} = \varepsilon u^\mu u^\nu + p P^{\mu\nu}$ and the $U(1)$ conservation equation is nothing but the continuity equation $\partial_\mu J^\mu = \partial_0 J^0 + \partial_i J^i = 0$. The condition (2.2) goes as

$$\begin{aligned}(\partial_\mu T_{J0}^{\mu\nu})u_\nu \\ &= -(\partial_\mu \varepsilon)u^\mu - (\varepsilon + p)(\partial_\mu u^\mu) \\ &= -(\partial_\mu \varepsilon)u^\mu - (Ts + \mu\rho)(\partial_\mu u^\mu) \qquad (2.43)\\ &= -T\partial_\mu(su^\mu) - \big(\partial_\mu \varepsilon - T(\partial_\mu s) - \mu(\partial_\mu \rho)\big)u^\mu - \mu\partial_\mu(\rho u^\mu) \\ &= 0, \end{aligned}$$

which gives the conservation of the entropy current $\partial_\mu(su^\mu) = 0$ and the thermodynamic relations $d\varepsilon = Tds + \mu d\rho$ for extensive variables. The latter also gives $dp = sdT + \rho d\mu$ by differentiating the constitute relation. Note we use the ideal conservation equation $\partial_\mu(\rho u^\mu) = 0$. Thus the $U(1)$ conservation equation plays role to establish the local second law of thermodynamics. This gives the explanation of the second term in (2.42).

At the first derivative order, we can write the stress tensor and the the current as

$$T_{J1}^{\mu\nu} = \varepsilon u^\mu u^\nu + pP^{\mu\nu} + \pi_S^{\mu\nu},$$
$$J^\mu = \rho u^\mu + \nu^\mu,$$
(2.44)

where the corrections are required to satisfy the conditions

$$\pi_S^{\mu\nu} u_\nu = 0, \qquad \nu^\mu u_\mu = 0,$$
(2.45)

to satisfy the Landau frame condition.

As mentioned above, there is another natural fluid frame we can consider, the Eckart frame, which is defined by the particle flow. To see the difference more clearly, we put the stress energy tensor in a different form, respecting the tensor structures explicitly with the time-like direction u^μ of the fluid and the transverse direction $P^{\mu\nu}$.

$$T^{\mu\nu} = \varepsilon_0 u^\mu u^\nu + p_0 P^{\mu\nu} + (k^\mu u^\nu + k^\nu u^\mu) + t^{\mu\nu},$$
$$J^\mu = \rho_0 u^\mu + j^\mu,$$
(2.46)

where ε_0, p_0, and ρ_0 are scalar quantities, k^μ and j^μ transverse vectors satisfying $k^\mu u_\mu = j^\mu u_\mu = 0$, and $t^{\mu\nu}$ transverse tensor satisfying $t^{\mu\nu} u_\nu = 0$. In this language, the Landau frame condition is nothing but $k^\mu = 0$. Two quantities, ε_0 and ρ_0, are fixed, while p_0, $t^{\mu\nu}$ and j^μ are modified by the derivative corrections. For the Eckart frame, we impose $j^\mu \equiv 0$. Thus there is no corrections to the current J^μ at the derivative orders, while the energy density ε gets corrected as $T^{\mu\nu} u_\nu = -\varepsilon_0 u^\mu - k^\mu$.

The condition (2.2) goes as (similar to the computation in (2.43))

$$(\partial_\mu T_{J1}^{\mu\nu}) u_\nu$$
$$= -(\partial_\mu \varepsilon) u^\mu - (\varepsilon + p)(\partial_\mu u^\mu) + (\partial_\mu \pi_S^{\mu\nu}) u_\nu$$
$$= -T \partial_\mu (s u^\mu) + \mu \partial_\mu (\nu^\mu) - \pi_S^{\mu\nu} (\partial_\mu u_\nu)$$
$$= -T \partial_\mu \left(s u^\mu - \frac{\mu}{T} \nu^\mu \right) - T \nu^\mu \partial_\mu \left(\frac{\mu}{T} \right) - \pi_S^{\mu\nu} (\partial_\mu u_\nu)$$
$$= 0,$$
(2.47)

where we use ideal order relation $d\varepsilon = Tds + \mu d\rho$ and the continuity equation $\mu\partial_\mu(\rho u^\mu) = -\mu\partial_\mu(\nu^\mu)$. The last line suggests that the following entropy current

$$s^\mu \equiv su^\mu - \frac{\mu}{T}\nu^\mu \tag{2.48}$$

satisfies the equation

$$\partial_\mu s^\mu = -\frac{1}{T}\pi_S^{\mu\nu}(\partial_\mu u_\nu) - \nu^\mu \partial_\mu\left(\frac{\mu}{T}\right) \geq 0. \tag{2.49}$$

The right-hand side is necessarily positive semidefinite. Thus we try to form complete squares for these two terms according to the given symmetries, which is the Lorentz symmetry. The first term on the right-hand side suggests that we rewrite $\partial_\mu u_\nu$ into two parts, a second rank symmetric traceless tensor $(1/2)(\partial_\mu u_\nu + \partial_\nu u_\mu - P_{\mu\nu}(\partial_\delta u^\delta))$ and a trace part $\partial_\delta u^\delta$. We already have encounter this form in (2.8). Thus the result for the stress energy tensor is the same as in (2.10) and $\eta \geq 0$ and $\zeta \geq 0$ as in (2.11).

For the current part in the second term of (2.49), we can set

$$\nu^\mu = -\kappa_2 TP^{\mu\nu}\partial_\nu\left(\frac{\mu}{T}\right) \tag{2.50}$$

with a new transport coefficient κ_2, which can be rescaled by other parameters such as temperature T as $\kappa_2 = \kappa_1/T \times (\cdots) = T\kappa_3 \times (\cdots)$ along with other constants represented by $\cdots$. For the particle number conservation, it is related to the thermal conductivity. There can be additional contribution, charge conductivity σ, if there are external gauge fields associated with a charge current. Thus,

$$-\nu^\mu \partial_\mu\left(\frac{\mu}{T}\right) = \kappa_2 TP^{\mu\nu}\partial_\mu\left(\frac{\mu}{T}\right)\partial_\nu\left(\frac{\mu}{T}\right) \geq 0, \tag{2.51}$$

where $\kappa_2 \geq 0$. Once we have the derivative corrections, the thermal conduction can be further refined. We can have the energy flux by a pure thermal conduction without particle flux, $J^i = \rho u^i + \nu^i = 0$. We come back to more sophisticated descriptions with a systematic approach for the various possible thermo-electromagnetic transport coefficients that include the thermal, electric, and magnetic properties in §2.6.

2.4.1 With broken parity

Here we revisit the charged hydrodynamics from a different angle. Hydrodynamics is advertised as a universal theory that respects the underlying symmetries. Thus if we are able to construct the possible scalar, vector, and tensor structures, they should come into play in view of (2.44) or (2.46).

The available parameters are u^μ, T, and μ for the charged case. Ideal order is already determined. In the first derivative order, there are three scalars, $u^\mu \partial_\mu T$, $u^\mu \partial_\mu(\mu/T)$, and $\partial_\mu u^\mu$. We have two scalar equations, $\partial_\mu J^\mu = 0$ and $(\partial_\mu T^{\mu\nu}) u_\nu = 0$. Thus there is only one independent scalar, that we choose $\partial_\mu u^\mu$. There are two additional pseudoscalars, $\tilde{B} = -(1/2)\epsilon^{\mu\nu\rho} u_\mu F_{\nu\rho}$ and $\tilde{\Omega} = -\epsilon^{\mu\nu\rho} u_\mu (\partial_\nu u_\rho)$, which are not constrained. As the energy is fixed in the Landau frame, these can enter into the pressure as

$$p \to p - \zeta(\partial_\delta u^\delta) - \tilde{\chi}_\Omega \tilde{\Omega} - \tilde{\chi}_B \tilde{B} . \tag{2.52}$$

For the second rank tensors, there are three candidates, $u_\mu \partial_\nu T$, $u_\mu \partial_\nu (\mu/T)$, and $\partial_\mu u_\nu$. The first two do not have any transverse projection components, for example $P^{\rho\mu} P^{\sigma\nu} u_\mu \partial_\nu T = 0$. Nontrivial transverse second rank tensors come from various traceless symmetric and antisymmetric projections of $\partial_\mu u_\nu$. They are

$$\pi_{\mu\nu} = \eta_{\mu\nu\rho\sigma}(\partial^\rho u^\sigma) + \eta^A_{\mu\nu\rho\sigma}(\partial^\rho u^\sigma)$$
$$= -\eta \sigma_{\mu\nu} - \eta_H \tilde{\sigma}_{\mu\nu} . \tag{2.53}$$

Here we use

$$\eta_{\mu\nu\rho\sigma} = \eta_{((\mu\nu),(\rho\sigma))} = -\eta\left(P_{\mu\rho} P_{\nu\sigma} + P_{\nu\rho} P_{\mu\sigma} - P_{\mu\nu} P_{\rho\sigma}\right) \tag{2.54}$$

for the Lorentz invariant case, and

$$\eta^A_{\mu\nu\rho\sigma} = \eta^A_{[(\mu\nu),(\rho\sigma)]}$$
$$= \frac{\eta_H}{2}\left(P_{\mu\rho} Q_{\nu\sigma} + P_{\nu\rho} Q_{\mu\sigma} + Q_{\mu\rho} P_{\nu\sigma} + Q_{\nu\rho} P_{\mu\sigma}\right) \tag{2.55}$$

for the systems with broken parity by introducing a transverse antisymmetric projection tensor $Q_{\mu\nu} = \epsilon_{\mu\nu\rho} u^\rho$.

For a vector, the first derivative order candidates are $\partial_\mu T, \partial_\mu(\mu/T), E^\mu = F^{\mu\nu}u_\nu$, and $u^\nu\partial_\nu u^\mu$. We can get four transverse vectors by contracting them with $P^{\mu\nu}$ and four pseudo vectors by contracting them with $Q^{\mu\nu}$. From the equation $\partial_\mu T^{\mu\nu} = F^{\nu\sigma}J_\sigma$, we can also construct a vector equation by contracting them with $P^{\mu\nu}$ and a pseudo vector one by contracting them with $Q^{\mu\nu}$. Thus there are 3 vectors and 3 pseudo vectors. We choose the first three candidates. Thus

$$\begin{aligned}\nu^\mu &= P^{\mu\nu}\left(\sigma_2 V_\nu + \chi_E E_\nu + \chi_T \partial_\nu T\right) \\ &\quad - Q^{\mu\nu}\left(\tilde\sigma_2 V_\nu + \tilde\chi_E E_\nu + \tilde\chi_T \partial_\nu T\right),\end{aligned} \quad (2.56)$$

where $V^\mu = E^\mu - TP^{\mu\nu}\partial_\nu(\mu/T)$ is a combination of two vectors that turns out to be useful. At the end of this subsection, we show that $\chi_E = \chi_T = 0$.

Combining all together, the results for the charged hydrodynamics with broken parity up to the first derivative order become

$$\begin{aligned}T^{\mu\nu} &= \varepsilon u^\mu u^\nu + \left(p - \zeta(\partial_\delta u^\delta) - \tilde\chi_\Omega \tilde\Omega - \tilde\chi_B \tilde B\right)P^{\mu\nu} \\ &\quad - \eta\sigma^{\mu\nu} - \eta_H \tilde\sigma^{\mu\nu}, \\ J^\mu &= \rho u^\mu + \tilde\chi_E \tilde E^\mu + \tilde\chi_T \epsilon^{\mu\nu\rho}u_\nu\partial_\rho T \\ &\quad + \sigma_2 V^\mu + \tilde\sigma_2 \tilde V^\mu,\end{aligned} \quad (2.57)$$

where $\tilde E^\mu = \epsilon^{\mu\nu\rho}u_\nu E_\rho$ and $\tilde V^\mu = \epsilon^{\mu\nu\rho}u_\nu V_\rho$. Here we put the expressions in a way to manifest the Landau frame condition, where the pressure is corrected as it is transverse to the energy flow and the current J^μ is also corrected at the derivative order.

Note that we treat the field strength tensor $F^{\mu\nu}$ as the first order as it is the derivative of the gauge field A_μ. This is true when the magnetic field B is small as well as the vorticity Ω is small. In general, thermodynamics can have non-zero static magnetic field B and vorticity Ω at the equilibrium as they are both parity-breaking effects and do not change the energy of the system. If their values are large, we need a separate treatment.

Here we adapt the so-called magnetovortical frame, where thermodynamic derivatives are evaluated at $B = 0$ and $\Omega = 0$. Then the constitute relations $\varepsilon + p = sT + \mu\rho$ and the corresponding exact differentials still hold. Detailed analysis reveal that the three transport coefficients must satisfy the conditions $\eta \geq 0$, $\zeta \geq 0, \sigma_2 \geq 0$. The Hall viscosity η_H and $\tilde{\sigma}_2$ are not constrained. The remaining parameters $\tilde{\chi}_\Omega$, $\tilde{\chi}_B$, $\tilde{\chi}_E$, $\tilde{\chi}_T$ are not independent, but specified by the thermodynamic functions exist in the system. Readers can refer to [24] for more details.

In 2+1 dimensions, the antisymmetric field strength tensor $F^{\mu\nu}$ is decomposed into a pseudo scalar, the magnetic field $B = -(1/2)\epsilon^{\mu\nu\rho}u_\mu F_{\nu\rho}$, and a transverse (two components) vector, the electric field $E^\mu = F^{\mu\nu}u_\nu$. Another possible combination $B^\mu = \epsilon^{\mu\nu\rho}F_{\nu\rho}$ has the following decomposition, $B^\mu = \tilde{E}^\mu + u^\mu B$ according to the transverse and the parallel directions of the velocity u_μ. Thus we include E^μ with parity invariance and $\tilde{E}^\mu$ and B for the cases with broken parity. As vectors cannot be used to correct the stress energy tensor for the Landau frame, only B can be used in $T^{\mu\nu}$. On the other hand, J^μ has the three different contributions in terms of E^μ and $\tilde{E}^\mu$ up to the first derivative order.

$$J^\mu \supset \sigma_2 E^\mu + (\tilde{\sigma}_2 + \tilde{\chi}_E)\epsilon^{\mu\nu\rho}u_\nu E_\rho, \tag{2.58}$$

The first contribution is the usual electric conductivity responsible for the Ohm's law. The second is the Hall conductivity which describes the generated electric current by the electric field perpendicular to the direction of current. $\tilde{\chi}_E$ is also present as $\tilde{V}^\mu$ and $\tilde{E}^\mu$ are independent combinations given in (2.57).

Before closing this subsection, we show that $\chi_E = \chi_T = 0$ using the constraint of the entropy current equation. For simplicity, we only consider the parts that preserve the parity symmetry. As advertised, all possible terms can contribute to the entropy current. Then (2.48) and (2.49) are modified with the presence of the background electric field and temperature gradient.

$$S^\mu = su^\mu - \frac{\mu}{T}\nu^\mu - \frac{u_\nu}{T}\tau^{\mu\nu}$$
$$+ b_0(\partial^\delta u_\delta)u^\mu + b_1(u^\delta \partial_\delta)u^\mu + b_2 E^\mu + b_3 \frac{V^\mu}{T}, \tag{2.59}$$

where the first line provides us the familiar expression for μ and $\tau^{\mu\nu}$, while the second line contains all the vector data. The divergence of this entropy current yields the first order data and also sum of the scalars with genuine second order derivatives, that cannot be decomposed into first order derivative terms. The genuine second order derivatives terms cannot be recast into complete squares and thus should vanish. Then,

$$\partial_\mu S^\mu = -\left(\partial_\mu\left(\frac{\mu}{T}\right) - \frac{E_\mu}{T}\right)\nu^\mu - \partial_\mu\left(\frac{u_\nu}{T}\right)\tau^{\mu\nu}$$
$$+ (b_0 + b_1)(u^\delta \partial_\delta \partial_\mu)u^\mu \qquad (2.60)$$
$$+ \left(b_2 - \frac{b_3}{T}\right)\partial_\mu E^\mu + b_3 P^{\mu\nu} \partial_\mu \partial_\nu\left(\frac{\mu}{T}\right) + \cdots,$$

we use the constitutive relations and the generalized local second law equation $(\partial_\mu T^{\mu\nu})u_\nu + E^\mu J_\mu + \mu \partial_\mu J^\mu = -sT(\partial_\mu u^\mu) - Tu^\mu \partial_\mu s + E^\mu \nu_\mu + u_\nu \partial_\mu \tau^{\mu\nu} + \mu \partial_\mu \nu^\mu = 0$, which is used to remove the combination of the terms of $\partial_\mu \nu^\mu$ and $\partial_\mu \tau^{\mu\nu}$. Here we see two combinations contribute to the first order transport coefficients. $V^\mu = E^\mu - TP^{\mu\nu} \partial_\nu(\mu/T)$ is responsible for the conductivity σ_2 in the current J^μ. The other one is $\partial_\mu u_\nu$ that is responsible for η, ζ in the stress energy $T^{\mu\nu}$. Note that the term with the derivative of temperature vanishes identically $\partial_\mu(u_\nu/T)\tau^{\mu\nu} = (1/T)(\partial_\mu u_\nu)\tau^{\mu\nu}$ as we impose the Landau frame condition, $\tau^{\mu\nu} u_\nu = 0$.

The second line in (2.60) tells us that $b_0 + b_1 = b_2 = b_3 = 0$. If we introduce the background metric, we can show that $b_1 = 0$ and thus $b_0 = 0$. The remaining terms $\cdots$ are proportional to these coefficients and thus vanish as well. These coefficients b_0, b_1, b_2, b_3 are related to the coefficients χ_E and χ_T. Thus $\chi_E = \chi_T = 0$ in (2.56). The connection between the coefficients b_0, b_1, b_2, b_3 and χ_E, χ_T can be more concrete, yet it requires to introduce more ingredients that we need here.

2.4.2 With broken boost

Here we also revisit the charged hydrodynamics for broken boost invariance. For simplicity, we set $F_{\mu\nu} = 0$. In addition to the stress energy tensor (2.14) given in §2.2.2, we add the $U(1)$ conserved current given in (2.44) along with the condition (2.45).

$$J^\mu = \rho u^\mu + \nu^\mu, \qquad \nu^\mu u_\mu = 0. \qquad (2.61)$$

We impose the local condition (2.42), whose combination is motivated with explicit computations above. Similar to the previous cases, we have

$$0 = (\partial_\mu T^{\mu\nu})u_\nu + \mu\partial_\mu J^\mu$$
$$= -T\partial_\mu s^\mu - \pi_A^{[\mu\nu]}(\partial_{[\mu}u_{\nu]} - u_{[\mu}u^\sigma\partial_\sigma u_{\nu]}) - T\nu^\mu\partial_\mu\left(\frac{\mu}{T}\right), \quad (2.62)$$

where the entropy current s^μ is given by (2.48), the middle term evaluated in (2.15), and the last term evaluated in (2.47). The condition for the broken boost, with rotation invariance for simplicity, is given by $P_{\rho\mu}\pi_A^{\mu\nu}P_{\nu\sigma} = 0$. The evaluations and discussions around (2.18) in the context of entropy current analysis point us that the contribution from the term $\partial_{[\mu}u_{\nu]}$ is the same as that of $-u_{[\mu}u^\sigma\partial_\sigma u_{\nu]} \equiv -u_{[\mu}a_{\nu]}$ with $a_\nu = u^\sigma\partial_\sigma u_\nu$. Thus we replace the former to the latter. The same condition $P_{\rho\mu}\pi_A^{[\mu\nu]}P_{\nu\sigma} = 0$ implies that π_A can be written in general as $\pi_A^{[\mu\nu]} = u^{[\mu}V_A^{\nu]}$. Here V_A^ν is a vector that can be fixed using the entropy current constructed below. Then, $-\pi_A^{[\mu\nu]}(\partial_{[\mu}u_{\nu]} - u_{[\mu}a_{\nu]}) = u^{[\mu}V_A^{\nu]}(u_{[\mu}a_{\nu]}) = -V_A^\mu a_\mu$, where we use $u^\mu a_\mu = 0$. The entropy current equation gives

$$\partial_\mu s^\mu = -\frac{1}{T}V_A^\mu a_\mu - \nu^\mu\partial_\mu\left(\frac{\mu}{T}\right). \quad (2.63)$$

To derive the general transport coefficients, we expand the dissipative terms as

$$\frac{1}{T}V_A^\mu = -\alpha_1 a^\mu - \alpha_2 P^{\mu\nu}\partial_\nu\left(\frac{\mu}{T}\right),$$
$$\nu^\mu = -\alpha_3 a^\mu - \alpha_4 P^{\mu\nu}\partial_\nu\left(\frac{\mu}{T}\right). \quad (2.64)$$

With the expression we can write the entropy current equation as

$$\partial_\mu s^\mu = \begin{pmatrix} a^\mu & P^{\mu\nu}\partial_\nu(\frac{\mu}{T}) \end{pmatrix}\begin{pmatrix} \alpha_1 & \alpha_2 \\ \alpha_3 & \alpha_4 \end{pmatrix}\begin{pmatrix} a^\mu \\ P^{\mu\nu}\partial_\nu(\frac{\mu}{T}) \end{pmatrix}. \quad (2.65)$$

Thus there are 4 independent transport coefficients as the off diagonal terms are not the same for the systems with broken parity. We write $\alpha_2 = C + \alpha'$ and $\alpha_3 = -C + \alpha'$. The dependence on C drops from the equation. Thus it corresponds to a dissipationless transport coefficient. α' is an another new dissipative transport coefficient. The other two dissipative coefficients are $\alpha_1 = \alpha/T$, that is the same as the neutral case in (2.18), and $\alpha_4 = \sigma_2 T$, that can be identified with the conductivity in (2.57). The positivity conditions on the coefficients are

$$\alpha\sigma_2 \geq (\alpha')^2, \quad \alpha \geq 0, \quad \sigma_2 \geq 0. \tag{2.66}$$

The transport coefficient C is dissipationless and can exist even at zero temperature similar to the Hall viscosity. Interested readers can refer to [26][27] for more details.

2.5 Green's function and Kubo formula

Various transport coefficients we have introduced above could be viewed as organizing the response of a fluid to hydrodynamic perturbations. When the perturbation is not strong enough, one can capture the essence of the responses at a linear level, which is referred as a linear response theory. A systematic description for the linear response theory can be developed by the Green's function of the quantum field theory using the general background field methods. The transport coefficients can be related to the retarded Green's functions through the Kubo formula.

2.5.1 Linear response theory

Here we introduce the linear response theory using the background field method, a useful and systematic way to capture all the known transport coefficients and thermodynamic quantities. Classical systems can be manipulated by controllable external forces. While there exist classical descriptions of the linear response theory, we provide a short review of its quantum description. Observables in quantum systems are described by operators $\mathcal{O}_\alpha$, whose dynamics is governed by the Hamiltonian $H(\mathcal{O})$, along with appropriate wave function $|\psi\rangle$.

2.5.1.1 Density matrix

We work with the quantum mechanical density matrix ρ. It is defined as the outer product of the wave function and its conjugate

$$\rho(t) = |\psi(t)\rangle\langle\psi(t)| \tag{2.67}$$

in the so-called Schrödinger picture, whose time dependence is carried by quantum states. The density matrix provides the quantum mechanical probability to find the system in a particular state $|\tilde{\psi}\rangle$ as $\langle\tilde{\psi}|\rho(t)|\tilde{\psi}\rangle = |\langle\tilde{\psi}|\psi(t)\rangle|^2$. The expectation value of an observable $\mathcal{O}$ is given by

$$\langle\mathcal{O}\rangle = \langle\psi(t)|\mathcal{O}|\psi(t)\rangle = \mathrm{Tr}[\rho(t)\mathcal{O}], \tag{2.68}$$

which traces over the product of two matrices. The wave function $|\psi(t)\rangle$ can be decomposed into a complete set $|n\rangle$, for example the energy eigenvectors, as $|\psi(t)\rangle = \sum_n c_n(t)|n\rangle$. Then the expectation value is $\langle\psi(t)|\mathcal{O}|\psi(t)\rangle = \sum_{mn} c_m^*(t)c_n(t)\langle m|\mathcal{O}|n\rangle = \sum_{mn} \rho_{nm}(t)\mathcal{O}_{mn} \equiv \mathrm{Tr}[\rho(t)\mathcal{O}]$, where we use

$$\begin{aligned}\rho(t) &= |\psi(t)\rangle\langle\psi(t)| \\ &= \sum_{mn} c_n(t)c_m^*(t)|n\rangle\langle m| = \sum_{mn} \rho_{nm}(t)|n\rangle\langle m|.\end{aligned} \tag{2.69}$$

Time evolution of the density matrix can be worked out as

$$\begin{aligned}\partial_t \rho(t) &= (\partial_t|\psi(t)\rangle)\langle\psi(t)| + |\psi(t)\rangle(\partial_t\langle\psi(t)|) \\ &= -(i/\hbar)[H,\rho]\end{aligned} \tag{2.70}$$

after using $\partial_t|\psi(t)\rangle = -i\hbar|\psi(t)\rangle$ and its conjugate. This is the same as the Heisenberg equation of motion, and the corresponding solution is given by $\rho(t;t_0) = U(t,t_0)\rho(t_0)U^\dagger(t;t_0)$ with time evolution operator $U(t,t_0) = e^{-(i/\hbar)\int_{t_0}^t H(t')dt'}$. Using the energy eigenstates $|n\rangle$, that satisfies $H|n\rangle = E_n|n\rangle$, the density matrix elements evolve as

$$\begin{aligned}\rho_{nm}(t;t_0) &= \langle n|\rho(t;t_0)|m\rangle = \langle n|U(t,t_0)\rho(t_0)U^\dagger(t;t_0)|m\rangle \\ &= \langle n|U(t,t_0)|\psi(t_0)\rangle\langle\psi(t_0)|U^\dagger(t;t_0)|m\rangle \\ &= e^{-i(E_n-E_m)(t-t_0)/\hbar}\rho_{nm}(t_0).\end{aligned} \tag{2.71}$$

The density matrix at thermal equilibrium is given by the equilibrium density matrix

$$\rho_0 = \frac{1}{Z} e^{-H/T}, \qquad (2.72)$$

with the quantum partition function $Z = \text{Tr}[e^{-H/T}]$ for the canonical ensemble (and more generally $\mathcal{Z} = \text{Tr}[e^{-(H-\mu N)/T}]$ for the grand canonical ensemble including the charge $N = \int d^d x \rho$). In terms of the energy eigenstates $|n\rangle$, $(\rho_0)_{nm} = \langle n|e^{-H/T}|m\rangle/Z = e^{-E_n/T}\delta_{nm}/Z = P_n \delta_{nm}$, where P_n is the probability. The thermal averaged expectation value of an operator $\mathcal{O}$ is

$$\langle \mathcal{O} \rangle = \frac{1}{Z} \sum_n e^{-E_n/T} \langle n|\mathcal{O}|n\rangle = \text{Tr}[\rho_0 \mathcal{O}]. \qquad (2.73)$$

2.5.1.2 Linear response

To properly set up the linear response theory, we consider the density matrix in the interaction picture that is useful for dealing with a small time-dependent perturbation $H_s(t)$ on top of a well established Hamiltonian H_0 that is time independent and has known solutions. The total Hamiltonian is separated as $H(t) = H_0 + H_s(t)$, where we treat $H_s(t)$ as an external potential for the system. In the interaction picture, the time evolution of the states is described by

$$|\psi(t)\rangle_I = U_0^\dagger |\psi(t)\rangle = U_s(t;t_0)|\psi(t_0)\rangle_I, \qquad (2.74)$$

where $U_0(t;t_0) = e^{-(i/\hbar)\int_{t_0}^t H_0(t')dt'}$, $U_s(t;t_0) = e^{-(i/\hbar)\int_{t_0}^t H_s(t')dt'}$ and $|\psi(t)\rangle$ is in Schrödinger picture.

The state and operator in interaction picture satisfy

$$\begin{aligned} i\hbar \frac{d}{dt}|\psi(t)\rangle_I &= H_s(t)|\psi(t)\rangle_I, \\ i\hbar \frac{d}{dt}\mathcal{O}_I(t) &= [\mathcal{O}_I(t), H_0], \end{aligned} \qquad (2.75)$$

where $i\hbar(d/dt)U_s(t,t_0) = H_s(t)U_s(t,t_0)$. The corresponding density matrix $\rho_I(t;t_0) = U_0^\dagger(t;t_0)\rho(t;t_0)U_0(t;t_0) = U_s(t;t_0)\rho(t_0)U_s^\dagger(t;t_0)$. The equation of motion becomes $\partial_t \rho_I(t) = -(i/\hbar)[H_s(t), \rho_I(t)]$. In the interaction picture, the operator $\mathcal{O}$ does not carry the dependence of H_s, $\mathcal{O}(t;t_0) = U_0(t;t_0)\mathcal{O}(t_0)U_0^\dagger(t;t_0)$.

An important example we consider here is an external source $\phi(t)$ that can be introduced by a term in the Hamiltonian as

$$H_s(t) = \mathcal{O}_\alpha(t)\phi_\alpha(t) \,. \tag{2.76}$$

This can be used to manipulate the quantum mechanical operators $\mathcal{O}_\alpha$. The source ϕ_α, with an index α, is the analogue of classical force, which can be checked by treating H_s as a classical Hamiltonian and computing the Euler-Lagrange equation. When the source is small enough, we can treat the effect of the source using a perturbation theory. Up to the first order, we have

$$\begin{aligned}\langle \mathcal{O}_\alpha(t)\rangle_\phi &= \mathrm{Tr}[\rho(t)\mathcal{O}_\alpha(t)] \\ &= \mathrm{Tr}[\rho(t_0)U_s^\dagger(t;t_0)\mathcal{O}_\alpha(t)U_s(t;t_0)] \\ &= \mathrm{Tr}[\rho(t_0)(\mathcal{O}_\alpha(t) + \frac{i}{\hbar}\int_{t_0}^{t}dt'[H_s(t'),\mathcal{O}_\alpha(t)] + \cdots)] \\ &= \langle\mathcal{O}_\alpha(t)\rangle_{\phi=0} + \frac{i}{\hbar}\int_{-\infty}^{\infty}\theta(t-t')dt'\langle[H_s(t'),\mathcal{O}_\alpha(t)]\rangle + \cdots .\end{aligned} \tag{2.77}$$

In the last line, we set $t_0 \to -\infty$ and extend the upper integration range to $t \to \infty$ by supplying the theta function to enforce the proper integration range.

Thus we have, after reintroducing the space index and (2.76), to the first order

$$\delta\langle\mathcal{O}_\alpha(t)\rangle_\phi = -\frac{i}{\hbar}\int_{-\infty}^{\infty}dt'\theta(t-t')\langle[\mathcal{O}_\alpha(t),\mathcal{O}_\beta(t')]\rangle\phi_\beta(t') \,. \tag{2.78}$$

This is nothing but the definition of the response function due to the change of the expectation value $\mathcal{O}_\alpha$ in the presence of the source ϕ_β, $\delta\langle\mathcal{O}_\alpha(t)\rangle_\phi = \int dt'\chi_{\alpha\beta}(t;t')\phi_\beta(t')$. The definition provides another equivalent definition for the response function as

$$\begin{aligned}\chi_{\alpha\beta}(t;t') &= \left.\frac{\partial\langle\mathcal{O}_\alpha(t)\rangle}{\partial\phi_\beta(t')}\right|_{\phi=0} \\ &= -i\theta(t-t')\langle[\mathcal{O}_\alpha(t),\mathcal{O}_\beta(t')]\rangle \,,\end{aligned} \tag{2.79}$$

where we set $\hbar = 1$. The equation (2.78) is nothing but the equation that lead to the retarded Green's function. Thus, we also use the term Green's function.

When the system is time translation invariant, the response function only depends on the time difference, $\chi_{\alpha\beta}(t;t') = \chi_{\alpha\beta}(t-t')$. It is sometimes more convenient to describe it in terms of frequency ω after taking the Fourier transformation $f(\omega) = \int dt e^{i\omega t} f(t)$ for arbitrary function $f(t)$. Then,

$$\delta\langle\mathcal{O}_\alpha(\omega)\rangle_\phi = \int dt' \int dt e^{i\omega(t-t')} \chi_{\alpha\beta}(t-t') e^{i\omega t'} \phi_\beta(t')$$
$$= \chi_{\alpha\beta}(\omega)\phi_\beta(\omega) \,. \tag{2.80}$$

If we work on real source ϕ and a hermitian operator $\mathcal{O}^\dagger = \mathcal{O}$, the response function $\chi(t)$ is real. We can see the consequences of them for $\chi(\omega)$ by decomposing $\chi(\omega)$ as

$$\chi_{\alpha\beta}(\omega) = \text{Re}\chi_{\alpha\beta}(\omega) + i\text{Im}\chi_{\alpha\beta}(\omega)$$
$$\equiv \chi'_{\alpha\beta}(\omega) + i\chi''_{\alpha\beta}(\omega) \,, \tag{2.81}$$

which is standard notation for the real and imaginary parts of the response function in momentum space. We can look into the imaginary part,

$$\chi''_{\alpha\beta}(\omega) = -(i/2)\left(\chi_{\alpha\beta}(\omega) - (\chi_{\alpha\beta}(\omega))^*\right)$$
$$= -(i/2)\int_{-\infty}^\infty dt\left(\chi_{\alpha\beta}(t)e^{i\omega t} - \chi_{\beta\alpha}(t)e^{-i\omega t}\right) \tag{2.82}$$
$$= -(i/2)\int_{-\infty}^\infty dt e^{i\omega t}\left(\chi_{\alpha\beta}(t) - \chi_{\beta\alpha}(-t)\right) \,.$$

This exercise also reveals the imaginary part is an odd function in ω as $\chi''_{\alpha\beta}(-\omega) = -\chi''_{\alpha\beta}(\omega)$. Thus the imaginary part $\chi''(\omega)$ survives only when the system is not invariant under the time reversal transformation $t \to -t$, which bears the information of time. Typically microscopic system is time reversal invariant, this contribution comes from dissipative effects.

2.5.2 Background field methods

We can generalize the linear response theory to its full scale by including all possible sources that we might be interested in depending on the symmetries. Here we only consider the conserved quantities related to the stress energy tensor $T^{\mu\nu}$ and a $U(1)$ current J^μ in the context of the quantum field theories. In particular, we look into the general properties of the retarded Green's functions defined by the variation of the one point function in terms of the source. The retarded Green's functions in the zero frequency limit coincide with the Euclidean time-ordered functions at zero frequency. The latter can be computed by the Euclidean functional integral in the grand canonical ensemble.

We consider the partition function $\mathcal{Z}[g_{\mu\nu}, A_\mu]$ that is coupled to the possible external sources, the background background metric $g_{\mu\nu}$ and gauge fields A_μ. Then the partition function can be used to compute the expectation values (or one point functions) of $\langle T^{\mu\nu}\rangle$ and $\langle J^\mu\rangle$ by evaluating the variations with respect to the corresponding sources.

$$\sqrt{-g}\langle J_\mu\rangle_{g,A} = \frac{\delta \log \mathcal{Z}}{\delta A^\mu},$$
$$\sqrt{-g}\langle T_{\mu\nu}\rangle_{g,A} = 2\frac{\delta \log \mathcal{Z}}{\delta g^{\mu\nu}}. \tag{2.83}$$

In the linear response theory, the sources are small and thus we can read off the response functions by setting the sources to vanish, $A_\mu \to 0$ and $g_{\mu\nu} = \eta_{\mu\nu} + h_{\mu\nu} \to \eta_{\mu\nu}$, after the computations.

Then, in the presence of the Lorentz invariance, the one point functions with the spatial indices to the first order are given by

$$\langle J_i\rangle = -\rho A_i - \sigma_{ij} E_j + \cdots,$$
$$\langle T_{ij}\rangle = p\delta_{ij} - \lambda_{ijkl} h_{kl} - \eta_{ijkl} \dot{h}_{kl} + \cdots, \tag{2.84}$$

where ρ is a charge density, $E_i = \partial_t A_i - \partial_i A_t$ is the electric field, σ_{ij} is the electric conductivity, p is the pressure, λ is the elastic modulus, and η is the viscosity we have defined previously. The first few term on the right side such as ρ, p, and λ do not have time dependence and are called the contact terms, that can be obtained by an additional variation with the sources. For example, the diamagnetic term ρ can be obtained by the second derivative of the partition function that has the A^2 term. These contact terms also play roles in the later parts of this book.

To find the Green's functions, we use (2.79). There are four different retarded Green's functions that are combinations for the background metric and gauge field.

$$G_R^{\mu,\nu}(x;x') = \left.\frac{\delta(\sqrt{-g}\langle J^\mu(x)\rangle)}{\delta A_\mu(x')}\right|_0 \approx -i\theta(t-t')\langle[J^\mu(x), J^\nu(x')]\rangle,$$

$$G_R^{\rho,\mu\nu}(x;x') = 2\left.\frac{\delta(\sqrt{-g}\langle J^\rho(x)\rangle)}{\delta h_{\mu\nu}(x')}\right|_0 \approx -i\theta(t-t')\langle[J^\rho(x), T^{\mu\nu}(x')]\rangle,$$

$$G_R^{\mu\nu,\rho}(x;x') = \left.\frac{\delta(\sqrt{-g}\langle T^{\mu\nu}(x)\rangle)}{\delta A_\rho(x')}\right|_0 \approx -i\theta(t-t')\langle[T^{\mu\nu}(x), J^\rho(x')]\rangle,$$

$$G_R^{\mu\nu,\rho\sigma}(x;x') = 2\left.\frac{\delta(\sqrt{-g}\langle T^{\mu\nu}(x)\rangle)}{\delta h_{\rho\sigma}(x')}\right|_0 \approx -i\theta(t-t')\langle[T^{\mu\nu}(x), T^{\rho\sigma}(x')]\rangle,$$

(2.85)

where $|_0$ indicates $h = A = 0$. $\approx$ means that the left-hand sides have the extra information about the contact terms, while the right-hand sides do not. On the right-hand side, the expectation values are evaluated by the grand canonical density matrix, for example $\langle[J^\mu(x), J^\nu(x')]\rangle = \text{Tr}[\rho_0[J^\mu(x), J^\nu(x')]]$, where $x = (t, \vec{x})$.

If there are translational invariance both for time and space, we can do the Fourier transformation to the momentum space $(\omega, \vec{q})$. Then (2.85) transform into the momentum space Green's function. For example, the current-current function has the form

$$G_R^{\mu,\nu}(\omega, \vec{q}; \omega', \vec{q}') = \left.\frac{\delta(\sqrt{-g}\langle J^\mu(\omega, \vec{q})\rangle)}{\delta A_\mu(\omega', \vec{q}')}\right|_{h=A=0}. \tag{2.86}$$

We apply this general formula to (2.84), then the spatial components Green's function become

$$G_R^{i,j}(\omega, \vec{q}) = -\rho\delta^{ij} + i\omega\sigma^{ij} + \cdots, \tag{2.87}$$

where $A_i(\omega, \vec{q}) = \int d^2x dt e^{i\omega t - i\vec{q}\cdot\vec{x}} A_i(t, \vec{x})$. The factors $i\omega$ come from the time derivative in E_i. We take the zero wave vector limit $\vec{q} = 0$ first as the Kubo formula is related to the infinitely slow processes. At finite wave vectors momentum relaxation is a fast process. Momentum relaxation is only infinitely slow at zero wave vector [31][32][33].

We decompose the Green's function into symmetric and antisymmetric parts using $\sigma^{ij} = \delta^{ij}\sigma + \epsilon^{ij}\sigma_H$, where σ and σ_H are the conductivity and Hall conductivity. Omitting the contact term, we have

$$\sigma = \lim_{\omega \to 0} \frac{\delta_{ij}}{2\omega} \mathrm{Im} G_R^{ij}(\omega, \vec{0}),$$
$$\sigma_H = \lim_{\omega \to 0} \frac{\epsilon_{ij}}{2\omega} \mathrm{Im} G_R^{ij}(\omega, \vec{0}),$$
(2.88)

where we use the fact that the imaginary part of the retarded Green's function is odd in ω. Here we also evaluate them at $\omega \to 0$ as the Green's function is evaluated at the equilibrium. This demonstrates that the Green's function evaluated using the background field methods can capture the contact terms not to mention the transport coefficients.

We move on to consider the stress energy – stress energy function that has the form in the momentum space with time and space translation invariance

$$G_R^{\mu\nu,\rho\sigma}(\omega, \vec{q}; \omega', \vec{q}') = \frac{\delta(\sqrt{-g}\langle T^{\mu\nu}(\omega, \vec{q})\rangle)}{\delta h_{\rho\sigma}(\omega', \vec{q}')}\bigg|_{h=A=0}.$$
(2.89)

We apply this general formula to (2.84) to consider the spatial components.

$$G_R^{ij,kl}(\omega, \vec{0}) = -\lambda_{ijkl} + i\omega\eta_{ijkl} + \cdots,$$
(2.90)

where we take the zero wave vector limit $\vec{q} = 0$ anticipating to capture the infinitely slow processes. As anticipated, the shear and bulk viscosities are parts of the shear tensor η_{ijkl}. To simply extract the shear viscosity, we can consider a specific component of the Green's function, $G_R^{xy,xy}(\omega, \vec{0}) = p - i\omega\eta + \mathcal{O}(\omega^2)$. We can find the covariant form of the shear and bulk viscosities as

$$\eta = \lim_{\omega \to 0} \frac{\delta_{ik}\delta_{jl} - \epsilon_{ik}\epsilon_{jl}}{8\omega} \mathrm{Im} G_R^{ij,kl}(\omega, \vec{0}),$$
$$\zeta = \lim_{\omega \to 0} \frac{\delta_{ij}\delta_{kl}}{4\omega} \mathrm{Im} G_R^{ij,kl}(\omega, \vec{0}).$$
(2.91)

We can check that the bulk viscosity ζ is the trace part with the tensor structure $\delta^{ij}\delta^{kl}\delta_{ij}\delta_{kl} = \delta_i^i \delta_k^k = 4$, while η is the traceless part that can be checked as $\delta^{ij}\delta^{kl}(\delta_{ik}\delta_{jl} - \epsilon_{ik}\epsilon_{jl}) = \delta^{ij}(\delta_i^l\delta_{jl} - \epsilon_i^l\epsilon_{jl}) = \delta^{ij}(\delta_{ij} - \delta_{ij}) = 0$.

Now we evaluate the Kubo formula for the Hall viscosity. It is sufficient to consider the local rest frame of the fluid at zero spatial momentum. Out of all the general metric perturbations $g_{\mu\nu} = \eta_{\mu\nu} + h_{\mu\nu}(t) + \mathcal{O}(h^2)$ in 2+1 dimensions, one can work with the following minimal set to ensure a consistency condition: $h = \{h_{xy}(t), h_{xx}(t), h_{yy}(t)\}$. Where $\mu, \nu = t, x, y$ and $\eta_{\mu\nu} = \mathrm{diag}(-1, 1, 1)$ is the flat metric with only the diagonal components.

We can work out a specific model that captures the essence of the process. We consider the tensor structures given in (2.38). Then, we get the energy momentum tensor for the component T^{xy} [34]

$$T^{xy} = -p h_{xy} - \eta \frac{\partial h_{xy}}{\partial t} + \frac{1}{2}\eta_H \frac{\partial (h_{xx} - h_{yy})}{\partial t} + \mathcal{O}(h^2).$$
(2.92)

Thus one finds

$$G_R^{xy,xx-yy}(\omega, \vec{0}) = 2i\omega\eta_H + \mathcal{O}(\omega^2).$$
(2.93)

One can put the expressions in a covariant form as

$$\eta_H = \lim_{\omega \to 0} \frac{\epsilon_{ik}\delta_{jl}}{4\omega} \mathrm{Im} G_R^{ij,kl}(\omega, \vec{0}),$$
(2.94)

The Kubo formula for the Hall viscosity has been worked out in [34].

Before moving forward, we also list the Kubo formula of α, the transport coefficient in the absence of boost invariance.

$$\alpha = \lim_{\omega \to 0} \frac{\delta^{ij}}{\omega c_b} \mathrm{Im} G^R_{0i,j0}(\omega, \vec{0}),$$
$$\alpha' = \lim_{\omega \to 0} \frac{-\delta^{ij}}{\omega c_b} \mathrm{Im} G^R_{i,j0}(\omega, \vec{0}),$$
(2.95)

where c_b is a constant associated with the thermodynamic quantities ϵ, p and the zero frequency two point functions [27]. We note that the stress energy tensors are not symmetric.

2.5.3 Thermodynamic response functions

In the previous sections, we have shown that the transport coefficients, the imaginary part of the response function, such as σ, η, ζ are constrained by the entropy current equation coming from the local second law of thermodynamics. New transport coefficients due to parity breaking effects, σ_H, η_H, are not constrained. This happens as forming a complete square with an antisymmetric coefficient vanishes, for example, $\eta^A_{ijkl} \xi^{ij} \xi^{kl} = 0$ with $\eta^A_{ijkl} = -\eta^A_{klij}$. Here we look into the symmetric part of the response function to constrain thermodynamics response parameters.

As mentioned above, practical computation can be done in equilibrium Euclidean functional integral in the local rest frame of the fluid $u^i = 0$. In the local rest frame, the stress energy tensor (2.44) becomes

$$T^{\mu\nu} = (\epsilon + p)\delta^\mu_0 \delta^\nu_0 u^0 u^0 + p\eta^{\mu\nu} + \pi^{\mu\nu},$$
$$J^\mu = \rho \delta^\mu_0 u^0 + \nu^\mu.$$
(2.96)

Here we remind that the independent parameters are the velocity u^μ, temperature T, and chemical potential μ. Without external sources, the partition function in the grand canonical ensemble is given by

$$\mathcal{Z}[T, \mu] = \mathrm{Tr}\left[\exp\left(-\frac{H}{T} + \frac{\mu N}{T}\right)\right],$$
(2.97)

where N is the $U(1)$ conserved charge associated with the chemical potential μ. The chemical potential can be defined as $\mu/T = i \int_0^\beta d\tau A_\tau$, where $\tau = it$ is the Euclidean time with the period $\beta = 1/T$. When a constant background gauge field, A_0, is turned on, the chemical potential is changed as $\mu'/T' = \mu/T + A_0/T$. Thus the effect of constant A_0 can be eliminated by rescaling the chemical potential.

Similarly, a constant metric component h_{00} can be absorbed by rescaling the Euclidean time τ. The length of the time circle does not change under the rescaling. Instead, the periodicity of the time circle is shifted $\beta' = \beta(1 - h_{00}/2)$ to the first order in h_{00}. This can be readily verified in the context of the black hole thermodynamics. Another metric component h_{0i} turns out to be eliminated by a coordinate transformation. The fluid velocity u^i does not change under this transformation to the first order in h_{0i}. These can be seen by checking the change of the invariant interval

$$\begin{aligned} ds^2 &= g_{\mu\nu} dx^\mu dx^\nu \\ &= (\eta_{00} + h_{00}) dx^0 dx^0 + 2 h_{0i} dx^0 dx^i + g_{ij} dx^i dx^j \\ &= (\eta_{00} + h_{00}) dx^0 dx^0 + g_{ij} d\tilde{x}^i d\tilde{x}^j + \mathcal{O}(h^2) \,, \end{aligned} \quad (2.98)$$

where $d\tilde{x}^i = dx^i + (h_{0j}/g_{ij}) dx^0$. Thus, the thermodynamic properties in the presence of constant and small background fields, A_0, h_{00}, and h_{0i}, are the same as those without the fields but appropriately shifted temperature, chemical potential, and normalized velocity field.

$$T' = T\left(1 + \frac{h_{00}}{2}\right), \quad \mu' = \mu\left(1 + \frac{h_{00}}{2}\right) + A_0 \,. \quad (2.99)$$

And the retarded Green's functions at zero frequency can be evaluated in the static equilibrium with $u^i = 0$.

Using the partition function $\mathcal{Z}[T', \mu']$ with shifted temperature and chemical potential, we reevaluate the time components of the retarded Green's functions, which provide the information of the constitute functions, ε and ρ (considered to be 'one point functions' compared to the sources T and μ).

$$\lim_{\vec{q}\to 0} G_R^{0,0}(0,\vec{q}) = \left(\frac{\partial \rho}{\partial \mu}\right)_T, \quad \lim_{\vec{q}\to 0} G_R^{0,00}(0,\vec{q}) = T\left(\frac{\partial \rho}{\partial T}\right)_{\mu/T},$$
$$\lim_{\vec{q}\to 0} G_R^{00,0}(0,\vec{q}) = \left(\frac{\partial \varepsilon}{\partial \mu}\right)_T, \quad \lim_{\vec{q}\to 0} G_R^{00,00}(0,\vec{q}) = T\left(\frac{\partial \varepsilon}{\partial T}\right)_{\mu/T}.$$
(2.100)

Note that the variation with A_0 is equivalent to that with μ for fixed T, while the variation with h_{00} is equivalent to that with T for fixed μ/T. We also note that only the real part of the Green's functions contributes as the imaginary part is odd in ω. These are called susceptibility conditions as there is no time dependence. We apply the results to (2.56) to check that $\chi_E = \chi_T = 0$.

2.6 Irreversible thermodynamics

In this section we expand our discussion to include details of multiple different transport phenomena such as the thermoelectric phenomena with heat current (stress energy) and electric current (an example of $U(1)$ current) at the same time. In general these two are not independent with each other. For example, temperature gradient can induce a charge current in addition to heat current, and vice versa. The electric current due to temperature gradient has a definite relation with the heat current in the presence of electric field when there are certain symmetries, such as the time reversal symmetry. The electric current and the heat current are in fact the same in appropriate units.

This is known as Onsager's reciprocal relations in the theory of irreversible thermodynamics [35][36]. Similar relations can be formulated for the systems with the broken parity symmetry with a little more care, for example in the presence of magnetic field [37][38]. When we consider the spin interactions in §4.5, there are more complicated relations among the heat, the electric, and the spin currents.

Before diving into this important topic, we hope to reveal some features that are counter-intuitive when dealing with non-zero off-diagonal components in multi-dimensional observables. This is even so for a single transport coefficient. For example, Ohm's law states that $\vec{J} = \sigma \vec{E}$, where $\vec{J}$ and $\vec{E}$ are the current and electric field, respectively. The conductivity tensor σ in 2 spatial dimensions (x,y) has the form

$$\sigma = \begin{pmatrix} \sigma_{xx} & \sigma_{xy} \\ -\sigma_{xy} & \sigma_{xx} \end{pmatrix}, \qquad (2.101)$$

for the rotationally invariant system in xy plane. The constraints $\sigma_{xx} = \sigma_{yy}$ and $\sigma_{yx} = -\sigma_{xy}$ can be verified by demanding a 2 by 2 matrix is invariant under a similarity transformation

$$\begin{pmatrix} \sigma_{xx} & \sigma_{xy} \\ \sigma_{yx} & \sigma_{yy} \end{pmatrix} = \begin{pmatrix} \cos\theta & \sin\theta \\ -\sin\theta & \cos\theta \end{pmatrix} \begin{pmatrix} \sigma_{xx} & \sigma_{xy} \\ \sigma_{yx} & \sigma_{yy} \end{pmatrix} \begin{pmatrix} \cos\theta & -\sin\theta \\ \sin\theta & \cos\theta \end{pmatrix}.$$

The off-diagonal component σ_{xy} is non-zero, for example, when electrons move in the presence of magnetic field perpendicular to the plane.

We also have an equivalent description with the resistivity tensor ρ that is the inverse of the conductivity tensor.

$$\rho = \frac{1}{\sigma_{xx}^2 + \sigma_{xy}^2} \begin{pmatrix} \sigma_{xx} & -\sigma_{xy} \\ \sigma_{xy} & \sigma_{xx} \end{pmatrix}. \qquad (2.102)$$

When the off-diagonal component of the conductivity tensor vanishes, $\sigma_{xy} = 0$, we have a familiar realtion. $\rho_{xx} = 1/\sigma_{xx}$. Thus, if $\sigma_{xx} = 0$, $\rho_{xx} = \infty$. These two quantities, $\sigma_{xx} = 0$, $\rho_{xx} = \infty$, describe the same object that is a perfect insulator. On the other hand, $\sigma_{xx} = \infty$ and $\rho_{xx} = 0$ does a perfect conductor.

Now we turn to the situation $\sigma_{xy} \neq 0$. If $\sigma_{xy} \neq 0$, $\rho_{xx} = \sigma_{xx}/(\sigma_{xx}^2 + \sigma_{xy}^2)$ and $\rho_{xy} = -\sigma_{xy}/(\sigma_{xx}^2 + \sigma_{xy}^2)$. When $\sigma_{xx} = 0$, we get $\rho_{xx} = 0$ and $\rho_{xy} = -1/\sigma_{xy}$, which is completely different from the case when $\sigma_{xy} = 0$. Here $\sigma_{xx} = 0$ and $\rho_{xx} = 0$ signify that there is no current flowing along the longitudinal direction parallel to the applied electric field, while $\sigma_{xy} \neq 0$ indicates that the current flows transverse to the field. This exercise warns us that we need to be careful for what we observe when different physical quantities act together.

Some of these features can be demonstrated in the Drude model for a particle with charge q and mass m, that is described by

$$m\partial_t v_i = q(E_i + \epsilon_{ijk} v_j B_k) - m\frac{v_i}{\tau_s}, \tag{2.103}$$

where E_i, B_i are the electric and magnetic fields and τ_s is the so-called scattering time that tells how long the particle is ballistically accelerated by the electric field until it collides with something that changes its direction. Let us consider a steady-state motion, $\partial_t v_i = 0$, in the xy plane with the magnetic field along perpendicular to it, $B_k = B\delta_{kz}$. Then the equation goes as

$$\left(\delta_{ij} - \frac{q\tau_s}{m}B\epsilon_{ij}\right)v_j = \frac{q\tau_s}{m}E_i. \tag{2.104}$$

This can be solved by multiplying $(\delta_{im} - (q\tau_s B/m)\epsilon_{im})$ both sides. After relabeling the indices and identifying the current (density) $J_i = nqv_i$, where n is the particle density, we get

$$J_i = \frac{nq^2\tau_s/m}{1+\omega_B^2\tau_s^2}(\delta_{ij} + \omega_B\tau_s\epsilon_{ij})E_j, \tag{2.105}$$

where $\omega_B = qB/m$ is the cyclotron frequency. In terms of the conductivity tensor,

$$\sigma = \frac{nq^2\tau_s/m}{1+\omega_B^2\tau_s^2}\begin{pmatrix} 1 & \omega_B\tau_s \\ -\omega_B\tau_s & 1 \end{pmatrix}. \tag{2.106}$$

Thus the magnetic field B produces the off-diagonal components of the conductivity tensor. It turns out that 2 dimensional electron systems under a strong magnetic field at low temperatures reveals the quantum Hall effects, where the Hall resistivity ρ_{xy} exhibits a plateau for a range of the magnetic field, while the longitudinal resistivity ρ_{xx} vanishes.

2.6.1 Onsager's reciprocal relation

To illustrate the Onsager's reciprocal relation, we consider the thermo-electric effect in the rest of this subsection and the thermo-electromagnetic effect in §2.6.3 that happen when simultaneous flow of electric current and heat in a system under the influence of magnetic field.

One can imagine, for example, a solid with electrons as charge carriers. We already encounter the constant volume thermodynamic relation $du = Tds + \mu dn$, where u is the local energy density, s the local entropy density, μ the electrochemical potential per electron (not to be confused with the magnetic dipole moment vector $\vec{\mu}$ or a component μ_z used in Chapter 4), n the number density of electrons.

$$ds = \frac{1}{T}du - \frac{\mu}{T}dn. \tag{2.107}$$

One can generalize this equation by including different charge carriers by introducing an index $k = 1, 2, \cdots$ as $\mu_k dn_k$ along with the corresponding relevant equations. The conserved current can be written as the continuity equation, $\partial_t n + \vec{\nabla} \cdot \vec{J}_N = 0$ or $\partial_t n + \partial_i J_N^i = 0$ using the index notation. The internal energy conservation also can be written as $\partial_t u + \partial_i J_U^i = 0$.

Similarly, these current densities are related as

$$\vec{J}_S = \frac{1}{T}\vec{J}_U - \frac{\mu}{T}\vec{J}_N, \tag{2.108}$$

where $\vec{J}_S$, $\vec{J}_U$, and $\vec{J}_N$ are the current densities of entropy, energy, and number of electrons, respectively. Here we assume the other components are immobile except electrons.

Now we are ready to compute the entropy production rate per volume by using the continuity equation as

$$\begin{aligned}\frac{ds}{dt} &= \frac{\partial s}{\partial t} + \vec{\nabla} \cdot \vec{J}_S = \frac{1}{T}\frac{\partial u}{\partial t} - \frac{\mu}{T}\frac{\partial n}{\partial t} + \vec{\nabla} \cdot (\frac{1}{T}\vec{J}_U - \frac{\mu}{T}\vec{J}_N) \\ &= \vec{\nabla}(\frac{1}{T}) \cdot \vec{J}_U - \vec{\nabla}(\frac{\mu}{T}) \cdot \vec{J}_N,\end{aligned} \tag{2.109}$$

where we use the continuity equation for the energy density $\partial u/\partial t + \vec{\nabla} \cdot \vec{J}_U = 0$ and number density $\partial n/\partial t + \vec{\nabla} \cdot \vec{J}_N = 0$. A little more convenient form is heat current $\vec{J}_Q = \vec{J}_U - \mu \vec{J}_N$ instead of the energy current. Thus we have

$$\frac{ds}{dt} = \vec{\nabla}\left(\frac{1}{T}\right) \cdot \vec{J}_Q - \frac{1}{T}\vec{\nabla}\mu \cdot \vec{J}_N. \tag{2.110}$$

The heat current in a steady-state flow shows something we usually see in thermodynamics. By taking the divergence of heat current and using the divergencelessness of $\vec{J}_U$ and $\vec{J}_N$, $\vec{\nabla} \cdot \vec{J}_U = \vec{\nabla} \cdot \vec{J}_N = 0$ for the steady-state flow, we get $\vec{\nabla} \cdot \vec{J}_Q = -\vec{\nabla}\mu \cdot \vec{J}_N$. Thus the rate of heat current increase is equal to the rate of decrease in the electro-chemical potential energy current. By plugging in (2.110), we get

$$\frac{ds}{dt} = \vec{\nabla}\left(\frac{1}{T}\right) \cdot \vec{J}_Q + \frac{1}{T}\vec{\nabla} \cdot \vec{J}_Q = \vec{\nabla} \cdot \left(\frac{\vec{J}_Q}{T}\right). \tag{2.111}$$

Thus in steady-state flow, the increase in entropy comes from the flow of heat and also appearance of heat current.

The significance of this equation (2.110) is the relation between the currents $\vec{J}_Q, \vec{J}_N$ and the corresponding generalized forces, called affinities, $\vec{\nabla}(1/T), (1/T)\vec{\nabla}\mu$, respectively. This can be viewed as generalizing the Ohm's equation in the presence of the heat and the electric currents. The dynamical equations that connect them are (in 1 dimension)

$$\begin{aligned} -\vec{J}_N &= L_{11}\frac{\vec{\nabla}\mu}{T} + L_{12}\vec{\nabla}\left(\frac{1}{T}\right), \\ \vec{J}_Q &= L_{12}\frac{\vec{\nabla}\mu}{T} + L_{22}\vec{\nabla}\left(\frac{1}{T}\right), \end{aligned} \tag{2.112}$$

where we use the Onsager relation $L_{12}(\vec{H}) = L_{21}(-\vec{H})$, meaning that number current flow is the same as the heat current flow (in the presence of magnetic field). Here we comment on μ. The chemical potential can include different contributions such as electric potential μ_e, the spin contribution, and other chemical portions. The charge of an electron is e, $\mu_e = e\phi$ with ϕ as an electrostatic potential. Thus, $\vec{\nabla}\mu_e = -e\vec{E}$. We will be back to the spin-up and spin-down chemical potentials $\mu_\pm$ later in §4.5.

Back to (2.112), these coefficients L are related to transport coefficients with no applied magnetic fields. For example, electric conductivity is the electric current density $\vec{J}_e = e\vec{J}_N$ per electric field $\vec{E}$ in an isothermal system $\vec{\nabla}T = 0$ (without controlling over the heat current). Thus

$$\sigma = \frac{\vec{J}_e}{\vec{E}} = \frac{e\vec{J}_N}{\vec{E}} = -\frac{e\vec{J}_N}{(\vec{\nabla}\mu_e/e)} = \frac{e^2 L_{11}}{T}. \tag{2.113}$$

Similarly, the heat conductivity κ is defined as the heat current density per unit temperature gradient when there is no electric current $\vec{J}_N = 0$ (without controlling the change of the chemical potential). To enforce the condition $\vec{J}_N = 0$, we solve the first equation in (2.112). Thus $\vec{\nabla}\mu = -(L_{12}/L_{11})T\vec{\nabla}(1/T)$. Plugging this into the second equation in (2.112),

$$\kappa = -\frac{\vec{J}_Q}{\vec{\nabla}T} = \frac{L_{11}L_{22} - L_{12}^2}{L_{11}T^2}. \tag{2.114}$$

There are three independent transport coefficients L_{11}, L_{22}, and L_{12}. In addition to the electric and thermal conductivities, one can define the thermoelectric power as the potential gradient (electric field) per unit temperature gradient when there is no electric current $\vec{J}_N = 0$ (without controlling the heat current).

$$\epsilon = \frac{\vec{E}}{\vec{\nabla}T} = -\frac{1}{e}\frac{\vec{\nabla}\mu}{\vec{\nabla}T} = -\frac{L_{12}}{eTL_{11}}. \tag{2.115}$$

Thus we have $L_{11} = T\sigma/e^2$, $L_{12} = -\epsilon\sigma T^2/e$, and $L_{22} = T^2\kappa + T^3\epsilon^2\sigma$. Thermoelectric power ϵ is another name of the Seebeck coefficient, that is sometimes denoted as S. (ϵ is not to be confused with ε that is used for energy density in §2.2 and $\epsilon_{\mu\nu}$ that is used as totally antisymmetric tensor with indices.)

Now the dynamic equation (2.112) can be rewritten in terms of the electric, thermal conductivities σ, κ and thermoelectric power ϵ

$$\begin{aligned}-\vec{J}_N &= \frac{\sigma}{e^2}\vec{\nabla}\mu + \frac{\epsilon\sigma}{e}\vec{\nabla}T, \\ \vec{J}_Q &= \frac{-T\epsilon\sigma}{e}\vec{\nabla}\mu - (\kappa + T\epsilon^2\sigma)\vec{\nabla}T,\end{aligned} \tag{2.116}$$

In terms of electric current, electric field, heat current, and temperature difference, we get

$$\begin{pmatrix} \vec{J}_e \\ \vec{Q} \end{pmatrix} = \begin{pmatrix} \sigma & \epsilon\sigma \\ T\epsilon\sigma & \kappa + T\epsilon^2\sigma \end{pmatrix} \begin{pmatrix} \vec{E} \\ -\vec{\nabla}T \end{pmatrix}. \qquad (2.117)$$

Thus we express all the transport coefficients in terms of the conductivities and thermoelectric power. Frequently, alternative form is used in the literature that change the relations as $\vec{E}(\vec{J})$. By converting the (2.117), we get

$$\begin{pmatrix} \vec{E} \\ \vec{Q} \end{pmatrix} = \begin{pmatrix} \rho & \epsilon \\ T\epsilon & -\kappa \end{pmatrix} \begin{pmatrix} \vec{J}_e \\ \vec{\nabla}T \end{pmatrix}, \qquad (2.118)$$

where the resistivity is inverse of the conductivity, $\rho = 1/\sigma$. As mentioned before, the thermoelectric power is also Seebeck coefficient $\epsilon = S$. In the coming section, we explore the physical properties of the transport coefficients in more detail.

We check that the off-diagonal components of (2.117) and (2.118) are the same with appropriate dimensionful conversion factor T. These are the explicit demonstrations of the Onsager's reciprocal relations.

2.6.2 Seebeck, Peltier, and Thompson effects

In this subsection, we continue to discuss one dimensional thermo-electric transport phenomena given in (2.112), (2.117), or (2.118) without magnetic fields. We briefly review the experimental setups for the Seebeck effect and its coefficient thermo-electric power, the Peltier effect, and the Thompson effect because they are frequently mentioned in the literature.

2.6.2.1 Seebeck effect
In general, Seebeck effect is the phenomenon that temperature gradient produces electric field in conducting materials. When one side of the conductor is hotter than the other, the mobile carriers in the hotter side are excited, diffused, and accumulated in the

 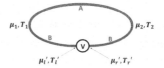

Figure 2.3 Seebeck effect. Left: Electric field is generated by a temperature gradient. Right: Illustration of the thermocouple. Two different materials A and B with different thermoelectric powers, ϵ_A and ϵ_B are connected at two different points with different temperatures and generate different electric potential difference. The potential difference can be measured by the voltmeter V.

colder side.[1] If the carriers have electric charges, electric field can be generated as illustrated in the left panel of Fig. 2.3. When the material is connected to a wire, electric current can flow.

In particular, the Seebeck effect refers to the production of an electromotive force in a thermocouple with vanishing electric current $\vec{J}_N = 0$. This is directly related to the definition of the thermoelectric power ϵ given in (2.115). Let us consider a thermocouple with junctions for two different materials A and B at temperature T_1 and $T_2 (T_2 > T_1)$ as in the right panel of Fig. 2.3. A voltmeter V is attached in the middle of the material B with a temperature T'. The voltmeter has a large resistance and does not allow electric current, while there is no resistance to the heat flow.

From the first equation in (2.116), we get $\vec{\nabla}\mu = -e\epsilon\vec{\nabla}T$. By integrating this function, one can compute

$$\mu_2 - \mu_1 = -e \int_1^2 \epsilon_A dT, \tag{2.119}$$

where ϵ_A is the thermoelectric power for the material A. The potential difference, Voltage, at the voltmeter can be computed as $V = (\mu'_r - \mu'_l)/e = [(\mu_2 - \mu_1) - (\mu_2 - \mu'_r) - (\mu'_l - \mu_1)]/e = -\int_1^2 (\epsilon_A - \epsilon_B) dT$, where we use the fact that the temperature

---

[1] This may not be true for electrons with a particular spin polarization in the in plane magnetic field. As we see in §4.5.1, electrons with two opposite polarization show opposite behavior, spin-up electrons accumulate in low temperature side, while the spin-down ones in the high temperature side. Overall, more electrons accumulate in the hotter side.

Figure 2.4 Peltier effect is generated when two conducting materials are connected with isothermal junction with electric current.

difference at the left and right side of the voltmeter vanishes. Thus the voltage on the voltmeter is integral of the difference of the thermoelectric power of the two materials.

The meaning of the thermoelectric power ϵ, or the Seebeck coefficient S, can be understood more clearly if we eliminate the terms proportional to $\vec{\nabla}\mu$ in (2.116), we get $\vec{J}_Q = Te e \vec{J}_N - \kappa \vec{\nabla} T$. The entropy current $\vec{J}_S = \vec{J}_Q/T = \epsilon e \vec{J}_N - \kappa \vec{\nabla} T/T$. Thus the thermoelectric power can be viewed as the entropy that is transported per coulomb by the electron flow.

2.6.2.2 Peltier effect

The Peltier effect is inverse of the Seebeck effect. Temperature gradient can be generated by the electric field due to the accumulation of the charge carriers. This happens because charge carriers are also heat carriers. In particular, the Peltier effect is the evolution of heat accompanying the flow of electric current $e\vec{J}_N$ across an isothermal junction (the middle point in Fig. 2.4) of two conducting materials A and B. The isothermal junction prevents heat exchange and thus creates the discontinuity of $\vec{J}_Q$. The total energy current, $\vec{J}_U = \vec{J}_Q + \mu \vec{J}_N$, is also discontinuous. Thus,

$$\vec{J}_U^B - \vec{J}_U^A = \vec{J}_Q^B - \vec{J}_Q^A = T(\epsilon_B - \epsilon_A)(e\vec{J}_N) \equiv \pi_{AB}(e\vec{J}_N) , \qquad (2.120)$$

where we use that the electric current and the chemical potential are continuous and use $\vec{J}_Q = T\epsilon e \vec{J}_N$ that is evaluated from (2.116) or (2.118) with the isothermal condition $\vec{\nabla} T = 0$.

The Peltier coefficient $\pi_{AB} = T(\epsilon_B - \epsilon_A)$ is the heat that needs to be supplied to the junction when unit electric current passes from A to B. As we see the Peltier effect is directly related to the Seebeck effect and thermoelectric power, which is called as second Kelvin relation.

A,T, x = 0	B, T + dT, x = dx	A,T, x = 0	B, T + dT, x = dx

Figure 2.5 Thompson effect. Left: heat current without electric current. Right: heat and electric currents.

2.6.2.3 Thompson effect

The Thompson effect concerns the heat absorbed per unit electric current and per unit temperature gradient. Let us consider a clearer setup [39]. Consider the left panel of Fig. 2.5 that there is a heat current without electric current due to the temperature difference $dT = \delta T$ at two ends A and B of a conducting material. The conductor are placed in heat reservoir so that there would be no heat interchange between the conductor and reservoir.

Now let the electric current flows as in the right panel of Fig. 2.5. Then the heat interchange will take place between the conductor and heat reservoir. The heat exchange has two parts: Joule heat and Thompson heat. The total energy flow can be computed by $\vec{\nabla} \cdot \vec{J}_U = \vec{\nabla} \cdot (\vec{J}_Q + \mu \vec{J}_N) = \vec{\nabla} \cdot \vec{J}_Q + \vec{\nabla}\mu \cdot \vec{J}_N$, where we use $\vec{\nabla} \cdot \vec{J}_N = 0$ due to current conservation. We are interested in the change of energy as a function of current $\vec{J}_N$ without changing the temperature profile. We replace $\vec{\nabla}\mu$ using the first equation of (2.116) and $\vec{J}_Q$ using $\vec{J}_Q = T\epsilon e \vec{J}_N - \kappa \vec{\nabla} T$ that is from the combination of the two equations of (2.116). Then

$$\vec{\nabla} \cdot \vec{J}_U = T\vec{\nabla}\epsilon \cdot (e\vec{J}_N) - \vec{\nabla} \cdot (\kappa \vec{\nabla} T) - \frac{e^2}{\sigma}\vec{J}_N^2. \tag{2.121}$$

The first term on the right-hand side is the desired Thompson term that represent the heat absorbed from the thermal reservoir when the current $e\vec{J}$ travels through the temperature gradient.

Using $\vec{\nabla}\epsilon = (d\epsilon/dT)\vec{\nabla}T$, the Thompson coefficient (defined as the Thompson heat absorbed per unit electric current and per unit temperature gradient) reads

$$\tau = \frac{\text{Thompson heat}}{\vec{\nabla}T \cdot (e\vec{J}_N)} = T\frac{d\epsilon}{dT}. \tag{2.122}$$

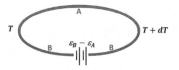

Figure 2.6 The first Kelvin relation among Seebeck, Peltier, and Thompson effects.

The second term on the right-hand side vanishes according to our setup. This can be checked with the condition $\vec{\nabla} \cdot \vec{J}_U = 0$ if $\vec{J}_N = 0$, meaning that the change of the energy vanished when there is no electric current. The last term is nothing but the Joule heating term.

2.6.2.4 First Kelvin relation

There exist an interesting relation between the three coefficients, Seebeck, Peltier, and Thompson coefficients. By taking a derivative of the Peltier coefficient, we get $d\pi_{AB}/dT + \tau_A - \tau_B = \epsilon_A - \epsilon_B$. This can be demonstrated by the thermocouple with the Voltmeter replaced by a battery that negates the Seebeck voltage so that there is no electric current in the thermocouple as in Fig. 2.6. Here the thermocouple is placed in the thermal bath that was considered for Thompson heating with the two junctions to the end points of the conductor. This is known as the first Kelvin relation.

2.6.3 Thermo-electromagnetic effects

We generalize the discussion of the thermoelectric transport coefficients to the two spatial dimensions with magnetic field $\vec{H}$ perpendicular to the plane. This is illustrated in Fig. 2.7. Following the discussion in §2.6, we consider the entropy change in terms of the currents and the corresponding generalized forces.

$$\frac{ds}{dt} = \vec{\nabla}\left(\frac{1}{T}\right) \cdot \vec{J}_Q - \frac{1}{T}\vec{\nabla}\mu \cdot \vec{J}_N = -\frac{1}{T^2}\vec{\nabla}T \cdot \vec{Q} + \frac{1}{T}\vec{E} \cdot \vec{J}_e . \quad (2.123)$$

Hereafter on we adapt the notation $J_{e,x} = eJ_{N,x}, J_{e,y} = eJ_{N,y}, Q_x = J_{Q,x}, Q_y = J_{Q,y}$, and $\vec{\nabla}\mu_e = -e\vec{E}$.

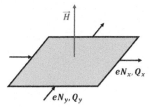

Figure 2.7 Setup for the thermo-electromagnetic transport coefficients in the presence of the magnetic field $\vec{H}$ as well as the charge $\vec{J}_e = e\vec{J}_N$ and heat $\vec{Q} = \vec{J}_Q$ currents.

The dynamical equation, the generalization of (2.116), becomes

$$-J_{e,x} = L'_{11}\frac{e^2}{T}E_x + L'_{12}\frac{e^2}{T}E_y + L'_{13}\frac{e}{T^2}(-\nabla_x T) + L'_{14}\frac{e}{T^2}(-\nabla_y T),$$

$$-J_{e,y} = -L'_{12}\frac{e^2}{T}E_x + L'_{11}\frac{e^2}{T}E_y - L'_{14}\frac{e}{T^2}(-\nabla_x T) + L'_{13}\frac{e}{T^2}(-\nabla_y T),$$

$$Q_x = L'_{13}\frac{e}{T}E_x + L'_{14}\frac{e}{T}E_y + L'_{33}\frac{1}{T^2}(-\nabla_x T) + L'_{34}\frac{1}{T^2}(-\nabla_y T),$$

$$Q_y = -L'_{14}\frac{e}{T}E_x + L'_{13}\frac{e}{T}E_y - L'_{34}\frac{1}{T^2}(-\nabla_x T) + L'_{33}\frac{1}{T^2}(-\nabla_y T).$$

(2.124)

Isotropy in x and y coordinates puts the 'diagonal components' to be the same, $L'_{11} = L'_{22}, L'_{13} = L'_{24}, L'_{31} = L'_{42}, L'_{33} = L'_{44}$. The presence of magnetic field breaks the parity symmetry and allows the off-diagonal components, $L'_{21} = -L'_{12}, L'_{23} = -L'_{14}, L'_{41} = -L'_{32}, L'_{43} = -L'_{34}$. The diagonal components are identified as even functions of the magnetic field, while the off-diagonal components are odd functions of the magnetic field. Moreover, Onsager relations $L'_{ij}(H) = L'_{ji}(-H)$ put further constraints as $L'_{13}(H) = L'_{31}(-H) = L'_{31}(H)$ and $L'_{14}(H) = L'_{41}(-H) = -L'_{32}(-H) = L'_{32}(H)$. Thus $L'_{13} = L'_{31}, L'_{14} = L'_{32}$. We identify all these related transport coefficients in (2.124).

This set of algebraic equations (2.124) can be converted into another set of equations with the electric currents and the temperature gradients on the right-hand side as these are experimentally controlled.

$$\begin{aligned}
E_x &= L_{11}J_{e,x} + L_{12}J_{e,y} - L_{13}\nabla_x T - L_{14}\nabla_y T, \\
E_y &= -L_{12}J_{e,x} + L_{11}J_{e,y} + L_{14}\nabla_x T - L_{13}\nabla_y T, \\
Q_x &= -TL_{13}E_x - TL_{14}E_y - L_{33}\nabla_x T - L_{34}\nabla_y T, \\
Q_y &= TL_{14}E_x - TL_{13}E_y + L_{34}\nabla_x T - L_{33}\nabla_y T.
\end{aligned} \quad (2.125)$$

There is a definite and clear relations between Ls and L's, even though they are a little complicated.

We present the results in terms of the transport coefficients and focus on summarizing some interesting parts.

$$\begin{pmatrix} E_x \\ E_y \\ Q_x \\ Q_y \end{pmatrix} = \begin{pmatrix} 1/\sigma_i & HR_i & -\epsilon & -H\eta_N \\ -HR_i & 1/\sigma_i & H\eta_N & -\epsilon \\ -T\epsilon & -TH\eta_N & -\kappa_i & -H\kappa_i \mathcal{L}_R \\ TH\eta_N & -T\epsilon & H\kappa_i \mathcal{L}_R & -\kappa_i \end{pmatrix} \begin{pmatrix} J_{e,x} \\ J_{e,y} \\ \nabla_x T \\ \nabla_y T \end{pmatrix} \quad (2.126)$$

where H, T are magnetic field and temperature, while $\sigma, R_H, \epsilon, \eta_N, \kappa, \mathcal{L}_R$ are isothermal electric conductivity $\sigma = \sigma_i$, isothermal Hall coefficient $R_H = R_i$, absolute thermoelectric power (Seebeck coefficient ϵ), isothermal Nernst effect $\eta_N = \eta_i$, isothermal heat conductivity $\kappa = \kappa_i$, and Leduk-Righi coefficient $\mathcal{L}_R$, respectively. (There are several different usages for η. This $\eta_N = \eta_i$ is used for the Nernst coefficient, while η_H, η_{ijkl}, $\eta_{\mu\nu}$, and η are used for the Hall viscosity, shear tensor, metric tensor, and shear viscosity, respectively, throughout this book.) In addition to these isothermal coefficients, there are corresponding transport coefficients for the adiabatic process that can be found in [38][40].

2.6.3.1 Hall coefficient

The form (2.125) and (2.126) suggest the following vector form for the electric field in terms of the temperature gradient in the presence of magnetic field $\vec{H}$

$$\vec{E} = \rho \vec{J}_e + R_H \vec{H} \times \vec{J}_e + \epsilon(-\vec{\nabla}T) + \eta_N \vec{H} \times (-\vec{\nabla}T). \quad (2.127)$$

In addition to the resistivity ρ and thermoelectric power ϵ, we have additional two transport coefficients, R_H and η_N. In the presence of

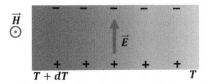

Figure 2.8 Illustration of the Nernst effect. Temperature gradient produce the electric field and current when connected to a closed circuit. In the presence of magnetic field, there is a transverse electric field and the corresponding transverse current when connected to a circuit. This can be understood as the Seebeck effect in the presence of magnetic field.

magnetic field, the Lorentz force change the direction of the electric current producing the transverse component of the current, which is measured by the Hall coefficient R_H. More specifically, the isothermal Hall effect is defined as $R_H = R_i = E_y/(HJ_{e,x})$ with $\nabla_x T = \nabla_y T = J_{e,y} = 0$.

$$R_H = \frac{E_y}{HJ_{e,x}} = \frac{L_{12}}{H}. \tag{2.128}$$

2.6.3.2 Nernst effect

Similarly, the electric field due to the temperature gradient (this is Seebeck effect with the thermoelectric power) develops transverse component in the presence of the magnetic field. This is called Nernst effect with the Nernst coefficient η_N. The isothermal Nernst effect, $\eta_N = \eta_i = E_y/(H\nabla_x T)$ with $\nabla_y T = J_{e,x} = J_{e,y} = 0$, measures the generation of transverse electric field upon applying a temperature gradient.

$$\eta_N = \frac{E_y}{H\nabla_x T} = \frac{L_{14}}{H}. \tag{2.129}$$

This is illustrated in Fig. 2.8.

2.6.3.3 Ettingshausen effect

We can think about the inverse of the Nernst effect, which is the generation of the temperature gradient due to the electric field in the presence of magnetic field. This is similar to the Peltier effect being inverse of the Seebeck effect. The Ettingshausen effect is given by $E_{tth} = -\nabla_y T/HJ_{e,x}$ with $\nabla_x T = Q_y = J_{e,y} = 0$. This coefficient is not in (2.126). It turns out that this coefficient can be rewritten in terms of other transport coefficients.

$$E_{tth} = -\frac{\nabla_y T}{HJ_{e,x}} = \frac{TL_{14}}{HL_{33}} = \frac{T\eta_N}{\kappa}. \tag{2.130}$$

2.6.3.4 Leduk-Righi effect

We comment one more transport coefficient $\mathcal{L}$ that is in (2.126). The Leduk-Righi effect is the generation of the transverse temperature gradient in the presence of a temperature gradient in the presence of magnetic field. It is defined as $\mathcal{L} = \frac{\nabla_y T}{H \nabla_x T}$ with $Q_y = J_{e,x} = J_{e,y} = 0$. Thus

$$\mathcal{L}_R = \frac{L_{34}}{HL_{33}} = \frac{L_{34}}{H\kappa}, \tag{2.131}$$

which is the part of L_{34} in (2.126). We discuss this Leduk-Righi effect in the context of magnon thermal Hall effect in §6.3.1. The details of exhaustive thermo-electromagnetic transport coefficients can be found in [38][40]. For example, the transport coefficients of the adiabatic process are listed there in detail along with the relations between L_{ij} and L'_{ij}.

Chapter 3

Hall Viscosity

After introducing the Hall viscosity in the context of universal hydrodynamics without parity symmetry, we would like to illustrate two independent computations of Hall viscosity done in quantum Hall systems. First, we compute it using the Berry phase [41] and the quantum mechanical adiabatic theory in quantum Hall fluid. This leads to a surprising relation between the viscosity and angular momentum of the system [42]. Second, we also demonstrate that the Hall viscosity is directly related to the Hall conductivity under the influence of electromagnetic fields [43]. These two different relations associated with the Hall viscosity are tied to the fact that there are two independent sets of mutually compatible symmetries discussed in §1.5. We start with the motion of a charged particle under the influence of electromagnetic fields, especially the magnetic field, that sets up the basis of quantum Hall systems.

3.1 Charged particles in electromagnetic fields

We consider the motion of a charged particle, with charge q and mass m, in the presence of external electric $\vec{E}$ and magnetic $\vec{B}$ fields. We utilize the Lagrangian formalism and correspondingly

---

Skyrmions and Hall Transport
Bom Soo Kim
Copyright © 2023 Jenny Stanford Publishing Pte. Ltd.
ISBN 978-981-4968-34-8 (Hardcover), 978-1-003-37253-0 (eBook)
www.jennystanford.com

the background gauge potentials $A^\mu(t, x^i(t)) = (\phi/c, \vec{A})^T$ where $\mu, \nu = 0, 1, 2, 3$ and $i, j, k = 1, 2, 3$ along with the metric signature $(-+++)$. We use t and x^0 interchangeably, and set $c = 1$. Then,

$$E_i = -\partial_i \phi - \dot{A}_i, \qquad B_i = \epsilon_{ijk} \partial_j A_k, \qquad (3.1)$$

where $\dot{}$ is a partial derivative with respect to t.

The action $\mathcal{S} = \int dt \mathcal{L}$ is given by

$$\mathcal{L} = \frac{m}{2} \dot{x}^\mu \dot{x}_\mu + q \dot{x}^\mu A_\mu, \qquad (3.2)$$

where repeated indices are summed over. The last term couples the particle's dynamics with the source A_μ in quantum field theory. This modifies the canonical momentum to $\vec{p} - q\vec{A}$ as we see below. If we develop the dynamics with the proper time τ as $x^\mu(\tau)$, this Lagrangian provides a fully relativistic description with manifest 'minimal coupling' scheme. Here, we single out time t as an independent parameter, and the spatial coordinates are functions of time, $x^i(t)$. The Lagrangian has a constant term $(m/2)\dot{x}^0 \dot{x}_0 = -m/2$, which is irrelevant and discarded. The Lagrangian is turned into

$$\mathcal{L} = \frac{m}{2} \dot{x}^i \dot{x}_i + q \dot{x}^i A_i - q\phi, \qquad (3.3)$$

where ϕ, A^i are the scalar and vector potentials.

A direct variation of action (3.3) with respect to the dependent variables x^i gives

$$\delta \mathcal{S} = \int dt \delta x^i \left[-m \ddot{x}_i - q \frac{dA_i}{dt} + q \dot{x}^\mu \partial_i A_\mu \right], \qquad (3.4)$$

up to total time derivative terms. Using $\delta \mathcal{S} = 0$ and the fact that the second term in (3.4) has two contributions as $(d/dt)A_i = \dot{x}^\nu \partial_\nu A_i = \dot{A}_i + \dot{x}^j \partial_j A_i$,

$$\begin{aligned} m\ddot{x}_i &= -q(\dot{A}_i + \dot{x}^j \partial_j A_i) + q \dot{x}^j \partial_i A_j - q \partial_i \phi \\ &= q(-\dot{A}_i - q\partial_i \phi) + q \dot{x}^j (\partial_i A_j - \partial_j A_i) \\ &= q \left(E_i + \epsilon_{ijk} \dot{x}^j B^k \right), \end{aligned} \qquad (3.5)$$

where we use $\dot{x}^0 = 1$, $A_\mu = (-\phi, A_i)$ and $\partial_i A_j - \partial_j A_i = \epsilon_{ijk} B^k$. In the vector notation, this gives the familiar Lorentz force law

$$\vec{F} = m\dot{\vec{v}} = q(\vec{E} + \vec{v} \times \vec{B}) . \tag{3.6}$$

One can check that the Euler-Lagrange equation for the coordinate x^i, $0 = (d/dt)(\partial \mathcal{L}/\partial \dot{x}^i) - \partial \mathcal{L}/\partial x^i = (d/dt)(m\dot{x}_i + qA_i) - q\dot{x}^\mu \partial_i A_\mu$, yields the same result.

With the canonical momentum defined by

$$p_i = \frac{\partial \mathcal{L}}{\partial \dot{x}^i} = m\dot{x}_i + qA_i , \tag{3.7}$$

one can compute the corresponding Hamiltonian

$$\begin{aligned} \mathcal{H} = p^i \dot{x}_i - \mathcal{L} &= \frac{m}{2} \dot{x}^i \dot{x}_i + q\phi + \frac{m}{2} \\ &= \frac{1}{2m}(\vec{p} - q\vec{A})^2 + q\phi + \frac{m}{2} , \end{aligned} \tag{3.8}$$

where the overall constant $m/2$ can be ignored. $q\phi$ is the electric potential energy. For example, $\phi = 0$ describes a free particle, $\phi = -Ex$ describes a particle under the influence of electric field $\vec{E} = E\hat{x}$, and $\phi = q_0/(4\pi\epsilon_0 r)$ describes a charged particle bounded by Coulomb potential with ϵ_0 as the permittivity of free space. The first term in the second line of (3.8) describes the motion of a charged particle under the influence of the external magnetic field. As it is the same as $m\dot{x}^i \dot{x}_i/2$, the energy is independent of the magnetic field, manifesting the fact the magnetic field does not work.

Under the so-called gauge transformation $A_\mu \to A_\mu + \partial_\mu \lambda$, or in a vector form

$$\vec{A} \to \vec{A} + \vec{\nabla}\lambda , \qquad \phi \to \phi - \partial_t \lambda , \tag{3.9}$$

the classical Lagrangian (3.2) is invariant up to a total derivative term. It is nothing but a surface term $q \int dt \dot{x}^\mu \partial_\mu \lambda = q \int d\lambda$ and does not change the equation of motion. Furthermore, the electric and magnetic fields are also invariant under the transformation. For a constant magnetic field, the vector potential can be written in the symmetric gauge

$$\vec{A} = \vec{B} \times \frac{\vec{x}}{2}, \tag{3.10}$$

as $(\vec{\nabla} \times \vec{A})_i = \epsilon_{ijk}\partial_j(\vec{B} \times \vec{x}/2)_k = \epsilon_{ijk}\epsilon_{klm}B_l(\partial_j x_m)/2 = \epsilon_{ijk}\epsilon_{klj}B_l/2 = B_i$, where $\partial_j x_m = \delta_{jm}$ and $\epsilon_{ijk}\epsilon_{klj} = 2\delta_{il}$ in 3 spatial dimensions as epsilon symbol is only a tensor when the indices fully cover the entire space-time dimensions. Then the Lagrangian can be recast to the form

$$\mathcal{L} = \frac{m}{2}\dot{x}^i\dot{x}_i + \frac{q}{2}\epsilon_{ijk}B_i x_j \dot{x}_k - q\phi. \tag{3.11}$$

We come back to this Lagrangian below.

It is interesting to consider the infinite magnetic field limit, $B \to \infty$. To have a nontrivial dynamics, we also keep the last term. These contributions can be achieved by taking a massless limit, $m \to 0$. Equation of motion $m\dot{\vec{v}} = q(\vec{E} + \vec{v} \times \vec{B})$ given in (3.5) reduces to

$$\vec{v} \times \vec{B} = -\vec{E}. \tag{3.12}$$

When the electric and magnetic fields are orthogonal to each other, $\vec{B} \perp \vec{E}$, we can get a simple equation for guiding center dynamics. By applying $\times \vec{B}$ on both sides of (3.12), we get $((\vec{v} \times \vec{B}) \times \vec{B})_i = \epsilon_{ijk}\epsilon_{jlm}v_l B_m B_k = v_k B_i B_k - v_i B_k B_k = -v_{\perp,i}B^2$. $v_{\perp,i} = v_i - B_i(\vec{v} \cdot \vec{B})/B^2$ is the velocity component perpendicular to the magnetic field. Then, (3.12) turns into

$$\vec{v}_\perp = \frac{\vec{E} \times \vec{B}}{\vec{B} \cdot \vec{B}}, \tag{3.13}$$

where $\vec{v}_\perp = \vec{v} - \vec{B}(\vec{v} \cdot \vec{B})/B^2$. The velocity component that is parallel to the magnetic field does not change. More detailed description of this guiding center dynamics is further described in §3.5.1.

In the massless limit, the system is described by a Lagrangian with a single time derivative. The action with the first time derivative can be recast to a familiar form

$$\mathcal{S} = q\int_{t_i}^{t_f} dt\,[\dot{x}^i A_i] = q\int_{x_i}^{x_f} dx^i A_i. \tag{3.14}$$

This action is independent of time as long as the path of the particle is the same. This is nothing but the Berry phase term. We come back to this below.

3.1.1 Quantum Hall fluid

In this subsection, we specialize in 2 spatial dimensions with coordinates (x,y), and carry on quantization in the presence of a magnetic field, $\vec{B} = B\hat{z}$, perpendicular to the plane. Here we provide some background materials for quantum Hall systems. For more information, you can find useful references in the literature. For example, [44] is available on arXiv.

The systems preserve the rotational and translational symmetries for a constant magnetic field. The corresponding classical Hamiltonian is given by (3.8).

$$\mathcal{H} = \frac{1}{2m}\vec{\pi}^2, \qquad \vec{\pi} = \vec{p} - q\vec{A}. \tag{3.15}$$

The canonical coordinates (x_i, p_i) with $i,j = x,y$ satisfy the Poisson relations.

$$\begin{aligned}\{x_i, p_j\} &= \delta_{ij}, \\ \{x_i, x_j\} &= \{p_i, p_j\} = 0,\end{aligned} \tag{3.16}$$

defined by the Poisson bracket $\{f,g\} = (\partial f/\partial x^i)(\partial g/\partial p^i) - (\partial g/\partial x^i)(\partial f/\partial p^i)$ with the repeated indices summed over. We see that $\vec{\pi} = m\dot{\vec{x}}$ is gauge invariant but not a canonical momentum as its Poisson brackets are given by

$$\begin{aligned}\{\pi_x, \pi_y\} &= qB, \\ \{\pi_x, \pi_x\} &= \{\pi_y, \pi_y\} = 0.\end{aligned} \tag{3.17}$$

The computation goes as

$$\begin{aligned}\{\pi_i, \pi_j\} &= \frac{\partial \pi_i}{\partial x^k}\frac{\partial \pi_j}{\partial p^k} - \frac{\partial \pi_j}{\partial x^k}\frac{\partial \pi_i}{\partial p^k} = -q\frac{\partial A_i}{\partial x^k}\delta_{jk} + q\frac{\partial A_j}{\partial x^k}\delta_{ik} \\ &= q(\partial_i A_j - \partial_j A_i) = q\epsilon_{ijk}B_k.\end{aligned} \tag{3.18}$$

On the other hand, p_i is a canonical momentum, but not a gauge invariant quantity (thus not physical observable) as it compensates the change of A_i under the gauge transformation.

We proceed to find the energy levels by quantizing the system. We do so by treating $\vec{p}$ and $\vec{A}$ as quantum operators and promoting the Poisson bracket to the canonical commutation relations.

$$[x_i, p_j] = i\hbar \delta_{ij} ,$$
$$[x_i, x_j] = [p_i, p_j] = 0 , \quad (3.19)$$

where $\hbar$ is the Planck constant and i the imaginary number satisfying $i^2 = -1$. Then $[\pi_x, \pi_y] = i\hbar qB$. We introduce the raising and lowering operators as

$$a = \frac{1}{\sqrt{2\hbar qB}}(\pi_x + i\pi_y) , \qquad a^\dagger = \frac{1}{\sqrt{2\hbar qB}}(\pi_x - i\pi_y) , \quad (3.20)$$

which satisfies

$$[a, a^\dagger] = 1 . \quad (3.21)$$

Using $\pi_x = \sqrt{\hbar qB/2}(a + a^\dagger), \pi_y = -i\sqrt{\hbar qB/2}(a - a^\dagger)$, the Hamiltonian is simply

$$\mathcal{H} = \frac{1}{2m}\vec{\pi}^2 = \hbar\left(\frac{qB}{m}\right)\left(a^\dagger a + \frac{1}{2}\right) . \quad (3.22)$$

This is similar to the Hamiltonian for a harmonic oscillator.

We construct the Hilbert space of the quantum Hall system by introducing a ground state $|0\rangle$ obeying $a|0\rangle = 0$. By applying the raising operator $a^\dagger$ repeatedly we build up the other states $|n\rangle$ with positive integer n. These states satisfy the orthogonal and normalized condition $\langle m|n\rangle = \delta_{mn}$. By defining the number operator $N = a^\dagger a$ that satisfies $N|n\rangle = n|n\rangle$, we get

$$a^\dagger|n\rangle = \sqrt{n+1}|n+1\rangle , \qquad a|n\rangle = \sqrt{n}|n-1\rangle . \quad (3.23)$$

From the definition $a^\dagger|n\rangle = c|n+1\rangle$ and its conjugate $\langle n|a = \langle n+1|c^*$, we can check $\langle n|aa^\dagger|n\rangle = \langle n|a^\dagger a + 1|n\rangle = n+1 = |c|^2\langle n+1|n+1\rangle = |c|^2$ that sets $c = \sqrt{n+1}$. Similarly, $a|n\rangle = \tilde{c}|n-1\rangle$ gives

$\tilde{c} = \sqrt{n}$. Thus, the spectrum of the Hamiltonian (3.15) is given by

$$E_n = \hbar\left(\frac{qB}{m}\right)\left(n + \frac{1}{2}\right), \tag{3.24}$$

for the states $|n\rangle = (a^{\dagger n}/\sqrt{n!})|0\rangle$ for $n = 0, 1, 2, 3 \cdots$. Thus the energy levels for the excited states are equally spaced by $\hbar qB/m$ that is proportional to the magnetic field B. They are called the Landau levels. Note that the ground state energy does not vanish and is given by $E_0 = \hbar qB/2m$. Notice that we do not need to specify the vector potential and the corresponding gauge choices to obtain the energy spectrum. Before going further, we note that the energy spectrum (3.24) is the same as that of one dimensional harmonic oscillator. Thus (3.24) does not fully capture the original 2 dimensional degrees of freedom. It turns out the energy levels are degenerate.

3.1.2 Wave function

Here we use the symmetric gauge introduced in (3.10) to explore the degeneracy and the wave function of the system. In 2 spatial dimensions,

$$\vec{A} = \vec{B} \times \frac{\vec{x}}{2} = B\left(-\frac{y}{2}\hat{x} + \frac{x}{2}\hat{y}\right), \tag{3.25}$$

which breaks the translational symmetry in x and y directions, while preserving the rotational symmetry, depicted in Fig. 3.1. This means the intermediate steps are not symmetric under the translation. Our final results are independent of the gauge choice, and thus restore all the symmetries.

If one goes through the detailed computations on π and the relations of a and $a^\dagger$ given in equations, (3.17), (3.20), and (3.21), it is straightforward to see that it works the same by replacing q to $-q$. Thus we define

$$\tilde{\vec{\pi}} = \vec{p} + q\vec{A}, \tag{3.26}$$

which is useful to illustrate the degeneracies. Note that it is not gauge invariant due to the relative sign, and thus does not contribute to any

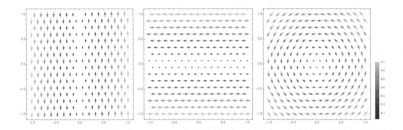

Figure 3.1 Illustration of the vector potentials in 2 spatial dimensions that provide $\vec{B} = B\hat{z}$. Left: $\vec{A} = xB\hat{y}$, Middle: $\vec{A} = -yB\hat{x}$, and Right: $\vec{A} = B(-y/2\hat{x} + x/2\hat{y})$ with $B = 1$. They show the corresponding symmetries, in particular the rotation symmetry in the right panel.

physical observables. Then,

$$[\tilde{\pi}_x, \tilde{\pi}_y] = -i\hbar qB,$$
$$[\tilde{\pi}_x, \tilde{\pi}_x] = [\tilde{\pi}_y, \tilde{\pi}_y] = 0. \tag{3.27}$$

Furthermore, one can show that

$$[\pi_i, \tilde{\pi}_j] = 0. \tag{3.28}$$

In general, this is not true in other gauge choices. Nevertheless, we can check it by deriving the general expressions, $[\pi_x, \tilde{\pi}_x] = -2i\hbar q \partial_x A_x$, $[\pi_y, \tilde{\pi}_y] = -2i\hbar q \partial_y A_y$, $[\pi_x, \tilde{\pi}_y] = [\pi_y, \tilde{\pi}_x] = -i\hbar q(\partial_x A_y + \partial_y A_x)$, which happen to vanish for the symmetric gauge. For other gauge choices, one can not simultaneously diagonalize $\tilde{\pi}$ together with π, and cannot provide information on the degeneracy.

Now we can define another set of ladder operators.

$$b = \frac{1}{\sqrt{2\hbar qB}} (\tilde{\pi}_x - i\tilde{\pi}_y), \qquad b^\dagger = \frac{1}{\sqrt{2\hbar qB}} (\tilde{\pi}_x + i\tilde{\pi}_y), \tag{3.29}$$

which satisfies

$$[b, b^\dagger] = 1. \tag{3.30}$$

As already worked out, the Hamiltonian is still the same as (3.22) with the same energy spectrum (3.24). As we construct another set of ladder operator $(b, b^\dagger)$ that commutes with $a, a^\dagger$, we can define a refined ground state $|0, 0\rangle$ that is annihilated by both a and b as $a|0, 0\rangle = b|0, 0\rangle = 0$. Then we construct the rest of the Hilbert space using both creation operators $a^\dagger$ and $b^\dagger$ as

$$|n, m\rangle = \frac{a^{\dagger n} b^{\dagger m}}{\sqrt{n! m!}} |0, 0\rangle . \tag{3.31}$$

The energy spectrum only depends on n and the degeneracy associated with the Landau level is described by the integer m.

With the information, we can construct the wave function of the Landau levels. For the ground state that occupies the lowest Landau level, the state satisfies $a|0, m\rangle = 0$. From this we can construct the ground state wave function. Explicitly, the operator a has the form

$$\begin{aligned} \sqrt{2\hbar qB} a &= (\pi_x + i\pi_y) = (p_x + ip_y) - q(A_x + iA_y) \\ &= -i\hbar(\partial_x + i\partial_y) - qB(-y + ix)/2 . \end{aligned} \tag{3.32}$$

To solve the equation $a|0, m\rangle = 0$, it is convenient to introduce holomorphic and anti-holomorphic coordinates

$$z = x + iy , \qquad \bar{z} = x - iy , \tag{3.33}$$

and the corresponding derivatives

$$\partial \equiv \partial_z = (\partial_x - i\partial_y)/2 , \qquad \bar\partial \equiv \partial_{\bar z} = (\partial_x + i\partial_y)/2 . \tag{3.34}$$

To derive ∂, one can define $\partial = c_x \partial_x + c_y \partial_y$ and demand $\partial z = 1$ and $\partial \bar z = 0$ that fix the coefficients as $c_x = -c_y = 1/2$. Similarly, $\bar\partial$ satisfies $\bar\partial z = 0$ and $\bar\partial \bar z = 1$. Now the operators a and $a^\dagger$ are given by

$$a = -i\sqrt{2} l_B \left(\bar\partial + \frac{z}{4 l_B^2} \right) , \qquad a^\dagger = -i\sqrt{2} l_B \left(\partial - \frac{\bar z}{4 l_B^2} \right) , \tag{3.35}$$

where $l_B = \sqrt{\hbar/qB}$ with $q = |e|$ the electric charge. We can see the Hermitian conjugate is different from the complex conjugate as the former picks up $-$ sign for a derivative in addition to exchange the holomorphic and anti-holomorphic coordinates.

The solution of the differential equation in the position basis $\langle z, \bar{z} | 0, m \rangle = a\psi(z, \bar{z}) = 0$ is given by

$$\psi_{LLL}^m(z, \bar{z}) = \psi_m(z) e^{-|z|^2/4l_B^2}, \tag{3.36}$$

where LLL indicates the lowest Landau level and $\psi_m(z)$ is a holomorphic function. The first order differential equation $(\bar{\partial} + z/(4l_B^2))\psi(z, \bar{z}) = 0$, with $\bar{\partial} = \partial/\partial \bar{z}$, is separable as $d\psi/\psi = z d\bar{z}/(4l_B^2)$, which can be solved by a direct integration. Thus, $\ln \psi = c(z) - z\bar{z}/(4l_B^2)$, where $\psi_m(z) = e^{c(z)}$ is an integration constant for $d\bar{z}$ integral. The wave functions for higher Landau levels can be constructed by acting the creation operators on the LLL states (3.36). They are

$$\psi_n^m = \frac{1}{\sqrt{n!}} (a^\dagger)^n \psi_{LLL}^m(z, \bar{z}), \tag{3.37}$$

where $a^\dagger$ is given by (3.34).

To determine $\psi_m(z)$, we note that the LLL state also satisfies $b|0, m = 0\rangle = 0$. The operator b can be written as

$$\sqrt{2\hbar q B} b = (\tilde{\pi}_x - i\tilde{\pi}_y) = (p_x - ip_y) + q(A_x - iA_y)$$
$$= -i\hbar(\partial_x - i\partial_y) - iqB\frac{(x-iy)}{2}. \tag{3.38}$$

Thus, using the holomorphic coordinates as before, we get

$$b = -i\sqrt{2} l_B \left(\partial + \frac{\bar{z}}{4l_B^2}\right), \qquad b^\dagger = -i\sqrt{2} l_B \left(\bar{\partial} - \frac{z}{4l_B^2}\right). \tag{3.39}$$

Using this, we see $\langle x|b|0, 0\rangle = \partial \psi_0(z) = 0$. Thus we get $\psi_0 = \mathcal{N}_0$, a constant. The rest of the LLL state can be constructed by using the creation operator $b^\dagger$ as

$$\psi_{LLL}^1(z, \bar{z}) = b^\dagger \mathcal{N}_0 e^{-|z|^2/4l_B^2}$$
$$= i\mathcal{N}_0 (z/\sqrt{2} l_B) e^{-|z|^2/4l_B^2}, \tag{3.40}$$

which shows that each time we apply $b^\dagger$ there is an additional factor $z/\sqrt{2} l_B$. As the wave function is not observable, overall i does not

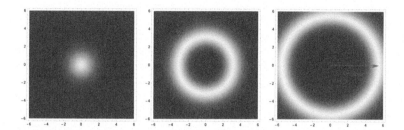

Figure 3.2 The LLL states $|\psi_m|^2$, $m = 0, 5, 15$ from the left, given in (3.41) for $l_B = 1$ for illustration purpose. $r_{m,peak} = \sqrt{2m}l_B$ is the peak radius of the state $|0, m\rangle$. $r_{5,peak} = 5.47723$.

make a difference. In the end, the general expression for the wave functions is given by

$$\psi_{LLL}^m(z, \bar{z}) = \mathcal{N}_m z^m e^{-|z|^2/4l_B^2},$$
$$\mathcal{N}_m^{-1} = \sqrt{\pi m!}\,(\sqrt{2}l_B)^{m+1}, \tag{3.41}$$

which contains $\sqrt{m!}$ according to (3.31). $\mathcal{N}_0$ and the general expression $\mathcal{N}_m$ can be determined by the condition $\int |\psi_{LLL}^m|^2 = 1$. The integral can be evaluated as $1 = |\mathcal{N}_m|^2 \int_0^\infty r dr \int_0^{2\pi} d\phi r^{2m} e^{-r^2/2l_B^2} = |\mathcal{N}_m|^2 \cdot \pi (2l_B^2)^{m+1} \Gamma(m+1)$ in polar coordinate with $z\bar{z} = r^2$. Some of the LLL states are illustrated as density plots in Fig. 3.2. The peak of the probability function $|\psi|^2$ is given by $r_{m,peak} = \sqrt{2m}l_B$ that can be obtained from $\partial_r |\psi|^2 = 0$.

To count the degeneracy of LLL state, we need a finite system that requires to specify a boundary condition. Here we roughly estimate it by computing the average size of the state with m.

$$\langle r^2 \rangle_m = \int d^2x r^2 |\psi|^2 = 2\pi |\mathcal{N}_m|^2 \int_0^\infty dr r^{2m+3} e^{-r^2/2l_B^2}$$
$$= 2l_B^2(m+1). \tag{3.42}$$

Thus when m gets bigger, the ring-shaped state is further away from the origin. For large m, both the average and peak radius grow as $r \sim \sqrt{2m}l_B$. This indicates the number of LLL states contained in area $A =$

πR^2, for a sufficiently large R, is

$$N = \frac{R^2}{2l_B^2} = \frac{qBA}{2\pi\hbar} = \frac{\Phi}{\Phi_0}. \qquad (3.43)$$

The number of ground states is given by the total magnetic flux going through the area $\Phi = BA$ divided by the flux quantum $\Phi_0 = 2\pi\hbar/q = B \cdot (2\pi l_B^2)$, the magnetic flux contained in the area $2\pi l_B^2$.

3.2 Landau Hamiltonian on torus

Let us consider a bunch of electrons moving in a 2 dimensional plane (with a periodic boundary condition) in the presence of an external magnetic field B, which is perpendicular to the plane. When the magnetic field is sufficiently strong, the electron spins are completely polarized. Thus, we can treat the electrons as spin-less particles. We utilize the Lagrangian formalism and correspondingly the gauge potentials $\vec{A}$. The aim of this section is to develop the wave function of quantum Hall system, which is crucial for computing the Hall viscosity. The reference [45] is useful for this section.

3.2.1 Flat torus

Let us consider a charged particle moving in a constant magnetic field B on a torus. The Hamiltonian is given by

$$\mathcal{H} = \frac{1}{2m}(\vec{p} - q\vec{A})^2 = -\frac{1}{2}\left((\partial_1 - iA_1)^2 + (\partial_2 - iA_2)^2\right), \qquad (3.44)$$

where we set $m = q = \hbar = 1$ and $B = \partial_1 A_2 - \partial_2 A_1$. Here we use the index $1, 2$ instead of x, y. There are many different choices for the vector potential that gives the same magnetic field. For example, The following three different choices, $(A_1, A_2) = (-Bx_2, 0)$, $(A_1, A_2) = (0, Bx_1)$, and $(A_1, A_2) = (-Bx_2/2, Bx_1/2)$, give the same magnetic field. We choose the first gauge, $A_1 = -Bx_2, A_2 = 0$, to proceed.

In 2 dimensions, it is convenient to employ the complex coordinates

$$z = x^1 + ix^2, \qquad \bar{z} = x^1 - ix^2, \qquad (3.45)$$

and the corresponding derivatives

$$\partial_z = \frac{1}{2}(\partial_1 - i\partial_2), \qquad \partial_{\bar{z}} = \frac{1}{2}(\partial_1 + i\partial_2), \qquad (3.46)$$

such that they satisfy $\partial_z z = \partial_{\bar{z}}\bar{z} = 1$ and $\partial_z \bar{z} = \partial_{\bar{z}} z = 0$. In this coordinates, the vector potentials are $A_z = (A_1 - iA_2)/2$, and $A_{\bar{z}} = (A_1 + iA_2)/2$. Then $A_z = A_{\bar{z}} = -B\operatorname{Im}(z)/2$ for $(A_1, A_2) = (-Bx_2, 0)$.

The torus can be represented by a parallelogram on the complex plane generated by the lattice vectors 1 and $\tau = \tau_1 + i\tau_2$ with $\tau_2 > 0$. τ is called moduli. The torus can be viewed as imposing periodic boundary conditions both on x^1 and x^2 coordinates. Now, changing moduli can be considered to change the boundary condition, which in turn corresponds to different deformation of torus. We remind that this is nothing but the small strain deformations that create stresses in the context of hydrodynamics.

We define the creation and annihilation operators

$$a = i\sqrt{\frac{2}{B}}(\partial_{\bar{z}} - iA_{\bar{z}}), \qquad a^\dagger = i\sqrt{\frac{2}{B}}(\partial_z - iA_z), \qquad (3.47)$$

which satisfy $[a, a^\dagger] = 1$. Then $\partial_1 - iA_1 = -i\sqrt{B/2}(a + a^\dagger), \partial_2 - iA_2 = -\sqrt{B/2}(a - a^\dagger)$. The Hamiltonian has the form

$$\mathcal{H}(B) = B\left(a^\dagger a + \frac{1}{2}\right). \qquad (3.48)$$

This is the Hamiltonian for a harmonic oscillator. Until now, the discussion is similar to the previous section except that we use the complex coordinates.

One important observation is that the flux B on torus has to be quantized. One way to see is that the usual translation generators $p_1 = -i\partial_1$ and $p_2 = -i\partial_2$ do not commute with the Landau Hamiltonian (3.44). On the other hand, the magnetic generators

$$T_1 = p_1, \qquad T_2 = p_2 + Bx_1, \tag{3.49}$$

commute with the Landau Hamiltonian. Moreover, it turns out that the two magnetic generators do not commute with each other. Now, we consider two finite translation operators, $G_1 = e^{ip_1}$ and $G_\tau = e^{i\tau_1 p_1 + i\tau_2(p_2 + Bx_1)}$, corresponding to the lattice vectors 1 and τ on torus. We compute $G_1 G_\tau = G_\tau G_1 e^{i\tau_2 B}$. We require the commutators for the finite translation operators to commute with each other. Then,

$$\text{Im}(\tau) B = 2\pi N, \tag{3.50}$$

where N is an integer. The integral for the torus is done over the parallelogram with the area τ_2. This relation tells that the modulus τ_2 is related to B, which in turn shows the Hamiltonian (3.48) depends on the modulus τ_2.

Another important observation associated with the Landau Hamiltonian on torus is the fact that the ground state is degenerate, which is mentioned in the previous subsection. The Dirac operator $D(B) = i\gamma^i(\partial_i - iA_i)$, with the Pauli matrices $\gamma^1 = \sigma^1, \gamma^2 = \sigma^2$, has the form

$$D(B) = \sqrt{2B}\begin{pmatrix} 0 & a^\dagger \\ a & 0 \end{pmatrix}, \quad D(B)^2 = 2B\begin{pmatrix} a^\dagger a & 0 \\ 0 & aa^\dagger \end{pmatrix}, \tag{3.51}$$

where we use (3.47). Here we choose $B > 0$ without loss of generality. Now we observe that

$$D(B)^2 \geq 0, \tag{3.52}$$

for the ground state $\varphi^{(0)}$. Thus the two conditions, $aa^\dagger \geq 0$ and $a^\dagger a = aa^\dagger - 1 \geq 0$, indicate that $a^\dagger \varphi^{(0)} \neq 0$. The dimension of function space for an operator A when satisfying $A^\dagger \varphi^{(0)} = 0$ is called dimension of Kernal. As $a^\dagger$ acting on the ground state does not vanish, dim Ker $a^\dagger = 0$. According to the Atiyah-Singer index theorem on compact manifolds

$$\text{index} D(B) = \dim \text{Ker } a - \dim \text{Ker } a^\dagger = \frac{1}{2\pi}\int F = N. \tag{3.53}$$

Thus the ground state, defined by $a\varphi_\alpha^{(0)} = 0$ with $\alpha = 1, 2, \cdots, N$, is N fold degenerate. As the spectra of $a^\dagger a$ and $a a^\dagger$ are the same, even the excited states are N fold degenerate as well.

The ground state $\varphi_\alpha^{(0)}$ can be constructed using $a\varphi_\alpha^{(0)} \propto (\partial_{\bar{z}} + iB\,\text{Im}(z)/2)\varphi_\alpha^{(0)} = 0$. We identify $B = 2\pi N/\text{Im}(\tau)$. Then

$$\varphi_\alpha^{(0)}(z|\tau) = \mathcal{N}_\alpha\, e^{-\frac{\pi N}{\text{Im}(\tau)}(\text{Im}(z))^2} f_\alpha(z|\tau) , \tag{3.54}$$

where f_α is a holomorphic function in z, $\partial_{\bar{z}} f(z) = 0$ and $\mathcal{N}_\alpha$ is a normalization factor. The excited states can be constructed using the creation operators, $\varphi_\alpha^{(n)} = (a^\dagger)^n \varphi_\alpha^{(0)}/\sqrt{n!}$, and the energy can be computed $E^{(n)} = (n + 1/2) 2\pi N/\text{Im}(\tau)$.

We determine the holomorphic function f by using the periodic boundary condition $G_1 \varphi_\alpha^{(0)} = \varphi_\alpha^{(0)}$ and $G_\tau \varphi_\alpha^{(0)} = \varphi_\alpha^{(0)}$ with the magnetic translation operators, T_1 and T_2 given in (3.49). Straightforward computations give

$$\begin{aligned} f_\alpha(z+1|\tau) &= f_\alpha(z|\tau) , \\ f_\alpha(z+\tau|\tau) &= f_\alpha(z|\tau) e^{-i\pi N(2\tau+z)} , \end{aligned} \tag{3.55}$$

where the second line is the result of the fact that two exponents $\tau_1 p_1$ of G_1 and $\tau_2(p_2 + Bx_1)$ of G_τ do not commute with each other. It turns out that Jacobi theta functions satisfy these periodicity conditions. Thus,

$$f_\alpha(z|\tau) = \vartheta \begin{bmatrix} \alpha/N \\ 0 \end{bmatrix}(Nz|N\tau) = \sum_{m \in \mathbb{Z}} e^{i\pi(m+\alpha/N)^2 N\tau + 2\pi i(m+\alpha/N)Nz} . \tag{3.56}$$

For our reference, we list some properties of the Jacobi theta function here.

$$\vartheta\begin{bmatrix} a \\ b \end{bmatrix}(z|\tau) = \sum_{m \in \mathbb{Z}} e^{i\pi(m+a)^2 \tau + 2\pi i(m+a)(z+b)} . \tag{3.57}$$

The theta function has the periodicity $\vartheta\begin{bmatrix}a+1\\b\end{bmatrix} = e^{-2\pi i a}\vartheta\begin{bmatrix}a\\b+1\end{bmatrix} = \vartheta\begin{bmatrix}a\\b\end{bmatrix}$. It also has the periodicity in $z \to z+1$ and $z \to z+\tau$ as

$$\vartheta\begin{bmatrix}a\\b\end{bmatrix}(z+1|\tau) = e^{2\pi i a}\vartheta\begin{bmatrix}a\\b\end{bmatrix}(z|\tau),$$

$$\vartheta\begin{bmatrix}a\\b\end{bmatrix}(z+\tau|\tau) = e^{-\pi i(\tau+2z+2b)}\vartheta\begin{bmatrix}a\\b\end{bmatrix}(z|\tau).$$

(3.58)

Here we can see that the periodic properties of $f_a(z|\tau)$ match with those of the theta function. Furthermore, the theta function has the modular transformation properties as

$$\vartheta\begin{bmatrix}a\\b\end{bmatrix}(z|\tau+1) = e^{-\pi i a(a+1)}\vartheta\begin{bmatrix}a\\b+a+1/2\end{bmatrix}(z|\tau),$$

$$\vartheta\begin{bmatrix}a\\b\end{bmatrix}\left(\frac{z}{\tau}\middle|-\frac{1}{\tau}\right) = (-i\tau)^{1/2} e^{\pi i z^2/\tau + 2\pi i a b}\vartheta\begin{bmatrix}b\\-a\end{bmatrix}(z|\tau).$$

(3.59)

3.2.2 Deformed torus

Let us consider a coordinate a transformation from the coordinates (x^1, x^2) to (θ^1, θ^2) with $0 \le \theta^1, \theta^2 \le 2\pi$ described by

$$z = x^1 + ix^2 = \frac{1}{2\pi}(\theta^1 + \tau\theta^2).$$

(3.60)

The line elements are given by $ds^2 = \delta_{ij}dx^i dx^j = g_{\mu\nu}d\theta^\mu d\theta^\nu$, where $i, j, \mu, \nu = 1, 2$ in this section. The moduli dependent metric has the form

$$g_{\mu\nu} = \frac{1}{4\pi^2}\begin{pmatrix}1 & \tau_1\\ \tau_1 & |\tau|^2\end{pmatrix}, \quad g^{\mu\nu} = \frac{4\pi^2}{\tau_2^2}\begin{pmatrix}|\tau|^2 & -\tau_1\\ -\tau_1 & 1\end{pmatrix}.$$

(3.61)

One can compute the volume, which is the area, as

$$V = \int_0^{2\pi} d\theta^1 \int_0^{2\pi} d\theta^2 \sqrt{\det(g)} = \tau_2.$$

(3.62)

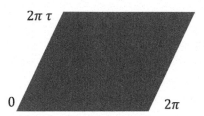

Figure 3.3 Deformed torus with the coordinates (θ^1, θ^2).

In the new coordinate, the Hamiltonian has the form

$$\mathcal{H} = -\frac{1}{2}g^{\mu\nu}D_\mu D_\nu$$
$$= -\frac{2\pi^2}{\tau_2^2}\left[|\tau|^2 D_1^2 - \tau_1(D_1 D_2 + D_2 D_1) + D_2^2\right], \quad (3.63)$$

where $D_\mu = \partial/\partial\theta^\mu - iA_\mu$. From the transformation, we check

$$x^1 = \frac{1}{2\pi}(\theta^1 + \tau_1\theta^2), \qquad x^2 = \frac{1}{2\pi}\tau_2\theta^2. \quad (3.64)$$

For the gauge field, $A = A_i dx^i = -Bx_2 dx^1 = -(N/2\pi)\theta^2 d\theta^1 - (N/2\pi)\tau_1\theta^2 d\theta^2$. Thus we see that the gauge field has an additional contribution in the new coordinate. This can be removed by a gauge transformation by a factor $(N/4\pi)\tau_1(\theta^2)^2$, which in turn change the ground state wave function as $\psi_\alpha^{(0)} \to e^{i(N/4\pi)\tau_1(\theta^2)^2}\varphi_\alpha^{(0)}$. Thus the deformed ground state wave function has the form

$$\psi_\alpha^{(0)} = \mathcal{N}_\alpha\, e^{i(N/4\pi)\tau(\theta^2)^2}\,\vartheta\begin{bmatrix}\alpha/N\\ 0\end{bmatrix}\left(\frac{N}{2\pi}(\theta^1 + \tau\theta^2)\Big|N\tau\right). \quad (3.65)$$

Note the important fact that this ground state, $\psi_\alpha^{(0)}$, becomes a holomorphic function of τ (except the normalization factor $\mathcal{N}_\alpha$) as the gauge transformation combines nicely with the exponential factor in (3.54). The theta function has the normalization property such that $\int_0^{2\pi} d\theta^1 \int_0^{2\pi} d\theta^2 \bar{\psi}_\alpha^{(0)}\psi_\beta^{(0)} = \text{const.} \cdot \mathcal{N}^2 \delta_{\alpha\beta}/\sqrt{\tau_2}$. Thus the normalization factor is independent of α, the index representing degenerate states,

$$\mathcal{N}_\alpha = \text{const.} \cdot \tau_2^{1/4} . \tag{3.66}$$

The ground state satisfies the boundary condition

$$\begin{aligned}\psi_\alpha^{(0)}(\theta^1 + 2\pi, \theta^2|\tau) &= \psi_\alpha^{(0)}(\theta^1, \theta^2|\tau) , \\ \psi_\alpha^{(0)}(\theta^1, \theta^2 + 2\pi|\tau) &= e^{-iN\theta^1 + 2\pi i N\tau} \psi_\alpha^{(0)}(\theta^1, \theta^2|\tau) .\end{aligned} \tag{3.67}$$

After the gauge transformation, the potential has a simple form, $A = -(N/2\pi)\theta^2 d\theta^1$. Then

$$D_1 = \partial_1^\theta + \frac{N}{2\pi}\theta^2 , \qquad D_2 = \partial_2^\theta . \tag{3.68}$$

These satisfy $[D_1, D_1] = [D_2, D_2] = 0$ and $[D_1, D_2] = -iN/2\pi$. With the differential operators, we can find the deformed creation and annihilation operators

$$\begin{aligned}b(\tau) &= -\sqrt{\frac{\pi}{N\text{Im}(\tau)}} \left(\partial_2^\theta - \tau(\partial_1^\theta + i\frac{N}{2\pi}\theta^2) \right) , \\ b^\dagger(\tau) &= \sqrt{\frac{\pi}{N\text{Im}(\tau)}} \left(\partial_2^\theta - \bar{\tau}(\partial_1^\theta + i\frac{N}{2\pi}\theta^2) \right) .\end{aligned} \tag{3.69}$$

The factor i in front of N in (3.69) ensures $[b, b^\dagger] = 1$, not $[b, b^\dagger] = i$. Thus, $D_1 = -(i/2)\sqrt{N/\pi\tau_2}(b + b^\dagger)$ and $D_2 = -(i/2)\sqrt{N/\pi\tau_2}(\bar{\tau}b + \tau b^\dagger)$. The deformed Hamiltonian has the same form as in (3.48) with the operators $b, b^\dagger$ instead of $a, a^\dagger$. The excited states can be obtained by applying the creation operator $b^\dagger$ repeatedly.

3.2.2.1 Gauge connection with deformation

Finally, we are ready to compute the gauge connection and curvature associated with the change of the ground state by slowly deforming the moduli τ of the torus. This depends only on the instantaneous eigenstate of the deformed Hamiltonian as explained below.

As the ground state is holomorphic function of τ except the normalization factor $\mathcal{N}$, we use the derivative in the form $d = d\tau \partial_\tau + d\bar{\tau}\partial_{\bar{\tau}}$. We perform the computation first. The gauge connection is given by

$$\begin{aligned}\mathcal{A}_{\alpha\beta}^{(0)} &= i\langle\psi_\alpha^{(0)}|d|\psi_\beta^{(0)}\rangle \\
&= i\langle\psi_\alpha^{(0)}|(d\tau\partial_\tau + d\bar\tau\partial_{\bar\tau})|\psi_\beta^{(0)}\rangle \\
&= id\tau\partial_\tau\langle\psi_\alpha^{(0)}|\psi_\beta^{(0)}\rangle - i(d\tau\partial_\tau\langle\psi_\alpha^{(0)}|)|\psi_\beta^{(0)}\rangle + i\langle\psi_\alpha^{(0)}|d\bar\tau\partial_{\bar\tau}|\psi_\beta^{(0)}\rangle \quad (3.70)\\
&= 0 - \frac{id\tau}{4(\tau-\bar\tau)}\delta_{\alpha\beta} - \frac{id\bar\tau}{4(\tau-\bar\tau)}\delta_{\alpha\beta} \\
&= -\frac{1}{4}\frac{d\tau_1}{\tau_2}\delta_{\alpha\beta},\end{aligned}$$

where we use $\langle\psi_\alpha^{(0)}|\psi_\beta^{(0)}\rangle = \delta_{\alpha\beta}$, $|\psi_\beta^{(0)}\rangle \propto \tau_2^{1/4}f_\beta(\tau)$, $\langle\psi_\alpha^{(0)}| \propto \tau_2^{1/4}f_\beta(\bar\tau)$, and $\tau_2 = (\tau-\bar\tau)/(2i)$. The corresponding curvature can be evaluated as

$$\mathcal{F}_{\alpha\beta}^{(0)} = d\mathcal{A}_{\alpha\beta}^{(0)} = -\frac{1}{4}\frac{d\tau_1 \wedge d\tau_2}{\tau_2^2}\delta_{\alpha\beta}. \quad (3.71)$$

This is the result of a single particle ground state. For the fully degenerate ground states, we sum over the indices α and β, which results in multiplying the factor N to the equations (3.70) and (3.71).

For the first excited states, we compute

$$\begin{aligned}\mathcal{A}_{\alpha\beta}^{(1)} &= i(\langle\psi_\alpha^{(0)}|b)d(b^\dagger|\psi_\beta^{(0)}\rangle) \\
&= i(\langle\psi_\alpha^{(0)}|b)(d\tau\partial_\tau + d\bar\tau\partial_{\bar\tau})(b^\dagger|\psi_\beta^{(0)}\rangle) \\
&= i\langle\psi_\alpha^{(0)}|(bb^\dagger d\tau\partial_\tau + d\bar\tau\partial_{\bar\tau}bb^\dagger + d\tau b(\partial_\tau b^\dagger) - d\bar\tau(\partial_{\bar\tau}b)b^\dagger)|\psi_\beta^{(0)}\rangle \\
&= \mathcal{A}_{\alpha\beta}^{(0)} - \frac{1}{2}\frac{d\tau_1}{\tau_2}\delta_{\alpha\beta},\end{aligned} \quad (3.72)$$

where we use the fact that the derivatives does not change the orthogonal condition except generating extra factors of τ_2, and thus $bb^\dagger = 1$.

From the relation $\mathcal{A}_{\alpha\beta}^{(0)}$ and $\mathcal{A}_{\alpha\beta}^{(1)}$, one can get the recursion relation $\mathcal{A}_{\alpha\beta}^{(n)} = \mathcal{A}_{\alpha\beta}^{(n-1)} - (1/2)(d\tau_1/\tau_2)\delta_{\alpha\beta}$. Thus,

$$\mathcal{A}_{\alpha\beta}^{(n)} = -\frac{1}{2}\left(n+\frac{1}{2}\right)\frac{d\tau_1}{\tau_2}\delta_{\alpha\beta}. \tag{3.73}$$

The corresponding curvature two form for the excited state is

$$\mathcal{F}_{\alpha\beta}^{(n)} = d\mathcal{A}_{\alpha\beta}^{(n)} = -\frac{1}{2}\left(n+\frac{1}{2}\right)\frac{d\tau_1 \wedge d\tau_2}{\tau_2^2}\delta_{\alpha\beta}. \tag{3.74}$$

This is the Berry curvature that we have been aiming for. In the following sections, we connect this to the Hall viscosity.

3.3 Berry phase and Hall viscosity

In the previous section, we compute the Berry connection and curvature by using the moduli of a torus. The change in the moduli deforms the shape of torus and thus can be used as a strain transformation. In this section we connect this to the Hall viscosity through the Kubo formula. Let us start with a general formulation of adiabatic process.

3.3.1 Adiabatic process and Berry phase

Berry's phase is a quantum mechanical phase arising in systems that undergo a slow evolution. It has far reaching consequences that have been used in many different context, dubbed as a geometric phase.

Under an adiabatic variation of a system's parameter space, for example a slow variation of the moduli space τ in §3.2.2, the environment and the corresponding Hamiltonian H are slowly changed. Due to the slow change, the system will be in an eigenstate of the instantaneous H at any instant of time. When the Hamiltonian is returned to its original form, the system will return to its original state with a modified phase factor. This phase factor can be observed by interference if the adiabatically changed system is recombined with an another system that was separated at an earlier time and whose Hamiltonian was unchanged. This phase factor contains a path-dependent component $e^{i\gamma}$ in addition to the familiar dynamical one e^{-iEt}, which describes the time evolution of any stationary state. We recommend to read the original Berry's paper [41] as it is clear and inspiring.

To be more specific, consider a system described by a state $|\phi(R)\rangle$ and a Hamiltonian $H(R)$ depending on a set of parameters $R(t)$. The state evolves according to the Schrödinger equation

$$H(R(t))|\phi(t)\rangle = i\hbar|\dot\phi(t)\rangle, \tag{3.75}$$

where $\dot{}$ is a time derivative. At any instant of time, there exist eigenstates of H, $|\psi(R)\rangle$, that satisfy $H(R)|\psi(R)\rangle = E(R)|\psi(R)\rangle$. Note that there is no relation between the phases of the eigenstates $|\psi(R)\rangle$ at different R. For example, it is the case for the deformed Hamiltonian $H(\tau)$ with the moduli parameter τ considered in (3.63) and the corresponding ground state $|\psi_\alpha^{(0)}(\tau)\rangle$ given by (3.65) in §3.2.2. The energy in the example is independent of the deformed parameter τ.

When the system is adiabatically evolved from an initial state $|\phi(t=0)\rangle$ to a final state $|\phi(t=T)\rangle$ through a closed path C in the parameter space $R(t)$ such that $R(T) = R(0)$, the total phase change of the state $|\phi\rangle$ is given by

$$|\phi(T)\rangle = e^{i\gamma(C)}e^{-i\int_0^T dt E(R(t))/\hbar}|\phi(0)\rangle, \tag{3.76}$$

where γ is the Berry phase. For the adiabatic approximation to hold, T must be large compared to the other relevant scales of the system. Note that there is another phase factor, proportional to the energy $E(R)$, that comes from the time evolution of the system through the Hamiltonian.

Let us look into an intermediate state at an instant of time t. If the initial state $|\phi(0)\rangle$ is prepared to be an eigenstate $|\psi(R)\rangle$, the intermediate state can be described as

$$|\phi(t)\rangle = e^{i\gamma(t)}e^{-i\int_0^t dt' E(R(t'))/\hbar}|\psi(R(t))\rangle. \tag{3.77}$$

We work out the Schrödinger equation (3.75) for the state $|\phi(t)\rangle$ to get $H|\psi(R(t))\rangle = i\hbar(i\dot\gamma(t) - iE(R)/\hbar)|\psi(R(t))\rangle + i\hbar|\dot\psi(R(t))\rangle$, where the exponential factors $e^{i\gamma(t)}e^{-i\int_0^t dt' E(R(t'))/\hbar}$ do not contribute as they are common for both sides of the equation. Then, after using the eigenvalue equation for Hamiltonian, we have

$$\dot{\gamma}(t)|\psi(R(t))\rangle = i|\dot{\psi}(R(t))\rangle . \tag{3.78}$$

By applying the bra vector $\langle\psi(R(t))|$, we get $\dot{\gamma}(t) = i\langle\psi(R(t))|\nabla_R\psi(R(t))\rangle \cdot \dot{R}(t)$. Upon integrating along the loop C,

$$\gamma(C) = i\oint_C \langle\psi(R(t))|\nabla_R\psi(R(t))\rangle \cdot dR(t) . \tag{3.79}$$

This Berry phase is called a geometric phase and given by a loop integral in parameter space R. It is independent of the way the system is transported. For a general state, the computation of the phase can be done conveniently using the Stokes's theorem in vector calculus by transforming the loop integral to an integral for a surface, whose boundary is C in the parameter space. For the example given in §3.2.2, the Berry connection and the corresponding curvature are computed already in (3.70) and (3.71).

3.3.2 Kubo formula for Hall viscosity

To facilitate the identification of the Hall viscosity with the Berry phase, we would like to introduce the "effective field theory with general background fields" such as metric or gauge fields, which is introduced in §2.5. This method has been used extensively in high energy theory and is useful in capturing all the known transport coefficients and thermodynamic quantities in a systematic way. It has been extended, in recent years, to the condensed matter community by incorporating the non-relativistic symmetry [46].

Specifically, we apply this to the quantum mechanical adiabatic theory in terms of the uniform geometric deformation called strain ξ_{ij}. Strain is given by a symmetric derivative of a displacement vector, $\xi_i = x'_i - x_i$ for a deformation from x_i to x'_i. Thus, $\xi_{ij} = (\partial_i\xi_j + \partial_j\xi_i)/2$. In adiabatic approximation, the geometric deformation is small, and the general metric can be expanded as $g_{ij} = \delta_{ij} + 2\xi_{ij}$. The factor 2 between the metric and deformation requires attentions to get a proper Hall viscosity below. At the end of our computations, we set $g_{ij} \to \delta_{ij}$ to find physical quantities such as conductivities and viscosities.

We already mentioned the hydrodynamic description of the stress energy tensor in §2.2, which has the following form:

$$T_{ij} = -\lambda_{ijkl}\xi_{kl} - \eta_{ijkl}\dot{\xi}_{kl} + \cdots, \tag{3.80}$$

where λ_{ijkl} is the elastic modulus measuring the inverse compressibility. The Hall viscosity is a part of the shear tensor η_{ijkl}, which is related to the retarded Green's function of the commutator with the energy momentum tensor as discussed in §2.5. The result is known as the Kubo formula. As it is associated with time derivative of variation, it is sufficient to consider the local rest frame of the fluid at zero spatial momentum. Due to the symmetry, all the time-dependent geometric deformations $\xi_{ij}(t) = \{\xi_{xy}(t), \xi_{xx}(t), \xi_{yy}(t)\}$ are related in 2 dimensions. This is more transparent in the language of metric perturbations.

It is sufficient to consider the stress energy tensor T^{xy} [34] as there is only one independent Hall viscosity component. Shear and Hall viscosities split themselves according to their tensor structures. Then

$$T^{xy} \supset -\eta \dot{\xi}_{xy} + \eta_H(\dot{\xi}_{xx} - \dot{\xi}_{yy}) + \mathcal{O}(\xi^2). \tag{3.81}$$

The transport coefficients η and η_H are coefficients of the linear terms in the stress energy tensor. Thus it is reformulated in terms of retarded Green's function as

$$\langle T^{ij}(x^\mu)\rangle_\xi = \langle T^{ij}(x^\mu)\rangle_{\xi=0} \\ - \frac{1}{2}\int d^3x'\, G_R^{ij,kl}(x^\mu;x'^\mu)\xi_{kl}(x'^\mu) + \mathcal{O}(\xi^2), \tag{3.82}$$

where $x^\mu = (t, x, y)$ and the retarded Green's function $G_R^{ij,kl}(x^\mu;x'^\mu)$

$$G_R^{ij,kl}(x^\mu;x'^\mu) = -i\theta(t - t')\langle[T^{ij}(x^\mu), T^{kl}(x'^\mu)]\rangle, \tag{3.83}$$

and the Fourier transform $\tilde{G}_R^{ij,kl}(\omega, \vec{k})$ with the frequency ω and momentum $\vec{k}$ have the forms

$$\tilde{G}_R^{ij,kl}(\omega, \vec{k}) = \int dt e^{i\omega t} \int d^2 x e^{-ik^l x_l} G_R^{ij,kl}(t, \vec{x}; 0, \vec{0}) . \qquad (3.84)$$

Thus one finds

$$\tilde{G}_R^{xy,xx-yy}(\omega, \vec{0}) = 2i\omega \eta_H + \mathcal{O}(\omega^2) . \qquad (3.85)$$

One can work out the general expression in covariant form as

$$\eta_H = \lim_{\omega \to 0} \frac{\epsilon_{ik}\delta_{jl}}{4i\omega} \tilde{G}_R^{ij,kl}(\omega, \vec{0}) . \qquad (3.86)$$

Thus we see that linear response theory connects the transport coefficients such as Hall viscosity to some components of the retarded Green's functions. From the Kubo formula (3.86), we see that the Hall viscosity is antisymmetric for exchanging the two sets of symmetric indices (ij) and (kl).

3.3.3 Hall viscosity of quantum Hall states

Now we see this linear response formulas from slightly different point of view. We obtain the stress energy tensor by applying the variations of the general background metric, g_{ij}, to the Hamiltonian H. (We note the Hamiltonian is a Legendre transformation of the Lagrangian, $\log \mathcal{Z} \propto \mathcal{L}$, and contains the same information in different variables.)

$$T_{ij} = \langle \frac{\delta H}{\delta \xi^{ij}} \rangle = \frac{\delta E}{\delta \xi^{ij}} + \Omega_{ijkl} \dot{\xi}_{kl} + \cdots , \qquad (3.87)$$

where $\langle \ \rangle$ is the expectation value for a normalized quantum state $|\psi\rangle$. For a generic fluid, the first term on the right-hand side contains a contact term, pressure, and the elastic modulus tensor, which measures the inverse compressibility and can be rewritten as $\frac{\delta E}{\delta \xi^{ij}} \supset (1/2)(\delta^2 E/\delta \xi^{ij} \delta \xi^{kl})\xi^{kl} = -\lambda_{ijkl}\xi^{kl}$ for the rotational invariant systems. This is in consistent with (3.81).

The last term Ω_{ijkl} in (3.87) is the adiabatic or Berry curvature term for a slow dynamics, and given by the variation of the state with respect to the strain ξ_{ij}.

$$\Omega_{ijkl} = \text{Im}\left[\frac{\partial}{\partial \xi^{ij}}\langle\psi|\frac{\partial\psi}{\partial\xi^{kl}}\rangle - \frac{\partial}{\partial\xi^{kl}}\langle\psi|\frac{\partial\psi}{\partial\xi^{ij}}\rangle\right]$$

$$= 2\,\text{Im}\langle\frac{\partial\psi}{\partial\xi^{ij}}|\frac{\partial\psi}{\partial\xi^{kl}}\rangle = 8\,\text{Im}\langle\frac{\partial\psi}{\partial g^{ij}}|\frac{\partial\psi}{\partial g^{kl}}\rangle\,. \tag{3.88}$$

This adiabatic curvature is non-trivial if the phase of the state ψ changes along a closed path through the geometric deformation. Note that this adiabatic term is antisymmetric $\Omega_{ijkl} = -\Omega_{klij}$ under the exchange of the first and second symmetric pairs of indices (ij) and (kl). Thus it is directly related to the Hall viscosity.

Now we apply these to the charged particle moving in a constant magnetic field B on a torus considered in §3.2. The variation of metric is equivalent with the changing of the moduli τ with a fixed area on torus. Here we use slightly generalized metric of deformed torus than (3.90) by including the volume factor and change the domain of the coordinates $0 \leq \theta^1, \theta^2 \leq 1$ to avoid the factors of 2π.

$$ds^2 = \frac{V}{\tau_2}\left(d(\theta^1)^2 + 2\tau_1 d(\theta^1)d(\theta^2) + |\tau|^2 d(\theta^2)^2\right), \tag{3.89}$$

then the lower and upper index metrics are

$$g_{\mu\nu} = \frac{V}{\tau_2}\begin{pmatrix} 1 & \tau_1 \\ \tau_1 & |\tau|^2 \end{pmatrix}, \quad g^{\mu\nu} = \frac{1}{V\tau_2}\begin{pmatrix} |\tau|^2 & -\tau_1 \\ -\tau_1 & 1 \end{pmatrix}. \tag{3.90}$$

We can rewrite (V, τ_1, τ_2) in terms of (g^{11}, g^{12}, g^{22}) as $V = 1/\sqrt{\text{Det}(g_{up})}$, $\tau_1 = -g^{xy}/g^{yy}$, and $\tau_2 = \sqrt{\text{Det}(g_{up})}/g^{yy}$, where $\text{Det}(g_{up}) = g^{xx}g^{yy} - (g^{xy})^2$. Then we work out the chain rules for the derivatives of metric in terms of the moduli.

$$\frac{\partial}{\partial g^{11}} = -\frac{1}{2}\frac{\partial}{\partial V} + \frac{1}{2}\frac{\partial}{\partial \tau_2},$$

$$\frac{\partial}{\partial g^{22}} = -\frac{1}{2}\frac{\partial}{\partial V} - \frac{1}{2}\frac{\partial}{\partial \tau_2}, \tag{3.91}$$

$$\frac{\partial}{\partial g^{12}} = \frac{\partial}{\partial g^{21}} = -\frac{1}{2}\frac{\partial}{\partial \tau_1},$$

where we evaluate the result at $V = \tau_2 = 1, \tau_1 = 0$ to have the flat metric on torus after taking the derivatives.

Then the Berry curvature can be recast as

$$\Omega_{1112} = -\Omega_{2212} = 4 \operatorname{Im}\left(\langle\frac{\partial\psi}{\partial g^{11}}| - \langle\frac{\partial\psi}{\partial g^{22}}|\right)|\frac{\partial\psi}{\partial g^{12}}\rangle$$

$$= 2\operatorname{Im}\langle\frac{\partial\psi}{\partial\tau_1}|\frac{\partial\psi}{\partial\tau_2}\rangle \quad (3.92)$$

$$= -4\operatorname{Im}(i\partial_\tau\partial_{\bar\tau}\ln\mathcal{N}) = -(\partial_{\tau_1}^2 + \partial_{\tau_2}^2)(\ln\mathcal{N}),$$

where we use $|\psi\rangle = |\mathcal{N}f(\tau)\rangle$ and the normalization factor is real and a function of τ_2, $\mathcal{N} = \mathcal{N}(\tau_2)$. We present some intermediate steps. From $\tau = \tau_1 + i\tau_2$, we get $\partial_\tau = (\partial_{\tau_1} - i\partial_{\tau_2})/2$, $\partial_{\bar\tau} = (\partial_{\tau_1} + i\partial_{\tau_2})/2$, and thus $\partial_{\tau_1} = \partial_\tau + \partial_{\bar\tau}$, $\partial_{\tau_2} = i\partial_\tau - i\partial_{\bar\tau}$. When the derivatives act on the state, we use the holomorphic property of the state except the normalization factor.

$$\operatorname{Im}\langle\frac{\partial\psi}{\partial\tau_1}|\frac{\partial\psi}{\partial\tau_2}\rangle = \operatorname{Im}\left[i((\partial_\tau + \partial_{\bar\tau})\langle\mathcal{N}f(\bar\tau)|)((\partial_\tau - \partial_{\bar\tau})|\mathcal{N}f(\tau)\rangle)\right]$$

$$= \operatorname{Im}\left[i([(\partial_\tau \ln(\mathcal{N})) - \partial_{\bar\tau}]\langle\mathcal{N}f(\bar\tau)|)([\partial_\tau - (\partial_{\bar\tau}\ln(\mathcal{N}))]|\mathcal{N}f(\tau)\rangle)\right] \quad (3.93)$$

$$= \operatorname{Im}\left[-i(\partial_\tau \ln(\mathcal{N}))(\partial_{\bar\tau}\ln(\mathcal{N})) + i(\partial_{\bar\tau}\langle\mathcal{N}f(\bar\tau)|)(\partial_\tau|\mathcal{N}f(\tau)\rangle)\right]$$

$$= -2\operatorname{Im}(i\partial_\tau\partial_{\bar\tau}\ln\mathcal{N}),$$

where the notation $\langle\mathcal{N}f(\bar\tau)|$ indicates that it is an anti-holomorphic function. In the third line, we omit the real part and use the orthonormal property of the wave function. In the last line, we use $(\partial_{\bar\tau}\langle\mathcal{N}f(\bar\tau)|)(\partial_\tau|\mathcal{N}f(\tau)\rangle) = -2\partial_\tau\partial_{\bar\tau}(\ln\mathcal{N}) + (\partial_\tau \ln(\mathcal{N}))(\partial_{\bar\tau}\ln(\mathcal{N}))$ as $\langle\bar\psi|\psi\rangle = \langle\mathcal{N}f(\bar\tau)|\mathcal{N}f(\tau)\rangle = 1$ and thus $\partial_\tau\partial_{\bar\tau}(\langle\mathcal{N}f(\bar\tau)|\mathcal{N}f(\tau)\rangle) = 0$.

Using the normalization factor $\mathcal{N} = const. \cdot \tau_2^{N/4}$ evaluated in (3.66), we compute the Hall viscosity

$$\eta_H = \eta_{1112} = \Omega_{1112} = \frac{1}{4}\frac{N}{V}\hbar = \frac{1}{2}\bar s \bar n \hbar, \quad (3.94)$$

where $\bar n = N/V = \nu/(2\pi\ell_B^2)$ (with filling factor ν and magnetic length ℓ_B) is the average density of particles and $\bar s = -\ell = 1/2$ is the average angular momentum, which is, for example, spin due to the cyclotron motion for non-interacting particles in a magnetic field.

This non-trivial relation between the Hall viscosity and angular momentum is interesting because the average angular momentum is related to a topological quantity called the shift, $\bar{S} = 2\bar{s}$, the change in magnetic flux when the ground state is formulated on a sphere. In general, the shift $\bar{S}$ is the offset $N_\phi = \nu^{-1}N - \bar{S}$ that is required when the system is put in a curved space with a different genus compared to that of the plane geometry $N_\phi = \nu^{-1}N$, where N_ϕ is the number of magnetic flux quanta piercing the surface, N the number of particle, and ν the filling factor [47]. Here we put the expression as a general form following [48][42] for quantum Hall systems with a gap. See also some earlier computations of the Hall viscosity in [45][49][50]. Here we omit the discussion of shift, a very interesting topological quantity. Interested readers are encouraged to read the reference [47].

We did not explicitly talk about the underlying symmetries that are relevant for the result (3.94) except the magentic translation generators (3.49). Here we can understand that spontaneously generated angular momentum also exists due to the broken parity as it carries a preferred direction. As explained in §1.5, this is not compatible with the translation symmetry. Thus, the result (3.94) is valid for the physical systems with rotation symmetry and angular momentum.

Until now in this chapter, we consider the relation between the Hall viscosity and angular momentum in the context of some gapped systems. Now we change our gear and consider the Hall conductivity of the quantum Hall systems and its relation to the Hall viscosity.

3.4 Quantum Hall systems and Hall conductivity

Quantum Hall systems show interesting phenomena, especially in the conductivity data as a function of an external magnetic field $B(T)$ in the unit of Tesla for integer quantum Hall systems [51] and for fractional quantum Hall systems [52]. These data are nicely combined in [53], which is depicted in Fig. 3.4.

The Hall resistivity data ρ_{xy}, upper line, reveals that there are prominent plateaus for various filling factors $\nu = 2/5, 3/7, \cdots, 3/5, 2/3$ and especially $\nu = 1, 2, 3, 4, \cdots$ that are

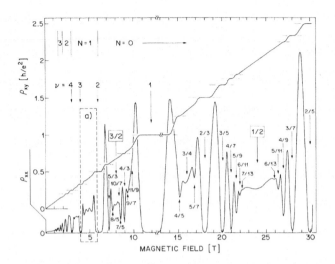

Figure 3.4 Resistivity ρ_{xx} and Hall resistivity ρ_{xy} data as a function of magnetic field for a quantum Hall system. N and ν indicate the Landau levels and filling factors respectively. Reproduced with permission from [53].

listed from the right side of Fig. 3.4. The corresponding longitudinal resistivity data points for ρ_{xx} turn out to vanish. Here we only consider the simpler case with the integer filling factors, $\nu =$ integer. This is the integer quantum Hall systems, where we can understand the system without taking into account the interactions between electrons.

As explained in §2.6, this can be understood for 2 dimensional conductivity σ or resistivity ρ formula as

$$\sigma = \begin{pmatrix} \sigma_{xx} & \sigma_{xy} \\ -\sigma_{xy} & \sigma_{xx} \end{pmatrix}, \quad \rho = \frac{1}{\sigma_{xx}^2 + \sigma_{xy}^2} \begin{pmatrix} \sigma_{xx} & -\sigma_{xy} \\ \sigma_{xy} & \sigma_{xx} \end{pmatrix}. \quad (3.95)$$

When the conductivity tensor has vanishing diagonal components $\sigma_{xx} = 0$ and non-vanishing off-diagonal components $\sigma_{xy} \neq 0$, $\rho_{xx} = \sigma_{xx}/(\sigma_{xx}^2 + \sigma_{xy}^2) = 0$, and $\rho_{xy} = -\sigma_{xy}/(\sigma_{xx}^2 + \sigma_{xy}^2) = -1/\sigma_{xy}$. The case $\sigma_{xx} = 0$ and $\rho_{xx} = 0$ signify that there is no current flowing along the longitudinal direction parallel to the applied electric field, while $\sigma_{xy} \neq 0$ and $\rho_{xy} \neq 0$ indicate that the current flows transverse to the field.

The classical computation for the conductivity using the Drude model for a particle with charge $q = -e$ and mass m have the result (2.106) in §2.6.

$$\sigma = \frac{ne^2\tau_s/m}{1+\omega_B^2\tau_s^2}\begin{pmatrix} 1 & \omega_B\tau_s \\ -\omega_B\tau_s & 1 \end{pmatrix}, \qquad (3.96)$$

where $\omega_B = -eB/m$ is the cyclotron frequency, n the particle density, and τ_s the scattering time. In terms of the resistivity tensor, we have

$$\rho = \frac{m}{ne^2\tau_s}\begin{pmatrix} 1 & -\omega_B\tau_s \\ \omega_B\tau_s & 1 \end{pmatrix}. \qquad (3.97)$$

The longitudinal and Hall resistivity have the values

$$\rho_{xx} = \frac{m}{ne^2\tau_s}, \qquad \rho_{xy} = \frac{B}{ne}. \qquad (3.98)$$

The Hall viscosity is independent of the scattering time τ_s and the particle mass m.

From the experimental data [51], the Hall viscosity has the values

$$\rho_{xy} = \frac{2\pi\hbar}{e^2}\frac{1}{\nu}. \qquad (3.99)$$

By comparing these two expressions we identify

$$n = \frac{eB}{2\pi\hbar}\nu = \frac{B}{\Phi_0}\nu, \qquad (3.100)$$

where $\Phi_0 = 2\pi\hbar/e$ is the density of electrons that is required to fill the ν-th Landau levels.

3.4.1 Conductivity for a single particle

We come up with the results (3.98) for the conductivity, or rather resistivity, by using the classical Drude model. We can compute the conductivity for a single free particle by applying an electric field and using the resulting wave function.

We have used the symmetric gauge as in (3.25) to construct the wave function with explicit holomorphic properties. Here we use the Landau gauge

$$\vec{A} = xB\hat{y},\tag{3.101}$$

which contains only x coordinate. The Hamiltonian (3.15) has the form

$$\mathcal{H} = \frac{1}{2m}\left(p_x^2 + (p_y + eBx)^2\right),\tag{3.102}$$

where we use $q = -e$, the charge for an electron.

As the Hamiltonian (3.102) has manifest translation invariance along the y coordinate, we try to construct the wave function as

$$\psi_k(x,y) = e^{iky}f_k(x).\tag{3.103}$$

Then the Hamiltonian acting on this wave function has the form

$$\mathcal{H}\psi_k(x,y) = \frac{1}{2m}\left(p_x^2 + (\hbar k + eBx)^2\right)\psi_k(x,y).\tag{3.104}$$

This is nothing but the Hamiltonian for a harmonic oscillator with a displaced center in x direction

$$\mathcal{H} = \frac{1}{2m}p_x^2 + \frac{m\omega_B^2}{2}(x + kl_B^2)^2,\tag{3.105}$$

where $\omega_B = eB/m$ and $l_B = \sqrt{\hbar/eB}$. The position of the harmonic oscillator is centered at $x = -kl_B^2$, which is the momentum along y direction.

From the experience with the harmonic oscillator, we know that the energy is quantized as (3.24) and the corresponding wave function has the form

$$\psi_{n,k}(x,y) \sim H_n(x + kl_B^2)e^{-(x+kl_B^2)^2/2l_B^2}e^{iky},\tag{3.106}$$

where H_n is the Hermite polynomial. We did not normalize the wave function. The wave function is extended along the y direction while exponentially localized around $x = -kl_B^2$, which is different from (3.36) for the symmetric gauge.

To compute the conductivity we turn on a constant electric field along x direction with a guage potential $\phi = -Ex$. Again, by anticipating the wave function to be a plane wave along the y coordinate, we have $p_y = \hbar k$. The Hamiltonian (3.102) is modified as

$$\mathcal{H} = \frac{1}{2m}\left(p_x^2 + (p_y + eBx)^2\right) - eEx$$

$$= \frac{1}{2m}p_x^2 + \frac{m\omega_B^2}{2}\left(x + eBp_y - \frac{mE}{eB^2}\right)^2 + \frac{Ep_y}{B} - \frac{mE^2}{2B^2}, \quad (3.107)$$

where we complete the square for the x coordinate to absorb the new term in the second line.

Thus the energies are given by

$$E_{n,k} = \hbar\omega_B\left(n + \frac{1}{2}\right) + \frac{E}{B}\hbar k - \frac{mE^2}{2B^2}. \quad (3.108)$$

This result contains $p_y = \hbar k$, and thus the degeneracy of the energy level discussed in §3.1.2 is lifted by the momentum in y direction. Thus, the states drift in y direction with the group velocity

$$v_y = \frac{1}{\hbar}\frac{\partial E_{n,k}}{\partial k} = \frac{E}{B}. \quad (3.109)$$

Thus, the cyclotron orbit of the electron drifts along y direction, that is not along the direction of electric field, but the direction of $\vec{E} \times \vec{B}$.

The wave function is also modified as

$$\psi(x,y) = \psi_{n,k}\left(x - \frac{mE}{eB^2}, y\right)$$

$$\sim H_n\left(x + kl_B^2 - \frac{mE}{eB^2}\right)e^{-(x+kl_B^2 - mE/eB^2)^2/2l_B^2}e^{iky}. \quad (3.110)$$

With this wave function we can compute the conductivity in an intuitive way [44].

From the velocity $m\dot{\vec{x}} = \vec{p} + e\vec{A}$ with the canonical momentum $\vec{p}$. The current is given by $\vec{I} = -e\dot{\vec{x}}$ that can be written in a quantum mechanical picture as

$$\vec{I} = -\frac{e}{m}\sum_{\text{filled states}}\langle\psi| -i\hbar\vec{\nabla} + e\vec{A}|\psi\rangle. \quad (3.111)$$

Here the filled states are given by the index n and wave vector $\vec{k}$. With the ν Landau levels filled and the gauge potential as $\vec{A} = xB\hat{y}$ in (3.101), we have the current in x direction as

$$I_x = -\frac{e}{m}\sum_{n=1}^{\nu}\sum_k \langle\psi_{n,k}|-i\hbar\partial_x|\psi_{n,k}\rangle = 0. \tag{3.112}$$

This vanishes as $p_x \propto a^\dagger - a$ with the creation and annihilation operators in harmonic oscillator (x) direction.

The current in y direction is more interesting.

$$\begin{aligned}
I_y &= -\frac{e}{m}\sum_{n=1}^{\nu}\sum_k \langle\psi_{n,k}|-i\hbar\partial_y + eBx|\psi_{n,k}\rangle \\
&= -\frac{e}{m}\sum_{n=1}^{\nu}\sum_k \langle\psi_{n,k}|\hbar k + eBx|\psi_{n,k}\rangle \\
&= -e\nu\sum_k \frac{E}{B} = -e\nu\frac{E}{\Phi_0}A.
\end{aligned} \tag{3.113}$$

Here we evaluate the expectation value of x as the shifted origin of the harmonic oscillator as $\langle\psi_{n,k}|x|\psi_{n,k}\rangle = -kl_B^2 + mE/eB^2 = -\hbar k/eB + mE/eB^2$. We also use the result (3.43), $\sum_k = N = \Phi/\Phi_0 = BA/\Phi_0$, which is the number of electrons with the area A. This gives the current (density) $\vec{I}/A = \vec{j} = \sigma \cdot \vec{E}$. Thus

$$\sigma_{xx} = \sigma_{yy} = 0, \quad \sigma_{xy} = -\sigma_{yx} = \frac{e\nu}{\Phi_0}. \tag{3.114}$$

In terms of the resistivity, we get

$$\rho_{xx} = \rho_{yy} = 0, \quad \rho_{xy} = -\sigma_{yx} = -\frac{e\nu}{\Phi_0}. \tag{3.115}$$

These conductivity and resistivity are the results of the quantum Hall plateaux. We recommend the readers to look into [44] for illuminating discussions on the roles of edge modes and the disorder in the context of the conductivity.

3.4.2 Kubo formula for Hall conductivity

Here we derive the Kubo formula for the Hall conductivity with a general multi-particle Hamiltonian H_0 in the context of linear responses theory discussed in §2.5.1. More systematic treatment can be found in [54]. The Hamiltonian H_0 is unperturbed before turning on an electric field $\vec{E}$. It has the energy eigenstates $|n\rangle$ with the eigenvalues E_n as $H_0|n\rangle = E_n|n\rangle$.

We add a background electric field $\vec{E} = -\vec{\nabla}\phi - \partial_t\vec{A} = -\partial_t\vec{A}$ by choosing the gauge $\phi = 0$. The full Hamiltonian is $H = H_0 + H_s$ with

$$H_s = -\vec{J} \cdot \vec{A}, \tag{3.116}$$

where $\vec{J}$ is the quantum operator associated with the electric current. Typically the interaction term is introduced as $\Delta L = J_\mu A^\mu$ as in Lagrangian (3.2) and thus $J^\mu = q\dot{x}^\mu$ in this case.

We want to compute the current $\langle \vec{J} \rangle$ due to the perturbation H_s by assuming that the electric field is small, and thus the perturbation theory is applicable. We are interested in computing the conductivity when the electric field is constant. To make things easier, we use the complex notation as $\vec{E}(t) = \vec{E}e^{-i\omega t}$ and thus

$$\vec{A} = \frac{\vec{E}}{i\omega}e^{-i\omega t}, \tag{3.117}$$

which is useful for evaluating the relation between the current and electric field. In the end, we take the DC limit $\omega \to 0$.

As in §2.5.1, we work with the interaction picture. The operators and states evolve as

$$\begin{aligned} \mathcal{O}(t) &= U_0^{-1}(t)\mathcal{O}U_0(t), \\ |\psi(t)\rangle_I &= U_s(t, t_0)|\psi(t_0)\rangle_I, \end{aligned} \tag{3.118}$$

where $U_0(t) = e^{-iH_0 t/\hbar}$ and $U_s(t, t_0) = T\exp\left(-\frac{i}{\hbar}\int_{t_0}^t H_s(t')dt'\right)$ with time ordering T. The operator $U_s(t, t_0)$ satisfies $i\hbar\, dU_s(t, t_0)/dt = H_s U_s(t, t_0)$. We use the notation $U_s(t, t_0 \to -\infty) = U_s(t)$.

For a many-body system, we prepare the system at $t \to -\infty$ in a specific many-body state, for example, the ground state $|0\rangle$. The expectation value of the current is given by

$$\langle \vec{J}(t) \rangle = \langle 0(t)|\vec{J}(t)|0(t)\rangle = \langle 0|U_s^{-1}(t)\vec{J}(t)U_s(t)|0\rangle$$
$$\approx \langle 0|\left(\vec{J}(t) + \frac{i}{\hbar}\int_{-\infty}^{t} dt'[H_s(t'),\vec{J}(t)]\right)|0\rangle \quad (3.119)$$

where we expand the evolution operator $U_s(t) \approx 1 - (i/\hbar)\int_{t_0}^{t} H_s(t')dt'$ and keep up to the first order in H_s. The first term in the final expression is current in the absence of electric field, which is assumed to vanish. The second term gives the current due to the applied electric field. Using (3.116) and (3.117), we get

$$\langle J_i(t)\rangle = \frac{1}{\hbar\omega}\int_{-\infty}^{t} dt'\, \langle 0|[J_j(t'),J_i(t)]|0\rangle\, E_j e^{-i\omega t'}$$
$$= \frac{1}{\hbar\omega}\left(\int_0^{\infty} dt''\, e^{i\omega t''}\langle 0|[J_j(0),J_i(t'')]|0\rangle\right) E_j e^{-i\omega t} \quad (3.120)$$

where, in the presence of the time translation invariance, the correlation function only depends on the time difference $t'' = t - t'$ between t and t'. We end up a suggestive result.

The expectation value of the current is proportional to the applied electric field with the same frequency. This is the essence of the linear response theory. The coefficient is the electric conductivity. As the same current components commute, only the Hall component contributes. Thus

$$\sigma_{xy}(\omega) = \frac{1}{\hbar\omega}\int_0^{\infty} dt\, e^{i\omega t}\langle 0|[J_y(0),J_x(t)]|0\rangle\,. \quad (3.121)$$

This is the Kubo formula for the Hall conductivity. This is consistent with the derivations in §2.5.1.

We recast (3.121) using time evolution of the operator $\vec{J}(t) = e^{iH_0t/\hbar}\vec{J}(0)e^{-iH_0t/\hbar}$ and by inserting a complete basis of energy eigenstates $\sum_n |n\rangle\langle n| = 1$.

$$\begin{aligned}\sigma_{xy}(\omega) &= \frac{1}{\hbar\omega}\int_0^\infty dt e^{i\omega t}\sum_n\Big[\langle 0|J_y(0)|n\rangle\langle n|J_x(0)|0\rangle e^{i(E_n-E_0)t/\hbar} \\ &\quad - \langle 0|J_x(0)|n\rangle\langle n|J_y(0)|0\rangle e^{i(E_0-E_n)t/\hbar}\Big] \\ &= -\frac{i}{\omega}\sum_{n\neq 0}\left[\frac{\langle 0|J_y(0)|n\rangle\langle n|J_x(0)|0\rangle}{\hbar\omega + E_n - E_0} - \frac{\langle 0|J_x(0)|n\rangle\langle n|J_y(0)|0\rangle}{\hbar\omega + E_0 - E_n}\right].\end{aligned}$$ (3.122)

In the last line, we perform the integral over t that requires $\omega \to \omega + i\epsilon$ with an infinitesimal parameter ϵ for the convergence of the integral. We explicitly see that the ground state $|0\rangle$ does not contribute as the contributions cancel out. We also assume that $E_n \neq E_0$.

Now we take the DC limit $\omega \to 0$ that we are interested in. By expanding the denominators in (3.122) as $1/(E_n - E_0 \pm \hbar\omega) \approx 1/(E_n - E_0) \mp \hbar\omega/(E_n - E_0)^2 + \mathcal{O}(\omega^2)$, we get

$$\sigma_{xy} = \lim_{\omega\to 0} -\frac{i}{\omega}\sum_{n\neq 0}\bigg[\frac{\langle 0|J_y(0)|n\rangle\langle n|J_x(0)|0\rangle + \langle 0|J_x(0)|n\rangle\langle n|J_y(0)|0\rangle}{E_n - E_0} \\ - \hbar\omega\frac{\langle 0|J_y(0)|n\rangle\langle n|J_x(0)|0\rangle - \langle 0|J_x(0)|n\rangle\langle n|J_y(0)|0\rangle}{(E_n - E_0)^2} + \mathcal{O}(\omega^2)\bigg].$$ (3.123)

The first term looks like diverging in the limit $\omega \to 0$. Indeed there exist such a divergent contribution for the longitudinal conductivity σ_{xx} due to the translation invariance. We can see that it actually vanishes for the Hall conductivity. The first term is symmetric under the exchange of x and y, while the second term is antisymmetric. From the observation $\sigma_{xy} = -\sigma_{yx}$ due to the rotational invariance, we know that the symmetric term should vanish. This can be also shown on general grounds from gauge invariance or from the conservation of current.

The second line in (3.123) is antisymmetric under the exchange of x and y, which survives. Thus, the Hall conductivity is actually finite.

$$\sigma_{xy} = i\hbar\sum_{n\neq 0}\frac{\langle 0|J_y(0)|n\rangle\langle n|J_x(0)|0\rangle - \langle 0|J_x(0)|n\rangle\langle n|J_y(0)|0\rangle}{(E_n - E_0)^2}.$$ (3.124)

This is the Kubo formula for the Hall conductivity. We did not explicitly carry along the area in the expression, yet the $\vec{J}$ is actually a current, not a current density. Thus an area factor is missing here. The formula is similar to that of the second order perturbation theory in quantum mechanics.

3.4.3 Topology and Hall conductivity

In this subsection, we put a system on a spatial torus T^2 and see the connection between the associated topology and the Hall conductivity. Interested readers can find more systematic discussions in [44].

The torus T^2 can be viewed as a rectangle with the lengths of two sides as L_x and L_y. Its opposite edges are identified, and we impose periodic boundary conditions for the wave function. It turns out that the obvious choice $\psi(x,y) = \psi(x+L_x,y) = \psi(x,y+L_y)$ is too restrictive in the presence of magnetic field as we already discussed as (more general) magnetic translation operators G_1 and G_τ and the generators around the equation (3.49). This happens the usual translation generators $p_x = -i\hbar\partial_x$ and $p_y = -i\hbar\partial_y$ do not commute with the Hamiltonian in the presence of magnetic field.

For a rectangle, we define the magnetic translation generators $G_i(\vec{L}) = e^{i\vec{L}\cdot(\vec{p}-q\vec{A})/\hbar}$. By choosing a Landau gauge $A_x = 0, A_y = Bx$, we get

$$G_x = e^{iL_x p_x/\hbar}, \qquad G_y = e^{iL_y(p_y + eBx)/\hbar}. \tag{3.125}$$

The translation generators act on the wave function around the cycles of torus to give

$$\begin{aligned} G_x\psi(x,y) &= \psi(x+L_x,y) = \psi(x,y), \\ G_y\psi(x,y) &= e^{ieBL_y x/\hbar}\psi(x,y+L_y) = \psi(x,y). \end{aligned} \tag{3.126}$$

The second equation indicates that the wave function is periodic with an additional gauge transformation.

As we already discussed in (3.50), there is a consistency condition to satisfy. We expect that the two magnetic translation operators commute. Simple computation shows

$$G_x G_y = e^{ieBL_xL_y/\hbar} G_x G_y \,. \tag{3.127}$$

It is only consistent when the following condition is satisfied.

$$eBL_xL_y = 2\pi\hbar\,\mathbb{Z}\,, \tag{3.128}$$

where $\mathbb{Z}$ is an integer.

Now we perturb the system to compute the conductivity. This can be done by threading two different fluxex Φ_x and Φ_y through the two cycles of torus. The gauge potential becomes

$$A_x = \frac{\Phi_x}{L_x}\,, \qquad A_y = Bx + \frac{\Phi_y}{L_y}\,. \tag{3.129}$$

Adding these fluxes can be reflected as extra terms in Hamiltonian as

$$H_s = -\frac{\Phi_x}{L_x} J_x - \frac{\Phi_y}{L_y} J_y\,. \tag{3.130}$$

We can treat this as a perturbation Hamiltonian similar to (3.116). From this we can compute the Hall conductivity using (3.124).

$$\sigma_{xy} = i\hbar L_x L_y$$
$$\times \sum_{n\neq 0} \frac{\langle\psi_0|J_y(0)|\psi_n\rangle\langle\psi_n|J_x(0)|\psi_0\rangle - \langle\psi_0|J_x(0)|\psi_n\rangle\langle\psi_n|J_y(0)|\psi_0\rangle}{(E_n - E_0)^2}\,, \tag{3.131}$$

where we change the notation for the energy eigenstates as ψ_n. We also assume that the ground state is non-degenerate and $E_n \neq E_0$.

To establish a connection between the conductivity formula and topological property, we discuss the quantum mechanical perturbation theory. The ground state $|\psi_0\rangle$ has the following form at the first order:

$$|\psi_0\rangle' = |\psi_0\rangle + \sum_{n\neq 0} \frac{\langle\psi_n|H_s|\psi_0\rangle}{E_n - E_0}|\psi_n\rangle\,. \tag{3.132}$$

By taking a variation of Φ_i, we get

$$|\frac{\partial \psi_0}{\partial \Phi_i}\rangle = -\frac{1}{L_i}\sum_{n\neq 0}\frac{\langle \psi_n|J_i|\psi_0\rangle}{E_n - E_0}|\psi_n\rangle . \tag{3.133}$$

Using this we can rewrite the conductivity formula as

$$\begin{aligned}\sigma_{xy} &= i\hbar\left[\langle\frac{\partial\psi_0}{\partial\Phi_y}|\frac{\partial\psi_0}{\partial\Phi_x}\rangle - \langle\frac{\partial\psi_0}{\partial\Phi_x}|\frac{\partial\psi_0}{\partial\Phi_y}\rangle\right]\\ &= i\hbar\left[\frac{\partial}{\partial\Phi_y}\langle\psi_0|\frac{\partial\psi_0}{\partial\Phi_x}\rangle - \frac{\partial}{\partial\Phi_x}\langle\psi_0|\frac{\partial\psi_0}{\partial\Phi_y}\rangle\right] .\end{aligned} \tag{3.134}$$

This expression is interesting. This is similar to the Berry curvature for the metric perturbations considered in (3.88). In the following we seek its connection to Berry curvature associated with the perturbation of gauge potentials

3.4.3.1 Berry connection and Chern number

We already discussed the Berry phase in the general context in (3.79), which can be recast into in the form

$$\gamma = -\oint_C \mathcal{A}_i(\lambda)d\lambda^i = -\int_S \mathcal{F}_{ij}dS^{ij} , \tag{3.135}$$

where λ is a parameter that controls the adiabatic process similar to $R(t)$ in (3.79). We use an identity in vector calculus to write the integral in a simpler form with a surface integral with S that has a closed boundary C. The Berry connection $\mathcal{A}$ and Berry curvature $\mathcal{F}$ have the forms.

$$\begin{aligned}\mathcal{A}_i(\lambda) &= -i\langle\psi_n|\partial_{\lambda^i}|\psi_n\rangle ,\\ \mathcal{F}_{ij}(\lambda) &= \frac{\partial \mathcal{A}_i}{\partial\lambda^j} - \frac{\partial \mathcal{A}_j}{\partial\lambda^i} .\end{aligned} \tag{3.136}$$

This form is suitable for our discussion with the gauge potential (3.129).

The fluxes Φ_x and Φ_y given in (3.129) are the parameters in the perturbation Hamiltonian. As the spectrum of the Hamiltonian only depends on Φ_i/Φ_0 with $\Phi_0 = 2\pi\hbar/e$, these parameters are also periodic. Thus the space of flux parameters are also a torus T_Φ^2. We introduce a dimensionless angular variables $0 \leq \theta_i < 2\pi$ as

$$\theta_i = \frac{2\pi \Phi_i}{\Phi_0} . \tag{3.137}$$

The Berry phase $\mathcal{A}_i(\Phi)$ that arise by varying these parameters θ_i naturally lives on the torus T_Φ^2. It is given by

$$\mathcal{A}_i(\Phi) = -i \langle \psi_0 | \partial_{\theta_i} | \psi_0 \rangle . \tag{3.138}$$

The corresponding Berry curvature is given by

$$\begin{aligned} \mathcal{F}_{xy}(\lambda) &= \frac{\partial \mathcal{A}_x}{\partial \theta_y} - \frac{\partial \mathcal{A}_y}{\partial \theta_x} \\ &= -i \left[\frac{\partial}{\partial \theta_y} \langle \psi_0 | \frac{\partial \psi_0}{\partial \theta_x} \rangle - \frac{\partial}{\partial \theta_x} \langle \psi_0 | \frac{\partial \psi_0}{\partial \theta_y} \rangle \right] . \end{aligned} \tag{3.139}$$

This is nothing but the Hall conductivity given in (3.134). To be more precise, we have

$$\sigma_{xy} = -\frac{e^2}{\hbar} \mathcal{F}_{xy} . \tag{3.140}$$

Thus we have a non-trivial connection between the Hall conductivity and Berry curvature.

Here comes a surprise! Let us average over all fluxes by integrating over the torus T_Φ^2. Then we get

$$\sigma_{xy} = -\frac{e^2}{\hbar} \int_{T_\Phi^2} \frac{d^2\theta}{(2\pi)^2} \mathcal{F}_{xy} = -\frac{e^2}{2\pi \hbar} C , \tag{3.141}$$

where C is always an integer called the first Chern number.

$$C = \int_{T_\Phi^2} \frac{d^2\theta}{2\pi} \mathcal{F}_{xy} \in \mathbb{Z} . \tag{3.142}$$

Thus the Hall conductivity is necessarily quantized when averaged over fluxes. This is referred as TKNN invariants after the authors of the paper [55]. The fact that the Hall conductivity is multiple of $e^2/2\pi\hbar$ is related to the integer quantum Hall effect.

3.4.4 Hall conductivity with momentum dependence

In this brief subsection we briefly mention the extension of the Hall conductivity to include the momentum k dependences, which has been computed for integer quantum Hall system in [56]. The computation is done by taking Fourier transform both in space and time coordinates for the current-current correlation function (3.121) and using the wave function (3.106).

The result normalized with the zero-momentum Hall conductivity is given by

$$\frac{\sigma_{xy}(k)}{\sigma_{xy}(0)} = 1 + \frac{\omega^2}{\omega_B^2} - \frac{3\nu}{4}(kl_B)^2 + \cdots, \qquad (3.143)$$

where $\sigma_{xy}(0) = \nu q^2/(2\pi\hbar)$, $\omega_B = eB/m$ with $\nu = N$. The last term is momentum dependent contribution, which is particularly simple. We return this formula at the end of this chapter to seek the relation to the Hall viscosity.

3.5 Hall conductivity and Hall viscosity

Here we change our gear from the previous sections and consider the relation between the Hall conductivity and Hall viscosity for quantum Hall states. We first consider the motion of a charged particle under the influence of a non-homogeneous electric field in an elementary fashion. Then we try to provide some intuitive picture on the momentum dependent Hall conductivity, a part of which is related to the Hall viscosity.

3.5.1 Non-homogeneous electric field

When a particle with a charge q and a mass m is placed in an electric field $\vec{E} = -E_0\hat{x}$, it shows a linear motion along the field direction. On the other hand, under an influence of a magnetic field $\vec{B} = B_0\hat{z}$, it shows a circular motion perpendicular to the field if the particle has a nonzero velocity perpendicular to the magnetic field. Note that these two fields are perpendicular to each other.

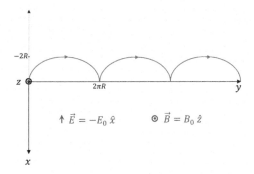

Figure 3.5 Motion of a charged particle under constant electric (along $-x$ direction) and magnetic (along $+\hat{z}$) fields.

The particle is pushed toward the negative x direction due to the electric field, and subsequently deflected toward the positive y direction because of the magnetic field. The motion is described by the Lorentz force $F_i = ma_i = q(E_i + \epsilon_{ijk}v_jB_k)$ in a vector form, $\vec{F} = q(\vec{E} + \vec{v} \times \vec{B}) = m\vec{a}$, where $\vec{v} = (\dot{x}, \dot{y}, 0)$ and $\vec{a} = \dot{\vec{v}}$. In components form, the motion is described by

$$m\ddot{x} = q(-E_0 + B_0\dot{y}),$$
$$m\ddot{y} = -qB_0\dot{x}. \tag{3.144}$$

To simplify the motion, we impose the initial conditions $x(t = 0) = y(0) = 0$ and $\dot{x}(0) = \dot{y}(0) = 0$. Then, we get

$$x = \frac{E_0}{\omega B_0}(1 - \cos(\omega t)),$$
$$y = \frac{E_0}{\omega B_0}(\omega t - \sin(\omega t)), \tag{3.145}$$

where $\omega = qB_0/m$.

To have a clear geometric description, we combine these equations using $\cos^2(\omega t) + \sin^2(\omega t) = 1$.

$$(x - R)^2 + (y - \omega R t)^2 = R^2, \tag{3.146}$$

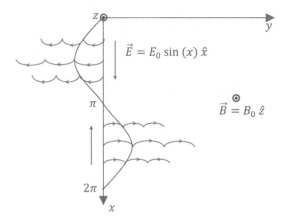

Figure 3.6 Motion of charged particles under a non-homogeneous electric field depending on x coordinate and a constant magnetic field pointing $+\hat{z}$ direction. Note that the shape of sin(x) function indicates the magnitude of the electric field and is not related to y-axis.

where $R = E_0/\omega B_0 = mE_0/qB_0^2$. It shows a cycloid motion, which can be represented by a rolling circle with a radius R depicted in Fig. 3.5. The center of the cycloid travels at a constant speed $\vec{v}_{center} = (0, \omega R, 0)$. We further note, for a later discussion, that the radius R gets smaller when the magnetic field B_0 gets stronger. This linear motion of the center in the presence of the strong magnetic field is nothing but the guiding center motion discussed in (3.12) and (3.13).

We generalize the elementary discussion with a non-homogeneous electric field $\vec{E}(x)$ as a function of the coordinate x, while the magnetic field is held fixed $\vec{B} = B_0\hat{z}$. In particular we choose $\vec{E}(x) = E_0 \sin(x)\hat{x}$ that changes its magnitude along the x direction and also the direction at $x = \pi$. This is depicted in Fig. 3.6. The sine graph in the figure represents the magnitude of the electric field and is independent of the coordinates y, z. The charged particles located at $0 < x < \pi$ are pushed downward (toward +x direction) and deflected to $-y$ direction as the electric field points +x direction. Similarly, the particles located at $\pi < x < 2\pi$ are pushed toward $-x$ direction and deflected to $+y$ direction as the electric field points +x direction.

Similar to the case in the constant electric field, the particle motion is described by the Lorentz force law as

$$m\ddot{x} = q(-E_0 \sin(x) + B_0 \dot{y}),$$
$$m\ddot{y} = -qB_0\dot{x}.$$
(3.147)

These coupled equations give a highly non-linear differential equation for $x(t)$, when decoupled, $\dddot{x} = (qE_0/m)\dot{x}\cos(x) - \omega^2 \dot{x}$. By integrating it once, we get $\ddot{x} = (qE_0/m)\sin(x) - \omega^2 x + c$.

In the following subsection, we outline a way to compute the Hall conductivity using the linear response theory.

3.5.2 Hall viscosity in terms of Hall conductivity

The relation between the Hall viscosity and the Hall conductivity has been established by an explicit computation utilizing the background field method based on symmetry arguments for a qantum field theory. Here we try to understand the results in an elementary fashion. The original reference [43] is useful to follow this subsection.

As already discussed in §3.4, the Hall conductivity is universal in the presence of constant electric field. This already indicates that the cycloid motion is nothing to do with the quantities we are going to discuss. We can turn on a small spatial variation of the electric field such as $\vec{E} = \hat{x}E_x(x)$, for example $E_x(x) = E\sin(kx)$. Then the guiding center motion (3.13) is space dependent as

$$v_y(x) = -\frac{E_x(x)}{B}.$$
(3.148)

This non-homogeneous electric field and space dependent velocity produces two independent contributions to the Hall conductivity by effectively changing the Lorentz force $\vec{F} = q(\vec{E} + \vec{v} \times \vec{B})$ through modifying the electric and magnetic fields.

Hall Viscosity

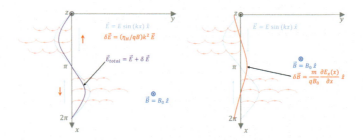

Figure 3.7 Corrections of the electric (left) and magnetic (right) fields due to the non-homogeneous electric field that depends on the wave vector k, where we take $k = 1$ for the illustration. The sine or cosine shapes (drawn for visualizations) indicate the qualitative magnitudes in appropriate directions.

From (3.80) and (3.81) in this chapter, we can get the strain rate and the corresponding stress tensor

$$\dot{\xi}_{xy} = \frac{1}{2}\left(\frac{\partial v_x}{\partial y} + \frac{\partial v_y}{\partial x}\right) = \frac{1}{2}\frac{\partial v_y(x)}{\partial x},$$

$$T_{xx} = -T_{yy} = 2\eta_{xxxy}\dot{\xi}_{xy} = \frac{\eta_H}{B}\frac{\partial E_x(x)}{\partial x},$$
(3.149)

where the factor 2 comes from the symmetric combination of the last two indices of $\eta_{xxxy} = \eta_{xxyx} = -\eta_H$. The diagonal component of the stress tensor is nothing but the pressure, force per unit area. Thus the force in the x-direction acting on the unit volume element is

$$f_x = -\frac{\partial T_{xx}}{\partial x} = -\frac{\eta_H}{B}\frac{\partial^2 E_x(x)}{\partial x^2},$$
(3.150)

where $-$ sign indicates the force acts on the volume. While the results depends on the specific form of the electric field, this can be interpreted as changing the electric field along the x-direction as

$$\delta E_x(x) = -\frac{\eta_H}{qB}\frac{\partial^2 E_x(x)}{\partial x^2} = \frac{\eta_H}{qB}k^2 E_x(x),$$
(3.151)

where we use, for example, $\vec{E}(x) = \hat{x}E_x(x) = \hat{x}E\sin(kx)$ with a wave vector k. This is illustrated in the left side in Fig. 3.7 for $(\eta_H/qB)k^2 < 0$.

The force along x-direction produces the Hall conductivity along y-direction due to the Lorentz force $\vec{f} = \vec{J} \times \vec{B}$ and in component form $f_x = BJ_y$, where we assume the charge carriers have positive charges (there is a relative − sign for a negative charge). Thus we arrive $J_y = \sigma_{yx} E_x$. When changed into the Fourier space, the x-derivative is replaced by a wave vector, say k. Thus we arrive the first correction to the conductivity after including $-k^2$ from $\partial_x = ik$ in the Fourier space and $\sigma_{xy} = -\sigma_{yx}$.

$$\sigma_{xy}^{(1)} = -\frac{\eta_H}{B^2} k^2 . \tag{3.152}$$

Note that this contribution is directly related to the Hall viscosity.

Now we search for the second contribution. In the presence of the velocity (3.148) due to the non-homogeneous electric field, the particles experience non-zero local angular momentum that is not relate to electromagnetic origin. In coordinate independent form, we have

$$\vec{w}_L = \frac{1}{2}(\vec{\nabla} \times \vec{v}) = \hat{z} \frac{1}{2} \frac{\partial v_y(x)}{\partial x} . \tag{3.153}$$

This local angular momentum is perpendicular to the two-dimensional plane, and can point $\pm \hat{z}$ depending on the sign of $\partial_x v_y(x)$.

In the local co-moving frame, this produces the Coriolis force $\vec{F}_C = 2m\vec{w}_L \times \vec{v}$, which has the form of the Lorentz force $\vec{F} = q\vec{v} \times \vec{B}$. Thus, this produces the effective magnetic field $\delta \vec{B}_L$ (anticipated as a small correction)

$$\delta \vec{B}_L = -\frac{2m}{q} \vec{w}_L = -\hat{z} \frac{m}{q} \frac{\partial v_y(x)}{\partial x} = \hat{z} \frac{m}{qB} \frac{\partial E_x(x)}{\partial x} . \tag{3.154}$$

Thus the second contribution contributes by changing the magnetic field. It is illustrated in the right panel of Fig. 3.7.

To proceed to find the corresponding effects on the Hall conductivity, we remind that the quantum Hall fluid is a diamagnetic material. From the energy expression $\epsilon = -\vec{M} \cdot \vec{B} = -MB$ by aligning the magnetic field along z direction, the corresponding magnetization can be obtained as $\vec{M} = -(\partial \epsilon / \partial B)\hat{B}$. As the small variation of the magnetic field is the same direction, we get

$$\delta \vec{M} = -\frac{\partial^2 \epsilon}{\partial B^2}\delta \vec{B}_L = -\hat{z}\epsilon''(B)\frac{m}{qB}\frac{\partial E_x(x)}{\partial x} \,. \tag{3.155}$$

The correction to the magnetization current can be computed from $\delta \vec{J}_L = \vec{\nabla} \times \delta \vec{M}$. For a diamagnetic material, $\delta \vec{M}(x,y) = -\hat{z}\delta M(x,y)$ with an opposite sign. Thus $\delta \vec{J}_L = -\vec{\nabla} \times (\hat{z}\delta M) = \hat{z} \times \vec{\nabla}\delta M = -\hat{y}\epsilon''(B)(m/qB)(\partial^2 E_x(x)/\partial x^2)$, which gives the Hall conductivity as

$$\sigma_{xy}^{(2)} = -\epsilon''(B)\frac{m}{qB}k^2 \,. \tag{3.156}$$

This is the second contribution that is proportional to k^2.

Combining the two corrections together, we arrive

$$\sigma_{xy} = \nu \frac{q^2}{2\pi\hbar} - \frac{\eta_H}{B^2}k^2 - \epsilon''(B)\frac{m}{qB}k^2 + \cdots \,. \tag{3.157}$$

This result is reported in [43] for a quantum Hall states with Galilean invariance in the presence of a background magnetic field B. Note that $q = -e$ is the charge for an electron and k is momentum. The first term is the Hall conductivity we obtained in (3.141). The quantity $\epsilon''(B)$ is the energy density, energy per unit area, as a function of external magnetic field B at a fixed filling factor. While the term with $\epsilon''(B)$ is not universal, its magnitude can be extracted independently.

The Hall conductivity has been computed for integer quantum Hall system in [56]. The result is listed in (3.143). In a slightly different form, it is

$$\sigma_{xy} = \nu\frac{q^2}{2\pi\hbar}\left(1 - \frac{3\nu}{4}(kl_B)^2 + \mathcal{O}(kl_B)^4\right), \tag{3.158}$$

where $\nu = N$. The energy density is given by $\epsilon(B) = (\nu^2/4\pi)(e^2B^2/m)$. Thus one can extract the Hall viscosity contributions η_H from this Hall conductivity data. There are further discussions for other interesting systems such as fractional quantum Hall states in [43].

After we digress to understand the spin dynamics in the following chapter, we come back to the Ward identity to discuss similar relations in the context of magnetic skyrmions.

Chapter 4

Spin Dynamics

Chiral magnetic skyrmion is made of a bunch of spins packed together in a beautiful way. Individual spins in the skyrmion are influenced by their environments, such as magnetic fields or other spins through spin interactions. We present this chapter in a systematic way through their interactions. By understanding the dynamics of each spin, we will be able to gain insights in the motion of domain walls and skyrmions. We start with this discussion of the basic spin motion, first introduced by Landau and Lifshitz and updated by Gilbert. And then we generalize the equation by various spin torques, including the spin transfer torque, spin orbit torque, spin Hall torque and the influence of emergent electromagnetic fields. In particular, the motion of a skyrmion center is governed by the Thiele equation, which also reveals the topological nature of spin configurations naturally. We also present the spin dynamics and associated transport properties driven by electric current, which is one of the major driving forces for skyrmions through the interactions between electron spins and skyrmion spins.

Skyrmions and Hall Transport
Bom Soo Kim
Copyright © 2023 Jenny Stanford Publishing Pte. Ltd.
ISBN 978-981-4968-34-8 (Hardcover), 978-1-003-37253-0 (eBook)
www.jennystanford.com

4.1 Landau-Lifshitz-Gilbert equation

Here we introduce the celebrated Landau-Lifshitz-Gilbert (LLG) equation by starting the discussion of ferromagnetism after introducing the exchange interaction. We look into the condition where we can treat the magnetization as a conserved quantity. We also treat the spin torque with an elementary description.

4.1.1 Ferromagnetism

We are interested in the magnetic properties of materials, which are fairly complicated in general. Thus we focus on primarily ferromagnetic materials that could exhibit a spontaneous magnetization and turn to anti-ferromagnets and ferrimagnets in Chapter 7. Contrast to the electronic crystal structure that is crucially related to the underlying lattice structure, the formation of magnetic structure is mainly due to the exchange interaction of the atom in the materials, which is quite independent of the direction of the total magnetic moment relative to the lattice [12].

Let us consider iron, Fe, with the atomic number 26. The electronic configuration is given by the following orbital structure

$$1s^2\ 2s^2\ 2p^6\ 3s^2\ 3p^6\ 3d^6\ 4s^2.$$

This is illustrated in the left panel of Fig. 4.1. The four unpaired electrons in $3d$ shell, which has higher energy level than $4s$, have the same spin (say, spin-up) due to the Hund's rule coupling between the electrons. As electrons are indistinguishable fermions, they are subject to the Pauli exclusion principle and their wave functions are antisymmetrized.

When two iron atoms are close each other, the unpaired electron wave functions $\psi_a(\vec{r}_i)$ from two different atoms overlap with each other. There are two cases we need to compare, spins of the neighboring atoms are parallel or anti-parallel. When the spins are parallel and anti-parallel, the spatial wave functions (without spin part) need to be antisymmetrized and symmetrized, respectively. Thus

$$\psi_{ab}^{\pm}(\vec{r}_1, \vec{r}_2) = \frac{1}{\sqrt{2}}\left(\psi_a(\vec{r}_1)\psi_b(\vec{r}_2) \pm \psi_b(\vec{r}_1)\psi_a(\vec{r}_2)\right), \quad (4.1)$$

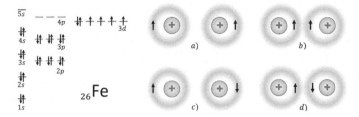

Figure 4.1 Left: Illustration of electronic configuration of *Fe*. Right: Simple model of two atoms with spin arrangements for exchange interactions. Minimizing the coulomb energy of two electrons excludes b) and d) as their average separation is small. In the main body, we discuss a) has lower coulomb energy compared to c) due to the exchange interaction in detail.

where $+(-)$ represent symmetrized (antisymmetrized) state, a, b represent orthogonal and normalized wave functions, and $\vec{r}_{1,2}$ the physical locations. $\sqrt{2}$ is added for a proper normalization. With the information, we can compute the energy difference between the two wave functions for the coulomb interaction between two electrons, which is $E_{coulomb} = 1/(4\pi\epsilon_0)(e^2/r_{12})$. Thus, the energy is minimized when the electrons are further apart. Thus, the configurations b) and d) are excluded as they have higher energy in the right panel of Fig. 4.1.

Computation can be simplified in one dimension, say x direction [57]. The separation can be evaluated by $\langle (x_1 - x_2)^2 \rangle = \langle x_1^2 \rangle + \langle x_2^2 \rangle - 2\langle x_1 x_2 \rangle$. A straightforward computation helps us understand it clearly. For an antisymmetric state $\overline{\psi_{ab}}$,

$$\langle x_1^2 \rangle_- = \frac{1}{2} \bigg[\int dx_1 x_1^2 |\psi_a(x_1)|^2 \int dx_2 |\psi_b(x_2)|^2 + (a \leftrightarrow b)$$
$$- \int dx_1 x_1^2 \psi_a^*(x_1) \psi_b(x_1) \int dx_2 \psi_b^*(x_2) \psi_a(x_2) - (a \leftrightarrow b) \bigg] \quad (4.2)$$
$$= \frac{1}{2} \big[\langle x^2 \rangle_a + \langle x^2 \rangle_b \big] ,$$

where $a \leftrightarrow b$ indicates the previous integral with two states a and b are exchanged. We also use the orthonormal properties $\int dx_2 \psi_b^*(x_2) \psi_a(x_2) = 0$ for two orthogonal states ψ_a and ψ_b. Similar computation gives $\langle x_2^2 \rangle = \langle x_1^2 \rangle = (\langle x^2 \rangle_a + \langle x^2 \rangle_b)/2$ as the coordinate x_1 and x_2 are integrated over. As the cross terms vanish, the result is the same for the symmetric state ψ_{ab}^+.

We can repeat the computation for the cross term $\langle x_1 x_2 \rangle$. Here we need to distinguish the computations for symmetric and antisymmetric states. For antisymmetric state

$$\langle x_1 x_2 \rangle_- = \frac{1}{2}\left[\int dx_1 x_1 |\psi_a(x_1)|^2 \int dx_2 x_2 |\psi_b(x_2)|^2 + (a \leftrightarrow b) \right.$$
$$\left. - \int dx_1 x_1 \psi_a^*(x_1)\psi_b(x_1) \int dx_2 x_2 \psi_b^*(x_2)\psi_a(x_2) - (a \leftrightarrow b) \right] \quad (4.3)$$
$$= \frac{1}{2}\left[\langle x \rangle_a \langle x \rangle_b + \langle x \rangle_b \langle x \rangle_a - \langle x \rangle_{ab} \langle x \rangle_{ba} - \langle x \rangle_{ba} \langle x \rangle_{ab} \right]$$
$$= \langle x \rangle_a \langle x \rangle_b - |\langle x \rangle_{ab}|^2 ,$$

where we use $\langle x \rangle_a^* = \langle x \rangle_a$ and $\langle x \rangle_{ab} = \int dx_i x_i \psi_a^*(x_i)\psi_b(x_i) = (\langle x \rangle_{ba})^*$ for $i = 1, 2$.

Putting them together, we get

$$\langle (x_1 - x_2)^2 \rangle_\pm = \langle x^2 \rangle_a + \langle x^2 \rangle_b - \langle x \rangle_a \langle x \rangle_b \mp |\langle x \rangle_{ab}|^2 . \quad (4.4)$$

Thus the electrons in the antisymmetrized state are further apart, $\langle (x_1 - x_2)^2 \rangle_- > \langle (x_1 - x_2)^2 \rangle_+$. For the coulomb interaction between the electrons in the $3d$ shell, the energy is smaller for the antisymmetric spatial wave function. The corresponding spin states are parallel, which is responsible for the ferromagnetic spin configuration.

Of course, there are other, less important, interactions that alter this magnetic structure. They are direct magnetic interactions between the magnetic moments of the atom and interactions between the magnetic moments and the electric fields of the crystal lattice. These interactions are relativistic effects and are suppressed by $\sim \mathcal{O}(1/c^2)$, where c is the speed of light. The ratio of the relativistic and exchange interactions is typically between 10^{-5} and 10^{-4}. If we are only concerned with the exchange interaction neglecting the other smaller interactions, the magnetization $\vec{M}$ can be considered as a conserved quantity and thus as an independent variable. The corresponding thermodynamic potential is a function of the magnitude M. This has been useful for understanding the Curie point of magnetic materials, where one can expand the thermodynamic potential as a function of M for small M. This approach provides a universal and simple understanding for the spontaneous magnetization and susceptibility [12].

It is well known that experimental hysteresis curves of ferromagnetic substances clearly show that beyond certain critical values of the applied magnetic field, the magnetization saturates, becomes uniform and aligns parallel to the magnetic field. In order to incorporate this experimental fact, from phenomenological grounds, Landau and Lifshitz [2] introduced the basic dynamical equation for the localized magnetization or spin $\vec{M}(\vec{r},t)$ vector in bulk materials, where the effect of relativistic interactions were also included as a damping term. In 1954, Gilbert [58] introduced a more convincing form with a damping term based on a Lagrangian approach. The combined form is now called the Landau-Lifshitz-Gilbert (LLG) equation, which is a fundamental dynamical system in applied magnetism. LLG equation can be understood from the view of various torque acting on the local magnetization vector.

4.1.2 Basic understanding of spin torque

The magnetization dynamics we consider with the Landau-Lifshitz-Gilbert (LLG) equation is the saturated (with a fixed magnitude $M_s = |\vec{M}|$) magnetization vector localized in a fixed location. Thus the main concerns is not the overall translation motion, but the rotation motion of the magnetization vector. The latter is described by the torque, the time derivative of the vector $\vec{T} = d\vec{M}/dt$. We call it as a spin torque for the magnetization vector is opposite to the spin vector $\vec{M} = -\langle \vec{S} \rangle$.

To understand the basic origin of the spin torque in an elementary fashion, we utilize a classical picture of angular momentum. Consider an atomic model described by an electron with charge e orbiting counterclockwise with radius r in xy plane viewed from $+\hat{z}$ axis. See the left panel of Fig. 4.2. We can easily visualize and compute the angular momentum, $\vec{L} = \vec{r} \times \vec{p} = rm_e v\hat{z}$,

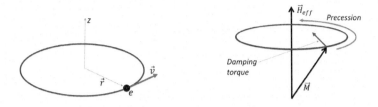

Figure 4.2 Left: A simple classical model of an atom provides an intuitive picture for gyromagnetic ratio. Right: Spin motion due to the precession and damping terms given in the LLG equation (4.6).

for the atomic orbit. The magnetic dipole moment of the orbit, $\vec{\mu} = \mu_z \hat{z} = -|e|vr/2\hat{z}$, is given by the product of the area (πr^2) and the current (charge over a period, $-|e|/(2\pi r/v)$). Here we define the gyromagnetic ratio γ_0, while negative for an electron, as magnitude of the magnetic moment over the angular momentum.

$$\gamma_0 = \left|\frac{\mu_z}{L_z}\right| = \frac{|e|}{2m_e} = \frac{\mu_B}{\hbar}, \qquad (4.5)$$

where μ_B is the Bohr magneton, $\hbar$ Planck constant, e the electric charge of an electron, and m_e the mass of an electron. For a quantum mechanical spin, we need an extra factor of 2 for the gyromagnetic ratio. The numerical value for an electron is $\gamma_0 = 1.76 \times 10^{11}$ $rad/(s \cdot T)$ with an SI unit radian per second times Tesla.

The definition of a torque for the angular momentum $\vec{L}$, and in turn for the dipole moment $\vec{\mu}$, is given by $\vec{T} = \partial \vec{L}/\partial t = -(1/\gamma_0)(\partial \vec{\mu}/\partial t)$. Thus, for a macroscopic volume that contains many atoms, the torque acting on a local magnetization $\vec{M}$ is given as $\partial \vec{M}/\partial t = -\gamma_0 \vec{T}$. There are several different torques action on the magnetization, which are described by the general form of the Landau-Lifshitz-Gilbert (LLG) equation.

$$\frac{\partial \vec{M}}{\partial t} = -\gamma_0 \vec{M} \times \vec{H}_{\text{eff}} + \frac{\alpha}{M_s} \vec{M} \times \frac{\partial \vec{M}}{\partial t} + \vec{T}_{STT} + \vec{T}_{SOT} + \vec{T}_{SHT}, \quad (4.6)$$

where M_s is the saturation magnetization. The first term in the right-hand side is the precessional torque that describes the precession of the magnetization around the local effective field $\vec{H}_{\text{eff}}$, which includes the external field, exchange interaction, anisotropy, and demagnetization. The second term is damping torques that describes the damping effect on the magnetization toward the equilibrium parallel to the effective field $\vec{H}_{\text{eff}}$. α is called the Gilbert damping parameter [2][58]. See the right panel of Fig. 4.2. The rest of the torques, $\vec{T}_{STT}$, $\vec{T}_{SOT}$, and $\vec{T}_{SHT}$ are so-called spin transfer torque, spin orbit torque, and spin Hall torque, respectively. We introduce them one by one in the following section.

Before considering these torque terms in detail, it is important to realize that the first two terms in (4.6) have the common structure with $\vec{M} \times \vec{A}$ with an unspecified vector $\vec{A}$. As mentioned in §4.1.1, we consider the conserved magnetization where the magnitude of the vector is constant

$$\vec{M} \cdot \vec{M} = M_s^2 = const. \tag{4.7}$$

By talking a time derivative, we can get $\vec{M} \cdot \partial_t \vec{M} = 0$, which sets the left-hand panel of (4.6) vanishes when dotted (scalar product) with $\vec{M}$. This is enforced by the special vector structure of the terms in the right-hand side. One can easily check this as

$$\vec{M} \cdot (\vec{M} \times \vec{A}) = \epsilon_{ijk} M_i M_j A_k = 0 \,, \tag{4.8}$$

which vanishes identically as the expression is the product of a totally antisymmetric expression ϵ_{ijk} and a totally symmetric one $M_i M_j$ for repeated indices i and j. One can see it easily by working with $k = 3$. Then the computation goes $(\epsilon_{ij3} M_i M_j) A_3 = (\epsilon_{123} M_1 M_2 + \epsilon_{213} M_2 M_1) A_3 = (M_1 M_2 - M_2 M_1) A_3 = 0$, where we use $\epsilon_{123} = -\epsilon_{213} = 1$.

Attentive readers might already suspect that the other torque terms would have the same vector structures. Yes, it is the case. All the torques $\vec{T}_{STT}$, $\vec{T}_{SOT}$, and $\vec{T}_{SHT}$ share the same structure. Similar discussions and structures will also play crucial role in understanding the so-called Thiele equation below.

4.1.3 Domain Wall illustration of LLG equation

Before introducing various spin torques, we study the original LLG equation

$$\frac{\partial \vec{M}}{\partial t} = -\gamma_0 \vec{M} \times \vec{H}_{\text{eff}} - \gamma_0 \frac{\alpha}{M_s} \vec{M} \times (\vec{M} \times \vec{H}_{\text{eff}}), \quad (4.9)$$

where the damping term is evaluated by iterating the equation (4.6). In particular, we explain in detail the precession and damping terms in the context of Domain Wall (DW) motion. This is depicted in Fig. 4.3. As the DW is driven by the external field $\vec{H}_{\text{eff}}$, it is called as a field-driven DW motion. *Basic theme: effective magnetic field drives the spin dynamics.* Throughout this chapter, we also use the DW motion to explain and contrast the other spin torques. See similar analysis on DWs dynamics using current [59].

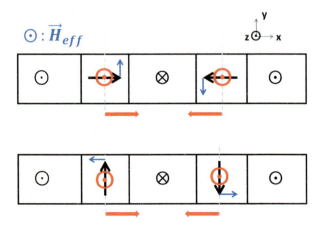

Figure 4.3 Top: Right-handed Néel DWs. Bottom: Right-handed Bloch DWs. Left-handed DWs have their in-plane magnetizations reversed. Effective field ($\vec{H}_{\text{eff}}$ with arrow head) expands the left and right blocks (with magnetizations pointing out) by turning the DW magnetization pointing out of the plane through the damping term (represented by red circles) in (4.9). Effectively the middle block with anti-parallel magnetization shrinks through the DW motion. The precession term changes the DWs structures, but cannot move them.

The left, middle, and right blocks in Fig. 4.3 are parts of a ferromagnetic layer with the uniform magnetization out of plane ($+\hat{z}$ illustrated with an arrow head $\odot$), into the plan ($-\hat{z}$, illustrated with an arrow tail $\otimes$), and out of plane ($+\hat{z}$, again an arrow head) directions, respectively. The external magnetic field $\vec{H}_{\text{eff}}$ does not have a torque on the local magnetizations of the three blocks as they are parallel or anti-parallel to each other. There are two DW regions sandwiched in the three blocks with the magnetization directions represented by the thick black arrows.

In the left DW of the top panel of Fig. 4.3, the magnetization of DW is along the $+\hat{x}$ direction, which is right-handed Néel DW. Upon applying the magnetic field out of plane (along the $\vec{H}_{\text{eff}} = H_0 \hat{z}$ direction), the precession term generate the torque $-\gamma_0 \vec{M} \times \vec{H}_{\text{eff}}$ that is represented as the short solid blue arrow, which is along $+\hat{y}$ for the magnetization in the top left DW. Thus the precession term forces the magnetization to rotate in the *xy* plane. This turns the right-handed Néel DWs (top figure) into right-handed Bloch DW (bottom figure), and proceed to left-handed Néel DWs and then to left-handed Bloch DWs, back to the original right-handed Néel DWs. More detailed structures of these right-handed DWs are illustrated in Fig. 4.4 using the side view. Thus the precession term in LLG equation does not move the DWs.

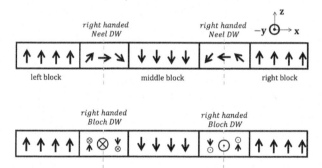

Figure 4.4 Side view of the DW configurations from $-\hat{y}$ axis given in Fig. 4.3. Top: Right-handed Néel DWs. Bottom: Right-handed Bloch DWs. Left-handed DWs have their magnetizations reversed. Thus the Néel DWs have no magnetization direction in $\hat{y}$, while the Bloch DWs have no magnetization direction in $\hat{x}$. If the two right-handed DWs are exchanged, they turn into the left-handed DWs.

The damping term proportional to α/M_s in (4.9) can be further evaluated by cross product of $\vec{M}$ and the precession term. Let us consider the left DW in the top panel of Fig. 4.3 again. The cross product of $-\gamma_0(\alpha/M_s)\vec{M} \times (\vec{M} \times \vec{H}_{\text{eff}}) = \gamma_0\alpha H_0 \vec{M} \times \hat{y} = \gamma_0\alpha H_0 M_s \hat{z}$, which points out of the plane. This is illustrated with the thick red circle with point inside the DW region $(+\hat{z})$. Thus, the direction of the magnetization vector of the top left DW turns to pointing out of the plane due to the damping term. With this, the left block with the magnetization out of plane expands. This is illustrated by the long red arrow outside (below) the ferromagnet layer.

Similarly, one can show that all the DWs illustrated in Fig. 4.3 has the damping term with the magnetization vector out of plane $\hat{z}$. We illustrate this with the right DW, with the magnetization pointing left $-\hat{x}$, in the top panel of Fig. 4.3. The precession term points $-\hat{y}$ direction as $-\gamma_0 \vec{M} \times \vec{H}_{\text{eff}} \propto -\hat{y}$. Thus the damping term points $\vec{M} \times (-\hat{y}) \propto \hat{z}$. Thus, the right block (with magnetization pointing out) expands.

We are going to adapt the effective magnetic field description to gain understanding the other spin torques. The message we carry over to these analysis is that *when the magnetization direction of a ferromagnetic block and the effective field direction on DWs are parallel, the ferromagnetic block expands*. This is one of the consequences of the celebrated LLG equation!

4.2 Spin torques

In this section we introduce the last three spin torques in the LLG equation (4.6) one by one. Understanding them is crucial as they play important roles in manipulating various spin textures. The third term in (4.6) describes the spin transfer torque (STT), $\vec{T}_{STT}$, whose contributions come from the spin-spin interaction between, for example, skyrmions and the conduction electrons flowing through the skyrmions. The fourth term describes the spin torque induced by the Rashba effect, whose contributions come from the current-induced spin-orbit interaction (SOT), $\vec{T}_{SOT}$. This Rashba effect can exist in the same layer of the local magnetization $\vec{M}$ or in a different layer. The last term is a torque due to spin Hall effects (SHE), $\vec{T}_{SHT}$. This Slonczewski-like spin torque is due to the spin Hall effect that is operating on the interface between a ferromagnet layer and a non-magnet layer.

4.2.1 Spin transfer torque

During the early 2000, surprising giant magneto-resistance effects were reported in magnetic multi-layers [60]. This stirred up activities to understand them theoretically [61][62] and to re-evaluate the Landau-Lifshitz-Gilbert equation that had the first two terms on the right-hand side of (4.6) [63][64]. Theoretical understanding of Zhang-Li [63] is highly insightful and has a significant impact on the further advancements on the field.

We considered two different types of electrons: the electrons near the Fermi surface that provide the spin-dependent transport phenomena and the electrons below the Fermi sea that involve the magnetization dynamics. The spin dynamics of the former, itinerant electrons, are described by the full quantum mechanical operators $\vec{s}$. The latter, localized electrons, have much slower dynamics and are described by the classical magnetization vector $\vec{S}$. Thus we set $\vec{S}/S = -\vec{M}(\vec{r},t)/M_s$ with $|\vec{M}| = M_s$, the saturation magnetization. The essential properties of their interactions turn out to be described by the following s-d Hamiltonian

$$\mathcal{H}_{sd} = -J_{ex}\vec{s}\cdot\vec{S} = \frac{SJ_{ex}}{M_s}\vec{s}\cdot\vec{M}(\vec{r},t). \tag{4.10}$$

This Hamiltonian (4.10) can be used to compute the induced spin density for a given $\vec{M}(\vec{r},t)$. In turn, the effects on the magnetization, in the language of the spin transfer torque, can be obtained by utilizing the induced spin density.

Using the generalized spin continuity equation by including the Hamiltonian (4.10) and spin relaxation $\Gamma(\vec{s})$ due to scattering with impurities and electrons, one can derive the equation of motion for the non-equilibrium conduction electron spin density $\vec{m}(\vec{r},t) = \langle\vec{s}\rangle$ and spin current density $\mathcal{J}(\vec{r},t) = \langle J\rangle$

$$\frac{\partial\vec{m}}{\partial t} + \vec{\nabla}\cdot\mathcal{J} = -\frac{SJ_{ex}}{\hbar M_s}\vec{m}\times\vec{M}(\vec{r},t) - \langle\Gamma(\vec{s})\rangle, \tag{4.11}$$

where $\langle\rangle$ represents the average over all occupied electronic states, and the third term comes from an explicit evaluation of $[\vec{s},H_{sd}]/i\hbar$. We note that the spin current $\mathcal{J}(\vec{r},t)$ is a tensor with charge current and the spin polarization of the current for the equation to make sense.[1]

To solve the equation (4.11) we separate the spin density and current density into two parts. One part, the slower component $\vec{M}$, satisfies the adiabatic approximation. The dynamics of magnetization is slow compared to that of conduction electrons, and thus the spins of conduction electrons approximately follow the direction of local magnetic moment $\vec{M}$. The other part $\delta\vec{m}$ captures the deviation from the adiabatic process, which is our main focus.

$$\vec{m}(\vec{r},t) = \vec{m}_0(\vec{r},t) + \delta\vec{m}(\vec{r},t) = n_0 \frac{\vec{M}(\vec{r},t)}{M_s} + \delta\vec{m}(\vec{r},t),$$

$$\mathcal{J}(\vec{r},t) = \mathcal{J}_0(\vec{r},t) + \delta\mathcal{J}(\vec{r},t) = -\frac{\mu_B P}{|e|}\vec{J}_e \otimes \frac{\vec{M}(\vec{r},t)}{M_s} + \delta\mathcal{J}(\vec{r},t),$$

(4.12)

where n_0 is the local equilibrium spin density parallel to the local magnetization, e the electron charge, J_e the electric current density, μ_B Bohr magneton, P spin current polarization of the ferromagnet. The term proportional to $\mu_B P$ is the spin current of the electron with the spin polarization parallel to the local magnetization vactor $\vec{M}(\vec{r},t)$.

To solve this analytically in a closed form, we assume $<\vec{\Gamma}(\vec{s})> = \delta\vec{m}(\vec{r},t)/\tau_{sf}$ with a spin-flip relaxation time τ_{sf}. We also keep only a linear term of $\delta\vec{m}(\vec{r},t)$ to the electric current J_e and time derivative of magnetization $\partial\vec{M}/\partial t$, discarding the higher order terms including $(\partial(\delta\vec{m})/\partial_t), j_e(\partial M/\partial t)$ and $\partial^2 M/\partial t^2$. Within this semi-classical approximation, we have the following linear relation $\delta\mathcal{J} = -D_0\vec{\nabla}\delta\vec{m}$ with a diffusion constant D_0 meaning that the non-adiabatic current is related to the non-equilibrium spin density. Then

---

[1] To see clearly, one can remind the continuity equation for particle transport given as $\frac{\partial\rho}{\partial t} + \vec{\nabla}\cdot\vec{J}_\rho = 0$ with states index i and spin index σ. The number density and corresponding current density are given by the wave function as $\rho = \psi^*_{i\sigma}\psi_{i\sigma}$ and $\vec{J}_\rho = -(i\hbar/m)\psi^*_{i\sigma}\vec{\nabla}\psi_{i\sigma}$, respectively. Here we assume the summation over the repeated indices and the reality of the current.
For spin degrees of freedom, the spin and spin current densities are $\vec{m} = \psi^*_{i\sigma}\vec{s}_{\sigma,\sigma'}\psi_{i\sigma'}$ and $\mathcal{J} = -(i\hbar/m)\psi^*_{i\sigma}\vec{s}_{\sigma,\sigma'} \otimes \vec{\nabla}\psi_{i\sigma'}$, where $\vec{s} = (\hbar/2)\vec{\sigma}$ and $\vec{\sigma}$ is a vector, whose Cartesian components are the three Pauli matrices. The indices of $\mathcal{J}$ are those in spin and real spaces. In general, spin is not conserved. Thus the continuity equation has extra contributions as in (4.11).

the non-equilibrium spin density satisfies

$$D_0 \vec{\nabla}^2 \delta \vec{m} - \frac{\delta \vec{m} \times \vec{M}}{\tau_{ex} M_s} - \frac{\delta \vec{m}}{\tau_{sf}} = \frac{n_0}{M_s} \partial_t \vec{M} + \frac{\mu_B P}{M_s} (\vec{v}_s \cdot \vec{\nabla}) \vec{M}(\vec{r}, t) , \quad (4.13)$$

where $\tau_{ex} = \frac{\hbar}{SJ_{ex}}$, the time scale of exchange interaction. Thus $\delta \vec{m}$ is determined by two different sources, time variation and spatial variation of magnetization. We discard the first term (with D_0) because we consider the slowly varying magnetization in space so that the domain wall width of magnetization is much larger than the transport length scale $\lambda = \sqrt{D_0 (1/\tau_{sf} + i/\tau_{ex})^{-1}}$. With the observation that the second and third terms in equation (4.13) are orthogonal each other, the equation can be solved algebraically as

$$\delta \vec{m} = \frac{-\tau_{ex}}{1 + \xi_0^2} \left[\frac{n_0 \xi_0}{M_s} \partial_t \vec{M} + \frac{n_0}{M_s^2} \vec{M} \times \partial_t \vec{M} \right.$$
$$\left. + \frac{\mu_B P \xi_0}{M_s} (v_s^i \nabla_i) \vec{M} + \frac{\mu_B P}{M_s^2} \vec{M} \times (v_s^i \nabla_i) \vec{M} \right] , \quad (4.14)$$

where $\xi_0 = \tau_{ex}/\tau_{sf}$ and $(\vec{v}_s \cdot \vec{\nabla}) = (v_s^i \nabla_i)$. This induced spin density affects the magnetization $\vec{M}$.

Now we can compute the spin transfer torque on the magnetization due to the induced conduction spin density $\delta \vec{m}$ in (4.14). The torque action on the slow magnetization $\vec{M}$ is given by the opposite of the torque listed in (4.11) as they are the action-reaction pairs. Thus, $\vec{T}_{sd} = -\vec{M} \times \vec{m}/(\tau_{ex} M_s) = -\vec{M} \times \delta \vec{m}/(\tau_{ex} M_s)$. *Basic theme: local magnetization and the spin of the conduction election exchange torques as action-reaction pair through the Hund's rule coupling.*

$$\vec{T}_{sd} = \frac{\tau_{ex}}{1 + \xi_0^2} \left[-\frac{n_0}{M_s} \frac{\partial \vec{M}}{\partial t} + \frac{n_0 \xi_0}{M_s^2} \vec{M} \times \frac{\partial \vec{M}}{\partial t} \right.$$
$$\left. + \frac{\mu_B P \xi_0}{M_s^2} \vec{M} \times (v_s^i \nabla_i) \vec{M} + \frac{\mu_B P}{M_s^3} \vec{M} \times [\vec{M} \times (v_s^i \nabla_i) \vec{M}] \right] . \quad (4.15)$$

The first two terms come from the magnetization variation in time. They are independent of the current and are already parts of LLG equation. On the other hand, the last two terms are due to the spatial variation of magnetization and are current-driven effect. It is important to know the regime of the validity of the result. This result (4.15) is valid when the spin transport length scale λ is much smaller than DW width or the scale of change of the magnetization. Thus this equation can not be simply applied to the multilayer magnetic materials where the effective domain wall length scale at the interface becomes zero.

We call the last two terms in (4.15) specifically as spin transfer torque (STT).

$$\vec{T}_{STT} = \frac{\tilde{\beta}}{M_s}\vec{M} \times (v_s^i \nabla_i)\vec{M} + \frac{\tilde{\beta}}{\xi_0 M_s^2}\vec{M} \times [\vec{M} \times (v_s^i \nabla_i)\vec{M}], \quad (4.16)$$

where $\tilde{\beta} = \tau_{ex}\mu_B P \xi_0 / M_s (1+\xi_0^2)$. These two terms are called non-adiabatic STT and adiabatic STT, respectively. The coefficients of the both terms are positive. We note that the terms in STT have similar mathematical structures as in (4.9), field-driven DW motion. In the literature, an alternative form is also used as $\vec{T}_{STT} = (\tilde{\beta}/M_s)\vec{M} \times (v_s^i \nabla_i)\vec{M} - (\tilde{\beta}/\xi_0)(v_s^i \nabla_i)\vec{M}$, where the adiabatic STT term is further simplified.[2] Note that $\vec{M}$ and its spatial derivatives are orthogonal to each other as $\nabla_i \vec{M}^2 = \vec{M} \cdot \nabla_i \vec{M} = \nabla_i(const.) = 0$. Thus LLG equation still satisfies the property $\vec{M} \cdot (\partial \vec{M}/\partial t) = 0$ after including this form of STT terms as discussed in §4.1.2. Moreover, when the magnetizations are uniform, these two STT contributions vanish due to the derivatives. Thus, the STT is only non-zero for the non-uniform regions in ferromagnetic materials.

4.2.2 Domain wall illustration of STT

A general motion of domain walls in the presence of electric current, so-called current-induced / current-driven DW motion, depends on various parameters such as the dimensions of nanowire (length, width, and thickness), the size of DW, relative size between the wire and DW, along with the spin polarization of the conduction electrons. For example, see [65] for extensive study for DWs in

---

[2] This can be easily derived by using
$\vec{M} \times (\vec{M} \times \vec{A}) = \hat{e}_i \epsilon_{ijk} M_j \epsilon_{klm} M_l A_m = \hat{e}_i(M_j M_i A_j - M_j M_j A_i) = -M_s^2 \vec{A}$, where one term vanishes as $M_j A_j \propto M_j(v_s^p \nabla_p M_j) = (1/2)v_s^p \nabla_p M^2 = 0$ where $M^2 = const.$

between opposite ferromagnetic spins aligned along the wire axis found in soft magnetic materials such as permalloy. Micromagnetic studies [64] found steady velocities of the transverse DW in nanowire is proportional to the electron velocity until vortices nucleate in the wire.

Here we would like to have a parallel study of the STT using the same DW configurations done in §4.1.3. STT is fundamentally different from the torque described in §4.1.3, as two players, local magnetization and electron spin, exchange torques to one another, while the external magnetic field dominates the magnetization dynamics. We take two different approaches, the effective field approach and the strong Hund's rule spin-spin interaction to explain STT. The strong Hund's rule coupling forces two adjacent spins to be parallel.

4.2.2.1 Using effective field approach (non-adiabatic STT)

There are two different terms in (4.16) that can serve as effective fields. We start with the non-adiabatic STT term using the form of LLG equation.

$$\frac{\partial \vec{M}}{\partial t} = -\gamma_0 \vec{M} \times \vec{H}_\beta + \frac{\alpha}{M_s} \vec{M} \times \frac{\partial \vec{M}}{\partial t},$$

$$\vec{H}_\beta = -\frac{1}{\gamma_0} \frac{\tilde{\beta}}{M_s} (v_s^i \nabla_i) \vec{M}.$$

(4.17)

Here we try to understand the DW dynamics using the effective field $\vec{H}_\beta$ illustrated in §4.1.3. The first term is precession term, which does not change the direction of the local magnetizations. It is the second damping term that derives the DW motion. The direction of damping term is along the effective field $\vec{H}_\beta$.

Let us examine the expression $(v_s^i \nabla_i)\vec{M}$ more closely. $\vec{v}_s$ is the velocity vector of the spin current due to the conduction electrons moving through the spin texture localized in ferromagnets, such as domain walls or skyrmions. The magnitude of the term $(v_s^i \nabla_i)\vec{M}$ depends on the magnitude of $\vec{v}_s$ and the variation of $\vec{M}$ along the direction of the current. If local magnetization is homogeneous along the current direction, the term vanishes.

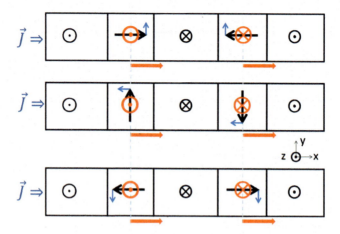

Figure 4.5 DW motion due to the non-adiabatic STT from the interaction with conduction electrons. Red vectors (⊙, ⊗) inside the blocks are the directions of $\vec{H}_\beta$, while the blue arrows are for the precession term in (4.17). The red arrows outside the blocks are the moving directions of the DWs. For the non-adiabatic STT-driven DWs motion, there is no difference whether the DWs are Néel (top and bottom figures) or Bloch (middle), right-handed (top and middle) or left-handed (bottom figure). Changing direction of current changes the direction of DW motion.

To understand the DW motion clearly, we choose the direction of the particle current $\vec{J} = J\hat{x}$ along the axis of the strip geometry, x direction in Fig. 4.5. The particle current is along the actual spin motion, which is opposite to the direction of the electric current. Note that we have $-$ sign in the term with $\vec{H}_\beta$ in (4.17). To figure out the direction of the vector $\vec{H}_\beta$, we examine the left (with ⊙) and middle (with ⊗) blocks along with the DW described by the thick right arrow ($\rightarrow$) in the top panel of Fig. 4.5. In this right-handed DW,

$$\vec{H}_\beta \propto -\frac{\partial \vec{M}}{\partial x} \approx -\frac{\Delta \vec{M}}{\Delta x} = -\frac{\vec{M}_\otimes - \vec{M}_\odot}{x_\otimes - x_\odot} = \odot = \hat{z}. \qquad (4.18)$$

This effective field $\vec{H}_\beta$ is depicted as the red circle in the top left DW. Using the lessen we learned in §4.1.3, we can conclude that the left ferromagnetic block (with ⊙) expands (the right side of the left block moves right) while the middle block (with ⊗) shrinks (the left side of the middle block moves right) due to $\vec{H}_\beta$ acting on the left DW. (Of course, one can carry on the analysis of the non-adiabatic SST term in (4.17), which produces the counter-clockwise in-plane torque (precession) depicted as small blue arrow in Fig. 4.5, followed by the damping term that forces the thick DW vector $\rightarrow$ to rotate toward out of plane, $+\hat{z}$.)

We repeat the analysis for the right DW (with $\leftarrow$) sandwiched between the middle (with $\otimes$) and right (with $\odot$) blocks in the top panel of Fig. 4.5. The direction of the effective magnetic field $\vec{H}_\beta$ can be evaluated as

$$\vec{H}_\beta \propto -\frac{\partial \vec{M}}{\partial x} \approx -\frac{\Delta \vec{M}}{\Delta x} = -\frac{\vec{M}_\odot - \vec{M}_\otimes}{x_\odot - x_\otimes} = \otimes = -\hat{z}. \qquad (4.19)$$

Thus the middle block with $\otimes$ expands as it has the same direction of the field, while the right block with $\odot$ shrinks. Combining the motion of the left and right DWs in the top panel of Fig. 4.5, we can conclude both the right-handed Néel DWs move to the right.

Now we consider the Bloch DWs in the middle panel of Fig. 4.5. The magnetization vectors in the left DW consistently change their directions (along the current direction) from out of plain ($+\hat{z}$) to into plane ($-\hat{z}$) in the yz plane with the rotational axis $-\hat{x}$ (right-handed convention). Thus $\vec{H}_\beta \propto +\hat{z}$. Thus the left DW moves to the right. Similarly for the right DW (depicted as $\downarrow$). We conclude that Bloch DWs have the same motion as Néel DWs under the non-adiabatic STT.

We also illustrate the left-handed Néel DWs in the bottom panel of Fig. 4.5. All the DWs move along the direction of applied spin current for the non-adiabatic STT. Changing the current direction changes the direction of DW motion.

4.2.2.2 Using effective field approach (adiabatic STT)

Let us turn to the adiabatic STT term. The term in (4.16) can be written as

$$\frac{\partial \vec{M}}{\partial t} = -\gamma_0 \vec{M} \times \vec{H}_{\beta'} + \frac{\alpha}{M_s} \vec{M} \times \frac{\partial \vec{M}}{\partial t},$$

$$\vec{H}_{\beta'} = -\frac{1}{\gamma_0} \frac{\tilde{\beta}}{\xi_0 M_s^2} [\vec{M} \times (v_s^i \nabla_i) \vec{M}]. \qquad (4.20)$$

The direction of the effective field $\vec{H}_{\beta'} = \vec{M} \times \vec{H}_{\beta}$ in the DWs is opposite to blue vectors inside the blocks in Fig. 4.5. It changes as the magnetization direction changes. Thus the adiabatic STT does not move the DWs, but changes the DW type and handedness.

4.2.2.3 Using Hund's rule approach

While the effective field approach, discussed in §4.2.2.1, works well for working with the mathematical expression in (4.16), there are also other intuitive explanations that give the same conclusion for STT. We can use the following basic theme: *local magnetization and the spin of conduction election exchange torques as action-reaction pair through the Hund's rule coupling that aligns the spins to be parallel.*

As the conduction electrons pass through the region of homogeneous local magnetization as in left, middle, and right blocks, there is no torque exchanged between them as the spins of the conduction electrons are aligned (anti-parallel) with the local magnetizations. When the conduction electrons pass the DW regions, for example the right-handed Néel DW depicted in the bottom left panel of Fig. 4.6, their spins are tilted toward $+\hat{x}$ direction following the local magnetization. The directions of the corresponding torque are depicted as green arrows. In turn, the local magnetization receives the opposite torque described by the blue arrows. Each individual electron has small effects, yet there are macroscopic number of conduction electrons. This process brings the local magnetization to $+\hat{z}$, which moves the DW to the right, $+\hat{x}$ direction.

We also consider the right-handed Bloch DW, depicted in the bottom right panel of Fig. 4.6. As the electrons pass through the DW, their spins are tilted into the page (toward $+\hat{y}$) in the first half, while they are tilted out of page (toward $-\hat{y}$) in the second half. These are depicted in green arrows. The local magnetization vectors receive the opposite torques, which are described as the blue vectors. Once again, the the right-handed Bloch DWs move toward right, $+\hat{x}$ direction.

Before move on, we would like to mention that the references [63][65][64] stress the importance of the non-adiabatic STT term for the current-driven DW motion. For example, one can also consider

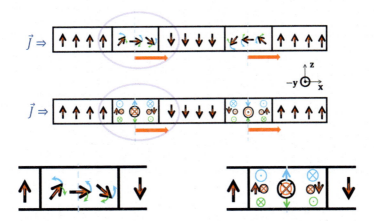

Figure 4.6 Illustration (side view) of DW motion by using the Hund's rule coupling between local magnetization and conduction electron spin to be parallel. The circled parts are enlarged in the bottom two figures. The black and red arrows are the directions of local magnetization and electron spins, respectively. Blue and green arrows are the STT acting on local magnetization and electron spins, respectively. Here $\otimes$ is into the page, $+\hat{y}$ direction.

a DW when the ferromagnetic blocks have in-plane magnetizations. It is also straight forward to analyze to see that the adiabatic STT term drives the DW along the current direction. There seems to be discrepancies, which is due to the assumption that we use the strong Hund's coupling limit. As mentioned in the beginning of this subsection, various other factors, such as polarized spin direction of conduction electrons, might play more important roles in the current-driven DW motion. We invite the readers to consult the references to uncover more on the subject.

4.2.3 Spin-orbit torque

The spin-orbit coupling of an electron spin can be obtained by coupling the relativistic Dirac equation with external electromagnetic fields and by taking the non-relativistic limit up to $\mathcal{O}(v^2/c^2)$. The corresponding Hamiltonian includes the term

$$\mathcal{H} = \frac{|e|\hbar}{4m_e^2 c^2} \vec{\sigma} \cdot (\vec{E} \times \vec{p}), \tag{4.21}$$

where $\vec{s} = \vec{\sigma}/2$ is the electron's spin angular momentum in terms of Pauli matrices $\vec{\sigma}$, $\vec{p}$ the linear momentum, $\vec{E} = -\vec{\nabla}\phi$ is the electric field in terms of a scalar potential ϕ. For a centrally symmetric electric field, $\vec{E} = -\vec{\nabla}\phi(\vec{r}) = -(\vec{r}/r)(d\phi/dr)$.

$$\mathcal{H}_{so} = -\frac{|e|\hbar^2}{2m_e^2 c^2} \frac{1}{r} \frac{d\phi}{dr} \left(\vec{L} \cdot \vec{s}\right), \tag{4.22}$$

where $\vec{L} = \vec{r} \times \vec{p}$ is the orbital angular momentum. Thus the Hamiltonian describes the spin-orbit coupling. Interested readers are encouraged to derive the Hamiltonian, which can be found in [66].

While mathematically oriented readers are satisfied with the formal derivation of the term (4.21), there is a more intuitive description for the term. A spin-orbit coupling (SOC) of an electron is present in general due to the atomic orbitals. Here we are interested in spin-orbit coupling that arises with a uniaxial symmetry without inversion symmetry, which are satisfied for hetero-structures grown from cubic crystals by breaking the cubic symmetry with asymmetric confinement of 2 dimensional electrons. The electron Hamiltonian associated with the spin-orbit coupling is known as the Rashba term [67].

$$\mathcal{H}_R = \frac{\alpha_R}{\hbar} \vec{\sigma} \cdot (\vec{p} \times \hat{z}) = -\vec{H}_{\text{eff}} \cdot \vec{\sigma}, \tag{4.23}$$

where α_R is the strength of SOC, $\hat{z}$ the unit vector along the symmetry axis, and $\vec{p}$ a linear momentum perpendicular to the symmetry axis. Note that the effective field is chosen such that the spin tends to align with the effective magnetic field $\vec{H}_{\text{eff}}$. The Hamiltonian $\mathcal{H}_R$ vanishes for $\langle\vec{p}\rangle = 0$ when the electrons do not move. We can see that this Hamiltonian $\mathcal{H}_R$ is the same as (4.21) after identifying α_R properly for the electric field along the symmetry axis, $\vec{E} = E\hat{z}$. The readers who are interested in the band structures due the Rashba term can read the references [68] for bulk material and [67] for a 2 dimensional electron gas.

To gain some physical understanding of the Rashba Hamiltonian, we consider an electron moving perpendicular to a net electric field $\vec{E} = E\hat{z}$. It is depicted in the left panel of Fig. 4.7. One can use the Lorentz transformation to an electron's rest frame, where the positive particles appear to move in opposite directions. There exists

Figure 4.7 Illustration of the Rashba field $\vec{H}_R$ in (4.24). Left: An electron moves perpendicular to the electric field $\vec{E} = E\hat{z}$. Right: In the electron's rest frame, Lorentz transform tells the existence of a magnetic field $\vec{H}_R$ in addition to an electric field (not in the figure).

a magnetic field $\vec{B}$ due to the motion of positive charges as in the right panel of Fig. 4.7. We call this Rashba field, $\vec{H}_R$.

$$\vec{H}_R = \frac{1}{c^2}\vec{E} \times \vec{v}. \qquad (4.24)$$

This Rashba field turns the electron spin to be anti-parallel to it. Now we apply this to an interesting application of manipulating local magnetization using conduction electrons with a completely different mechanism compared to the STT.

Recently, with the spin-orbit interactions introduced above, it has been proposed to use the electric current to generate a non-equilibrium spin density that gives rise to a spin torque on the local magnetization in a single nanomagnet [69]. Moreover, this spin torque does not involve spin transfer mechanism. We consider the following Hamiltonian for a single ferromagnetic 2 dimensional electron gas sandwiched between two dissimilar materials, so that the electric potential is highly asymmetric, leading to the presence of the Rashba interaction at the interfaces or within the ferromagnetic layer.

$$\mathcal{H} = \frac{\vec{p}^2}{2m_e} + \frac{\alpha_R}{\hbar}\vec{\sigma} \cdot (\vec{p} \times \hat{z}) - J_{sd}\vec{\sigma} \cdot \vec{M}. \qquad (4.25)$$

Here the electrons moving in asymmetric crystal-field potential experiences a net electric field $\vec{E} = E\hat{z}$, which can be transformed to a magnetic field in electron's rest frame as described by the second term, Rashba Hamiltonian, on the right-hand side. The corresponding

Rashba field makes the conduction electron's spin titled as in Fig. 4.8. The last term is the s-d Hamiltonian that describe the interaction between the conduction electron spin and local magnetization (assuming to be Hund's rule coupling to make them parallel) with the parameter J_{sd}, which is similar to J_{ex} used in (4.10). Thus an electric current passing through a uniformly magnetized ferromagnet exerts an exchange-mediated effective field $\vec{H}_{sd}$ on the local magnetization $\vec{M}$ that is parallel to $\vec{H}_R$. See the left panel of Fig. 4.8. As $\vec{p}$ is proportional to the particle current $\vec{J}_s$, this spin current- (the electron spins are polarized) driven effect can be used to switch the direction of the local magnetization.

Interested readers are recommended for explicit computations for deriving the non-equilibrium spin density and the spin torque in the presence of the current in [69].

$$\vec{T}_{SOT}^{DL} = -\gamma_0 \vec{M} \times \vec{H}_{sd},$$
$$\vec{H}_{sd} = \frac{2\alpha_R m_e P}{\hbar |e| M_s}(\hat{z} \times \vec{J}_s),$$
(4.26)

where P is the polarization, $\hat{z}$ is the direction of the local electric field, and $\vec{J}_s$ the spin current. This is referred as damping-like spin orbit torque (DL SOT). *Note that the direction of $\vec{H}_{sd}$ is same regardless of the directions of the local magnetizations. It tends to destroy the localized magnetization textures.* The domain wall motion associated with DL SOT has been investigated in [70][71], which is further discussed further in §4.2.4.

Once the torque (4.26) is accepted, it is natural to introduce another contribution to the spin torque, called field-like spin orbit torque (FL SOT) [72]. We can look into the LLG equation given in (4.6). The right-hand side has also the time derivative of the local magnetization. Thus we can re-express LLG equation with iteration as in (4.9). The new FL SOT term is nothing but the analogous Gilbert term for the DL SOT. Thus,

$$\vec{T}_{SOT}^{FL} = -\gamma_0 \vec{M} \times \vec{H}_{sot},$$
$$\vec{H}_{sot} = \frac{\hbar}{2\mu_0 |e|} \frac{\alpha_H}{M_s^2 t}(\vec{M} \times (\hat{z} \times \vec{J}_s)),$$
(4.27)

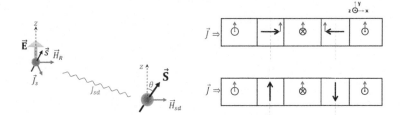

Figure 4.8 Left: Illustration of the interaction between the conduction electron and local magnetization through s-d Hamiltonian (with Hund's rule coupling). Right: The effect of effective field $\vec{H}_{sd}$ given in (4.26) on the DW structure. The field $\vec{H}_{sd}$ does not generate a DW motion.

where t is the thickness of the ferromagnet layer. It turns out that this FL SOT term (4.27) is much more efficient to move Néel-type *skyrmions*, while it does not help the domain wall motion as in Fig. 4.8.

Combining these two contributions, we have

$$\vec{T}_{SOT} = -\gamma_0 a_{SOT} \left[\tilde{\eta}\vec{M} \times (\hat{z} \times \vec{J}_s) + \vec{M} \times (\vec{M} \times (\hat{z} \times \vec{J}_s)) \right], \quad (4.28)$$

where $a_{SOT} = (\hbar/2\mu_0|e|)(\alpha_H/M_s^2 t)$ and $\tilde{\eta}$ a parameter to distinguish the strength of field-like SOT compared to that of damping-like SOT. We note that these terms also induce similar terms in the LLG equation through the Gilbert damping term, turning a field-like SOT term into a damping-like SOT term and vice versa. These four combined terms has been used in [72] to have a better fit for the skyrmion motion.

4.2.4 Domain Wall illustration of SOT

While SOT shares some common features with STT because of the conduction electrons, they are fundamentally different. Due to the electric field and the corresponding Rashba magnetic field, the conduction electron spins are anchored to particular direction which is assumed to be fixed while interacting with local magnetizations here. Then Hund's rule coupling forces the local magnetization to align to be parallel to the electron spins. As discussed above, STT

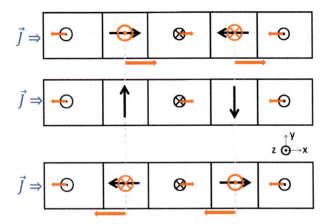

Figure 4.9 Illustration of FL SOT using an effective field $\vec{H}_{sot}$ given in (4.27). The red arrows inside the blocks, $\vec{H}_{sot}$, cannot flip all the local homogeneous magnetizations in ferromagnetic blocks except at the boundaries with DWs. The large red arrows outside the blocks indicates that the right-handed Néel DWs move to the right (top panel), while the left-handed Néel DWs move to the left (bottom panel). Bloch DWs illustrated in the middle panel are not affected by FL SOT.

transfers spin angular momentum directly to DW without any effects on uniform magnetization. On the other hand, SOT can act on local magnetization uniformly regardless of its structure.

4.2.4.1 For DL SOT

Let us consider DW motion due to DL SOT by using the effective field approach. By rewriting (4.28) into the form (4.9), we have

$$\frac{\partial \vec{M}}{\partial t} = -\gamma_0 \vec{M} \times \vec{H}_{sd} + \frac{\alpha}{M_s} \vec{M} \times \frac{\partial \vec{M}}{\partial t}, \quad (4.29)$$
$$\vec{H}_{sd} \propto (\hat{z} \times \vec{J}_s).$$

Once we figure out the direction of the effective magnetic field $\vec{H}_{sd}$, we can use the results given in §4.1.3. We already have discussed the direction of $\vec{H}_{sd}$, which is depicted as the blue arrows in the right

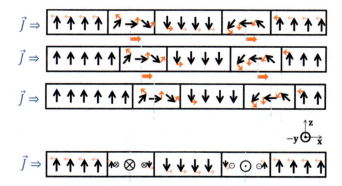

Figure 4.10 Side view of Fig. 4.9. Top panel corresponds to the top one in Fig. 4.9. Bright red arrows inside the blocks (representing the direction of $\vec{H}_{sot}$) changes the local magnetizations, while transparent red arrows do not have effects. Second and third panels are the results of FL SOT that show the motion of DWs. The process goes on. The bottom panel is for the Bloch DWs, which do not move under FL SOT.

panel of Fig. 4.8. Thus the effective field push the local magnetizations toward $+\hat{y}$. It is not useful for DW motion.

4.2.4.2 For FL SOT

Let us turn to DW motion due to FL SOT. The LLG equation turns into

$$\frac{\partial \vec{M}}{\partial t} = -\gamma_0 \vec{M} \times \vec{H}_{sot} + \frac{\alpha}{M_s} \vec{M} \times \frac{\partial \vec{M}}{\partial t},$$

(4.30)

$$\vec{H}_{sot} \propto \vec{M} \times (\hat{z} \times \vec{J}_s).$$

The directions of $\vec{H}_{sot}$ acting on the local magnetizations are depicted in Fig. 4.9. As described in §4.1.3, torque due to the effective fields pointing out of plane helps the block with the same magnetization expand. Thus right-handed Néel DWs move toward right, while the left-handed Néel DWs move to the left. We also illustrate the side view in Fig. 4.10 to clearly show the torques in various parts of the magnet.

From the discussions, we see that FL SOT is efficient to move Néel type DWs. This can be used to distinguish Néel DWs from Bloch DWs.

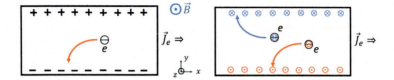

Figure 4.11 Analogy between the charge Hall effect (left panel) and the spin Hall effect (right panel). Note that the current $\vec{J}_e$ is electric current. Electrons move toward $-\hat{x}$. The charge Hall effects in the presence of magnetic field are from the Lorentz force law and independent of the spin polarization. Right: The spin Hall effects depend on the spins of the electrons without a magnetic field. Spin polarization of an electron along $\hat{z}$ behaves similar to the electron with magnetic field in the same direction. Thus spin-up and -down electrons move towards the opposite directions. The red circle with a dot represents an electron with up-spin, while the blue circle with a cross represents an electron with down-spin. The result is a spin imbalance between the top and the bottom of the 2 dimensional film.

It turns out that this FL SOT term is also much more efficient to move Néel type skyrmions as demonstrated [72].

4.2.5 Spin Hall torque

Spin Hall torque is very fruitful approach to move only a certain types of DWs, actually a Néel DW. As the spin Hall effect has wide variety of applications, we introduce the concept of the spin Hall effect [73][74]. We also consider the Slonczewski mechanism for transferring spin angular momentum into a different layer [75][76]. Then we consider the spin Hall torque that is generated on a ferromagnet through the interface with a heavy metal, whose interface produces a large spin-orbit coupling.

4.2.5.1 Spin Hall effect

The well known Hall effect is the result of magnetic field acting on a moving charged particle through the Lorentz force law $\vec{F} = q\vec{v} \times \vec{B}$ with an electric charge $q = -|e|$, velocity of the charged particle $\vec{v}$ and the magnetic field $\vec{B}$ as illustrated in the left panel of Fig. 4.11.

The spin Hall effect is the result of the motion of an electron's intrinsic spin polarization, depicted as red circle with a dot ($\odot$) or

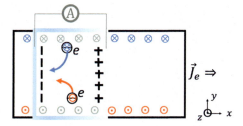

Figure 4.12 A metal such as *Pt* is overlaid to provide a pathway for the conduction electrons to move to reduce the spin imbalance. The inverse spin Hall effect results in a measurable electric potential or electric current if the two ends are connected.

blue circle with a cross (⊗) depending on the spin directions [74]. See the right panel of Fig. 4.11. These spin polarizations effectively play the similar role of the magnetic field. Of course, the spin polarization is random unless one polarizes the spins of the electrons. In a thin film, spin polarizations parallel to the film ($\pm\hat{y}$) produce the force out of the plane resulting in no visible effects. On the other hand, electrons with spin polarization perpendicular to the film ($\pm\hat{z}$) produces net spin accumulations, down-spins along the $+y$ side and up-spins along the $-y$ side of the thin material. This spin deflection can be described by

$$\vec{F}_{SH} \propto \theta_{SH} \vec{J} \times \sigma , \qquad (4.31)$$

where the particle current $\vec{J}$ is opposite to the electric current and σ is a spin polarization vector. θ_{SH} is so-called spin Hall angle which is defined by the dimensionless ratio between spin current and electric current. In right panel of Fig. 4.11, the spin deflections indicates that $\theta_{SH} < 0$. The sign of spin Hall angle depends on the materials, for example $\theta_{SH} > 0$ for Platinum (*Pt*) [77] and $\theta_{SH} < 0$ for Tantalum (*Ta*) [78]. Thus electron flow produces the net spin excess.

When there exists a pathway, the polarized spins accumulated in opposite ends move toward each other to remove the spin imbalance and produce spin current. In Fig. 4.12, a metal layer is overlaid on top of the setup of the right panel of Fig. 4.11. In the metal layer, the spin-up (⊙) electrons move up toward $+\hat{y}$, while the

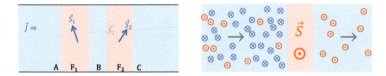

Figure 4.13 The middle conducting ferromagnetic layer, with up direction magnetization, serves as a spin polarizer by the spin transport of the majority spins, up-spins with arrow head. Accumulation of the minority spins, down-spins with arrow tail, on the left side of the ferromagnetic layer and that of the majority spins on the right side.

spin-down ($\otimes$) electrons move down toward $-\hat{y}$. Accordingly, these spins are deflected toward $-x$ direction due to the spin Hall effect. Thus electrons accumulate in the left side of the metal layer, which produces a charge imbalance. This is so-called the inverse spin Hall effect (iSHE) [74]. The resulting electric potential or electric field can be measured by using Voltmeter or Ammeter as in Fig. 4.12.

4.2.5.2 Slonczewski spin torque

The spin Hall effect can be used to produce spin torque when combined with the so-called Slonczewski mechanism [75][76] by transferring spin polarization from one layer into another layer. We explain Slonczewski mechanism [75].

Let us consider two different ferromagnetic layers F_1 and F_2 with their spin (magnetization) vectors $\vec{S}_1$ and $\vec{S}_2$, respectively in the left panel of Fig. 4.13. There are three other metallic layers, A, B, and C with the thickness of the spacer layer B less than the spin diffusion length. When electric currents are applied to the left (electrons move to the right with particle current $\vec{j}$), the spin polarization $\vec{S}_1$ in F_1 is present on the layer F_2. Slonczewski solved the quantum mechanical tunneling problem of the five layers to obtain the spin torques

$$\frac{\partial \vec{S}_{1,2}}{\partial t} \propto \vec{S}_{1,2} \times (\vec{S}_1 \times \vec{S}_2), \tag{4.32}$$

where angular momentum conservation has the key role as local spin polarization $\vec{S}_2$ reacts to the electron spins (influenced by $\vec{S}_1$) passing through. Note that the polarization $\vec{S}_1$ is also affected by $\vec{S}_2$ as not all the electrons started from layer A can pass through to the layer C. If we focus on three layers, B, F_2, and C, we can think that the electrons polarized with $\vec{S}_1$ in B interacts with the local spins $\vec{S}_2$ in the ferromagnetic layer F_2. We adapt this picture below in the context of spin Hall effect.

We consider an extreme case in the right panel of Fig. 4.13. In ferromagnetic conductors, the majority spins (electrons with the same polarization as the polarization of the ferromagnet layer) have large conductivity, while the minority spins (electrons with the opposite polarization compared to the polarization of the ferromagnet layer) have poor conductivity. In the interface between a normal metal and a ferromagnetic metal, majority spins are preferentially transmitted. Thus the ferromagnet sitting in the middle panel of Fig. 4.13 acts as a spin polarizer. With the electrons injected from left, there are accumulated spins on both sides of the ferromagnet layer, the minority (down) spins on the left side and the majority (up) spins on the right side of the ferromagnet with the up-direction magnetization. These collective spin accumulations (magnetization) can provide Slonczewski torque with appropriate arrangements of ferromagnet and metallic layers. As the process is quantum mechanical, the actual transmission of the electrons are not necessary for the torque to influence the magnetization in the ferromagnetic layer. Similar illustrations can be found in [59].

4.2.5.3 Spin Hall torque

Now let us put these two ingredients, spin Hall effect and Slonczewski mechanism, together to realize the spin Hall torque (SHT).

We consider the particle current toward right ($+\hat{x}$) in Pt layer that produces spin excess according to the spin Hall effect. The electrons with down-spin ($\otimes$, here we consider all the spin directions to be opposite to those of magnetic moments) move up toward the interface with the ferromagnetic layer. See the left panel of Fig. 4.14. The Pt layer and the spin polarization serve as the metallic layer B and the direction $\vec{S}_2$ in the left panel of Fig. 4.13, respectively. At the interface, we apply the Slonczewski torque (4.32). The spin torque acting on $\vec{M}$ in the ferromagnetic layer is described by

$$\vec{T}_{Slonczewski} = \frac{J_e \hbar \theta_{SH}}{2|e|M_s^2 t_F} \vec{M} \times (\vec{M} \times \vec{m}_f), \qquad (4.33)$$

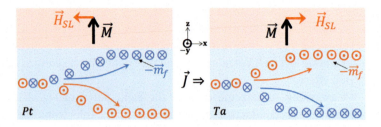

Figure 4.14 The middle conducting ferromagnetic layer, with up direction magnetization, serves as a spin polarizer by the spin transport of the majority spins, up-spins with arrow head. Accumulation of the minority spins, down-spins with arrow tail, on the left side of the ferromagnetic layer and that of the majority spins on the right side.

where $\vec{J}_e$ is the electric current, θ_{SH} the spin Hall angle, t_F the thickness of ferromagnetic layer with the magnetization $\vec{M}$. $\vec{m}_f$ is the magnetization of another ferromagnetic layer $\vec{S}_1$ in the left panel of Fig. 4.13, with which we can identify as the magnetization of accumulated electrons near the interface. As mentioned, this is quantum mechanical phenomena and does not depends on whether the spins really penetrate to the other layer or not.

Upon applying this Slonczewski torque for the bottom heavy metal layer with the top ferromagnetic layer, it is more convenient to use the electron magnetization (spin polarization) as $\vec{m}_f = \hat{z} \times \vec{J}_{HM}$. The spin Hall angle θ_{SH} also include various efficiencies of transferring the torque. Then, we can use the effective field form.

$$\vec{T}_{SHT} = -\gamma_0 \vec{M} \times \vec{H}_{SL},$$
$$\vec{H}_{SL} = -\frac{\hbar \theta_{SH}}{2|e|M_s^2 t_F}(\vec{M} \times (\hat{z} \times \vec{J}_{HM})), \quad (4.34)$$

where $\theta_{SH} > 0$ for Pt layer, while $\theta_{SH} < 0$ for Ta layer. See Fig. 4.14. Directions of the effective fields are also illustrated in the figure.

Before discuss the DW motion due to SHT, it is beneficial to distinguish the field $\vec{H}_{SL}$ with the $\vec{H}_{sot}$ that is associated with the FL SOT. SOT actually can apply torque even though the electric current is in a different layer as the interaction is mediated indirectly by s-d interaction between the conduction electron spin and the

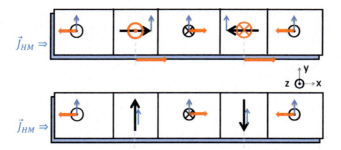

Figure 4.15 DW motion under the influence of SHT (4.34) and (4.35). The current $\vec{J}_{HM}$ flows in the heavy metal (HM) layer (blue layer below the ferromagnet layer) and the blue arrows are the direction of $\hat{z} \times \vec{J}_{HM}$. The red arrows are the direction of $\vec{H}_{SL}$ with $\theta_{SH} > 0$, which is the case for Pt layer. If $\theta_{SH} < 0$, the DW moves to the opposite direction. Top figure: right-handed Néel DWs, which move to the right. Bottom figure: right-handed Bloch DWs. There is no spin Hall torque acting on Bloch DWs. Note that the deriving mechanism is completely different compared to FL SOT, while the direction of $\vec{H}_{SL}$ that derives the DW motion is the same as the $\vec{H}_{sot}$ illustrated in Fig. 4.9. In the current setup, the effects of FL SOT is negligible compared to SHT [59].

local magnetization. Regardless the detailed Rashba mechanism, $\vec{H}_{sd}$ associated with the DL SOT forces the local magnetization to be antiparallel to the conduction electron spin. In the Pt layer illustrated in Fig. 4.14, magnetization vector is forced to point $-\hat{y}$ (out of page) due to the spin polarization $\otimes$ near the layers interface. First of all, the distance between the electron and the local magnetization here is larger than that we considered in §4.2.4 where they are in the same layer. Moreover, the field $\vec{H}_{sd}$ due to the accumulated electrons near the interface is expected to be partially canceled by the other electrons (with $\odot$) at the bottom of the Pt layer. Similarly, the effects of the FL SOT are expected to be small compared to SHT by $\vec{H}_{SL}$, as the effective fields $\vec{H}_{sot}$ is smaller due to the distance and partial cancellations as explained with $\vec{H}_{sd}$. These are experimental evidences that the SOT in Fig. 4.14 is negligible compared to SHT [59].

4.2.6 Domain Wall illustration of SHT

Similar to the other spin torques, we use the effective field description to understand the DW motion due to SHT, by utilizing

(4.9) in §4.1.3.

$$\frac{\partial \vec{M}}{\partial t} = -\gamma_0 \vec{M} \times \vec{H}_{SL} + \frac{\alpha}{M_s}\vec{M} \times \frac{\partial \vec{M}}{\partial t},$$

$$\vec{H}_{SL} \propto -\theta_{SH}\vec{M} \times (\hat{z} \times \vec{J}_{HM}).$$

(4.35)

Here we explicitly include the parameter θ_{SH} as its sign is crucial for the DW motion. The effective field $\vec{H}_{SL}$ (depicted as red arrows in Fig. 4.15) forces the magnetization vector $\vec{M}$ to precess and the corresponding damping force directs the vector to the direction $\vec{H}_{SL}$. We add explanations in the caption of Fig. 4.15. We note that the directions of $\vec{H}_{SL}$ and $\vec{H}_{sot}$ are the same. Nevertheless, the main driving mechanism of Néel DW motion is due to the SHT with $\vec{H}_{SL}$ as explained in the previous subsection [59]. We also add the detailed illustrations of the side view to clarify the whole picture of the SHT-driven motion of DWs in Fig. 4.16.

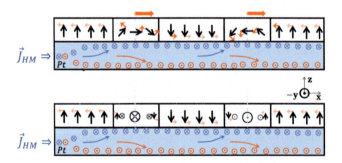

Figure 4.16 Side view for the DW motion under the influence of SHT (4.34) and (4.35) described in Fig. 4.15. Top and bottom layers are the ferromagnetic and heavy metal (HM) layers, respectively. The think black arrows are magnetization vectors, while the red arrows are the directions of $\vec{H}_{SL}$ in the top layers. In the HM layers, the red and blue arrows are the electron spins, which are opposite to their magnetic moments.

4.2.7 Emergent electromagnetic field

Before moving on, we would like to introduce the effect of the topological emergent electro-magnetic fields to the spin torque and the LLG equation. The emergent electromagnetic field has been first proposed in [79] and [80][81].

$$e_i = E_i^e = \vec{n} \cdot (\partial_i \vec{n} \times \partial_t \vec{n}) ,$$
$$b_i = B_i^e = \frac{1}{2} \epsilon_{ijk} \vec{n} \cdot (\partial_j \vec{n} \times \partial_k \vec{n}) , \qquad (4.36)$$

where $\vec{n} = \vec{M}/M_s$. Attentive readers might recognize the emergent magnetic field is nothing but the topological skyrmion charge. We also note that the emergent electric field only exists for time dependent spin configurations. Thus, the spin structure of the skyrmions provides the emergent electromagnetic fields (EEMF).

This EEMF interact with the conduction electrons due to the Hund's rule coupling, strong ferromagnetic coupling, between the electron spins and the spins of the Skyrmions. In turn, skyrmion motion can be influenced by the interaction, which has been considered in the context of skyrmion motion [80][81].

$$\vec{T}_{EM} = -\frac{\alpha'}{M_s^3} \left[\vec{M} \cdot \left(\partial_i \vec{M} \times (\partial_t + \frac{\beta'}{\alpha'} v_s^k \nabla_k) \vec{M} \right) \right] \partial_i \vec{M}$$
$$= -\left(\alpha' E_i^e + \beta' (\vec{v}_s \times \vec{B}^e)_i \right) \partial_i \vec{M} . \qquad (4.37)$$

Note that, while it is similar to the STT with derivative, there are two derivatives. They are higher order terms in derivatives. Nevertheless, these contributions can be significant in some metallic materials.

4.3 Generalized LLG equation

The Landau-Lifshitz-Gilbert (LLG) equation (4.6) consists of a precessional term due to an effective field and a phenomenological damping term as described in §4.1.2. We introduced four more torques, $\vec{T}_{STT}$, $\vec{T}_{SOT}$, $\vec{T}_{SHT}$, and $\vec{T}_{EM}$, one by one in the previous sections. We include them to the LLG equation.

Let us include the spin transfer torque $\vec{T}_{sd}$ we derived in (4.15) to the LLG equation (4.6). Note that the first term in (4.15) simply renormalizes the gyromagnetic ratio γ_0, while the second term renormalizes the Gilbert damping parameter α. Thus the first two temporal spin terms in (4.15) can be absorbed by the redefinition $\gamma_0' = \gamma_0(1+\eta')^{-1}$ and $\gamma_0'\alpha' = \gamma_0(\alpha + \xi_0 \eta')$ with $\eta' = (n_0/M_s)/(1+\xi_0^2)$. To make the notation uncluttered, we omit the prime for these parameters. Thus, we have two new contributions to the LLG equation, the last two spatially non-uniform magnetization vector terms in (4.15) or the two terms in (4.16), in the presence of conduction electrons. Then the LLG equation with the spin transfer torque has the form

$$\partial_t \vec{M} = -\gamma_0 \vec{M} \times \vec{H}_{\text{eff}} + \frac{\alpha}{M_s} \vec{M} \times \partial_t \vec{M}$$
$$+ \frac{\beta}{M_s} \vec{M} \times (v_s^i \nabla_i) \vec{M} + \frac{\beta}{\xi_0 M_s^2} \vec{M} \times [\vec{M} \times (v_s^i \nabla_i) \vec{M}], \quad (4.38)$$

where $\beta = P\mu_B \xi_0/M_s(1+\xi_0^2)$.

When we consider the skyrmions, we use the approximation $|\vec{M}| = const.$ as we have done in this chapter. This is justified because $|\vec{M}|$ varies only weakly in the skyrmion phase. The last term in (4.38) is simplified as $(\beta/\xi_0 M_s^2)\vec{M} \times [\vec{M} \times (v_s^i \nabla_i)\vec{M}] = -(\beta/\xi_0)(v_s^i \nabla_i)\vec{M}$ for $\vec{M}$ and its spatial derivatives are orthogonal each other. After normalizing the space scale compared to the time scale so that $\beta/\xi_0 \to 1$, the LLG equation becomes

$$(\partial_t + v_s^i \nabla_i)\vec{M} = -\gamma \vec{M} \times \vec{H}_{\text{eff}} + \frac{\alpha}{M_s} \vec{M} \times (\partial_t + \frac{\beta}{\alpha} v_s^i \nabla_i)\vec{M}. \quad (4.39)$$

Here the time and spatial derivatives come along together. The time derivative becomes the drift velocity $\vec{v}_d$ of the skyrmions when we consider their motion later, and $\vec{v}_s$ is the velocity of the conduction electrons passing through the skyrmions. The combination $\vec{v}_d - \vec{v}_s$ is the relative velocity between the center of skyrmion and conduction electrons. We can also include the emergent electromagnetic fields, effectively, by modifying the parameters α and β. This is further explained when we discuss the Thiele equation in §4.4.2.1.

We also generalize the LLG equation (4.6) by including the spin-orbit torque $\vec{T}_{SOT}$ and also the Hall torque $\vec{T}_{SHT}$ in a unified fashion.

$$\frac{\partial \vec{M}}{\partial t} = -\gamma_0 \vec{M} \times (\vec{H}_{\text{eff}} + \vec{H}_{sd} + \vec{H}_{SOT} + \vec{H}_{SL}) + \frac{\alpha}{M_s} \vec{M} \times \frac{\partial \vec{M}}{\partial t}, \quad (4.40)$$

where $\vec{H}_{sd}$, $\vec{H}_{SOT}$, and $\vec{H}_{SL}$ are given in (4.26), (4.27), and (4.34), respectively. Combining $\vec{H}_{sd}$, $\vec{H}_{SOT}$, and $\vec{H}_{SL}$ with $\vec{H}_{\text{eff}}$ is convenient and actually reasonable because all these acts on the ferromagnetic layer uniformly. All the spin torques can be combined to control the motion of DW or skyrmions. The STT term $\vec{T}_{STT}$, which has the form (4.40) is rather distinct from those of field like spin torques $\vec{T}_{SOT}$ and $\vec{T}_{SHT}$ given in (4.39). Of course, we can include all the STT terms as a part of the effective field so that we can write all the spin torques as

$$\frac{\partial \vec{M}}{\partial t} = -\gamma_0 \vec{M} \times \vec{H}_{tot} + \frac{\alpha}{M_s} \vec{M} \times \frac{\partial \vec{M}}{\partial t},$$
$$\vec{H}_{tot} = \vec{H}_{\text{eff}} + \vec{H}_\beta + \vec{H}_{\beta'} + \vec{H}_{sd} + \vec{H}_{SOT} + \vec{H}_{SL}, \quad (4.41)$$

where $\vec{H}_\beta$ and $\vec{H}_{\beta'}$ are given in (4.17) and (4.20), respectively. Depending on physical situations, we can choose different descriptions by including $\vec{T}_{STT}$ and $\vec{T}_{EM}$ in terms of velocity along with the skyrmion drift velocity or by separating them as force terms.

4.4 Thiele equation

In this section we derive the celebrated Thiele equation that describes the skyrmion center without internal deformations. The Thiele's original derivation [82] is highly inspirational. We encourage readers the original paper [82]. We also present a modern derivation and also generalize the Thiele equation with various spin torques that we have introduced above in §4.2.

4.4.1 Thiele's original derivation

On 1973, Thiele published a paper that describes the steady-state motion of magnetic domains, which can be also directly applied to the topological magnetic skyrmions as well. His paper [82] is beautifully written and captures deep insights.

The i-th component of the LLG equation (4.6) without the three torque terms (or (4.38) without the last two terms) has the form

$$-\frac{1}{\gamma_0}\partial_t M_i + \frac{\alpha}{\gamma_0 M_s}\epsilon_{ijk}M_j\partial_t M_k - \epsilon_{ijk}M_j H_k^{\text{eff}} = 0 \,. \tag{4.42}$$

We consider the effects of the other torque terms, $\vec{T}_{STT}$, $\vec{T}_{SOT}$, $\vec{T}_{SHT}$, and $\vec{T}_{EM}$, below in this section.

We focus on studying the spatially constant saturation magnetization, $M_s^2 = M_i M_i = $ constant. By taking a derivative $\partial_j(M_i M_i) = 0$, we see that $M_i(\partial M_i/\partial x^j) = 0$, meaning that the magnetization vector and its spatial derivatives are orthogonal to each other. Furthermore, contracting M_i with (4.42) followed by summing over the index i reveals that $M_i \partial_t M_i = 0$. They are summarized as

$$M_i \partial_t M_i = 0 \,,$$

$$M_i \frac{\partial M_i}{\partial x^j} = 0 \,, \tag{4.43}$$

$$M_s^2 = M_i M_i = \text{const.}$$

These relations tells that $\vec{M}$ is orthogonal to $\partial_t \vec{M}$ and also to $\vec{\nabla}\vec{M}$.

Thiele realized that the following equation is equivalent to (4.42) with the conditions (4.43)

$$\tilde{\beta}M_j + \frac{-1}{\gamma_0 M_s^2}\epsilon_{jkl}M_k \partial_t M_l + \frac{-\alpha}{\gamma_0 M_s}\partial_t M_j + H_j^{\text{eff}} = 0 \,. \tag{4.44}$$

One can verify the equivalence by multiplying $-\epsilon_{jik}M_k$ to (4.44), summing over j and renaming indices. Multiplying M_j to (4.44) seems to produce a new result $\tilde{\beta}M_s^2 + M_j H_j = 0$, yet it just fixes the value $\tilde{\beta}$, which does not enter into the physical quantities we consider. Note that the first three terms are mutually orthogonal among themselves.

Thus, with proper identifications of $\tilde{\beta}$, one can derive (4.44) directly from (4.42) by acting $\epsilon_{ijk}M_k$ on it.

To derive the Thiele's equation, we consider the steady motion of spin structure such as a domain wall or skyrmion lattice with the center position X_i that moves with a drift velocity v_i^d, while field position is denoted by x_i.

$$M_i = M_i(x_j - X_j), \qquad X_j = v_j^d t, \tag{4.45}$$

and thus $\partial_t M_i = -v_j^d \partial M_i / \partial x_j$. A crucial step to get the Thiele equation is multiplying $-\partial M_j / \partial x_i$ on (4.44) to project the LLG equation into the translation modes. Thiele called the resulting terms as the force densities. The first term in (4.44) vanishes. The second term gives $g_{ij} v_j = (1/M_s^2 \gamma_0') \epsilon_{lmn} M_l (\partial M_m / \partial x_i)(\partial M_n / \partial x_j) v_j$ that can be rewritten in a more convenient form as the epsilon symbol makes the two derivative terms antisymmetric.

$$\begin{aligned} f_i^g &= \epsilon_{ijk} g_j v_k, \\ g_j &= \frac{-1}{2M_s^2 \gamma_0'} \epsilon_{jpq} \epsilon_{lmn} M_l \frac{\partial M_m}{\partial x_p} \frac{\partial M_n}{\partial x_q}. \end{aligned} \tag{4.46}$$

The integral of this f_i^g over a total volume is the total Magnus force with the gyromagnetic coupling vector $\vec{g}$. Even without considering the topological object such as skyrmions, this gyromagnetic term contains the topological charge density (1.97) in §1.3.4 that can be identified with $\vec{n} = \vec{M}/M_s$ in two spatial dimensions. Thus we can readily apply the result for the systems with skyrmions. How exciting it is!

The third term proportional to α gives

$$\begin{aligned} f_i^\alpha &= \alpha' d_{ij} v_j, \\ d_{ij} &= \frac{-1}{M_s \gamma_0'} \frac{\partial M_k}{\partial x_i} \frac{\partial M_k}{\partial x_j}. \end{aligned} \tag{4.47}$$

Here d_{ij} is a second rank tensor. The integral of this f_i^α over a total volume is the total dissipative drag force. This term contains a second rank tensor d_{ij} that is similar to the non-uniform energy term considered in Landau-Lifshitz [12].

Multiplying $-\partial M_j/\partial x_i$ to the last term in (4.44) followed by integrating over the volume captures the total externally applied force $\vec{F}$. Thus we have

$$\vec{\mathcal{G}} \times \vec{v}_d + \alpha \mathcal{D} \cdot \vec{v}_d + \vec{F}' = 0, \tag{4.48}$$

where $\vec{\mathcal{G}} = \int_V \vec{g} d$, $\mathcal{D} = \int_V d\, dV$ with $\mathcal{D}$ and d as second rank tensors. This is the original form developed by Thiele [82], which describes the translational motion of the center of a domain wall or skyrmion without macroscopic deformation of their structures. While the skyrmion is extended object and the spin configuration is complex, the quantity $\vec{\mathcal{G}}$ is able to capture the topological charge of the skyrmions. It turns out that $\vec{\mathcal{G}}$ is also directly related to the emergent magnetic field that we discussed already in §4.2.7.

4.4.2 Generalizations with various spin torques

There are several different generalizations of the Thiele equation. We introduce them one by one here. They include the cases with the spin transfer torque terms and the emergent electromagnetic fields in §4.4.2.1, with the spin orbit torque in §4.4.2.2, and with the spin Hall torque in §4.4.2.3.

4.4.2.1 Spin transfer torque and emergent fields

As the original Thiele's derivation is a little complicated, we revisit the derivation by generalizing the Thiele equation with the spin transfer torque terms §4.2.1 and also the emergent electromagnetic fields contributions introduced in §4.2.7. To start, let us consider the LLG equation for the unit vector $\vec{m} = \vec{M}/M_s$,

$$\begin{aligned}(\partial_t + \vec{v}_s \cdot \vec{\nabla})\vec{m} = &-\gamma \vec{m} \times \vec{H}_{\text{eff}} \\ &+ \vec{m} \times (\alpha \partial_t + \beta \vec{v}_s \cdot \vec{\nabla})\vec{m} \\ &- \vec{m} \cdot \left[\partial_i \vec{m} \times (\alpha' \partial_t + \beta' \vec{v}_s \cdot \vec{\nabla})\vec{m}\right] \partial_i \vec{m},\end{aligned} \tag{4.49}$$

where the second line contains the usual spin transfer torque [63] and the third line the emergent fields [81][80]. We recover the previous case by setting $\vec{v}_s = \beta = \alpha' = \beta' = 0$.

By taking the cross product $\vec{m}\times$, the equation can be rewritten as by taking $\vec{m}\times$ to it.

$$\begin{aligned}\gamma M_s \vec{H}_{\text{eff}} = &\, M_s \vec{m} \times (\partial_t + \vec{v}_s \cdot \vec{\nabla})\vec{m} \\ &+ M_s(\alpha \partial_t + \beta \vec{v}_s \cdot \vec{\nabla})\vec{m} \\ &+ M_s \vec{m} \times \left[\alpha' \boldsymbol{e}_i + \beta'(\vec{v}_s \times \boldsymbol{b})_i\right]\partial_i \vec{m} \,.\end{aligned} \qquad (4.50)$$

To get the Thiele equation, we use the parametrization of the center of skyrmion motion as $\vec{m}(\vec{x} - \vec{v}_d t)$, which turns time derivative into spatial one with velocity $\vec{v}_d$ and thus the emergent electric field turns into emergent magnetic field multiplied by the drift velocity $\vec{v}_d$. Then we project the LLG equation (4.50) to the translation mode by taking a dot product $\partial_i \vec{m} \cdot$ and integrate over a unit cell (UC). Then we get the Thiele equation extended with the emergent electromagnetic fields [83].

$$\vec{\mathcal{G}} \times (\vec{v}_s - \vec{v}_d) + \mathcal{D} \cdot (\tilde{\beta}\vec{v}_s - \tilde{\alpha}\vec{v}_d) + \vec{F}_{pin} = 0 \,, \qquad (4.51)$$

where

$$\begin{aligned}\mathcal{G}_i &= \int_{UC} d^2 x \, M_s \vec{b}_i = 4\pi Q M_s \hat{b}_i \,, \\ \mathcal{D}_{ij} &= \int_{UC} d^2 x \, M_s \partial_i \vec{m} \partial_j \vec{m} = D \mathcal{P}_{ij} \,.\end{aligned} \qquad (4.52)$$

Here $\hat{b}_i$ is the unit vector along the direction of emergent magnetic field. It is chosen to be along $\hat{z}$. In this form, it is clear that the first term $\vec{\mathcal{G}}$ is the multiplication of two different contributions, the topological charge Q and the net spin density $M_s \hat{b}$. If the net spin density vanishes as in the anti-ferromagnetic materials, the skyrmion Hall effect is expected to vanish. We come back to this later in §7.1.2.

The projection operator is given by $\mathcal{P}_{ij} = \mathbb{I} - \hat{B} \cdot \hat{B}^T$ with the unit vector $\hat{B}$ along the external magnetic field direction. This is also $\hat{B} = \hat{z}$ for our case. Ground state Néel skyrmion is evaluated for this to get $D = \mathcal{D}_{xx} = \mathcal{D}_{yy}$ and $\mathcal{D}_{xy} = \mathcal{D}_{yx} = 0$. Now

$$\tilde{\alpha} = \alpha + \alpha' \frac{D'}{D} \,, \qquad \tilde{\beta} = \beta + \beta' \frac{D'}{D} \,, \qquad (4.53)$$

with

$$D' = \int_{UC} d^2x\, M_s(\vec{b})^2 . \tag{4.54}$$

The last term $\vec{F}_{pin}$ represents the pinning effects, that cannot be obtained by following the Thiele's procedure. It depends on the materials that we consider. Some materials have large pinning effects, while the others have negligible effects. When $\vec{F}_{pin} = 0$, we get

$$\vec{\mathcal{G}} \times (\vec{v}_s - \vec{v}_d) + \mathcal{D} \cdot (\beta \vec{v}_s - \tilde{\alpha}\vec{v}_d) = 0 . \tag{4.55}$$

This equation can be useful for the skyrmions without pinning or impurity effect [84][72]. Here the contribution of the conduction electron current with the velocity $\vec{v}_s$ can be important in conducting materials. The equation (4.38) has the combination $(\partial_t + v_s^i \nabla_i)\vec{M}$ in two different places. This combination has a suggestive form, $(\partial_t + v_s^i \nabla_i)\vec{m} = (v_s^i - v_d^i)\nabla_i \vec{m}$ upon using (4.45).

For the contribution of the topological emergent electromagnetic fields, we multiply $(\vec{M}/M_s^2)\times$ to (4.37), take j th component of it, contract with $\partial m_j/\partial x_i$, and perform the integral over the unit cell. The results are

$$\int_{UC} d^2x\, M_s \left(B_e^2 (\beta \vec{v}_s - \alpha \vec{v}_d) - \vec{B}_e B_e^i (\beta v_i^s - \alpha v_i^d) \right) . \tag{4.56}$$

The first term can be directly added to the contribution to the third term in (4.51) or (4.55). The second term vanishes when the magnetic field is perpendicular to the velocities, which is typically the case for the 2 dimensional film geometry.

4.4.2.2 Spin orbit torque

Here we also generalize the Thiele equation with the spin-orbit torque terms given in (4.28). We follow the original Thiele's paper described in §4.4.1. We multiply $-1/\gamma_0$ and $-\epsilon_{jlk}M_k$ to the l the component of (4.28), followed by contracting the index l, and finally multiply $-\partial M_j/\partial x_i$ to the equation followed by summing over the index j. After some algebra, we get

$$f_i^{SOT} = a_{SOT} \vec{M} \cdot \left(\frac{\partial \vec{M}}{\partial x_i} \times \left[(-\hat{z} \times \vec{J_e}) + \eta \vec{M} \times (\hat{z} \times \vec{J_e}) \right] \right), \quad (4.57)$$

where we put the force term to the other side to match the form (4.51). This can be rewritten as

$$f_i^{SOT} = T_{ij} J_{e,j} = (T_{ij}^1 + T_{ij}^2) J_{e,j},$$

$$T_{ij}^1 = a_{SOT} \epsilon_{zlj} \epsilon_{lmn} \frac{\partial M_m}{\partial x_i} M_n, \quad (4.58)$$

$$T_{ij}^2 = -a_{SOT} \eta \left[\epsilon_{zmn} M_m \frac{\partial M_n}{\partial x_i} M_j - \epsilon_{jmn} M_m \frac{\partial M_n}{\partial x_i} M_z \right].$$

Here T_{ij} is a second rank tensor. The Thiele equation of the magnetization $\vec{M}$ in the presence of an in-plane electric current $\vec{J_e}$ has multiple contributions.

$$\vec{\mathcal{G}} \times (\vec{v}_s - \vec{v}_d) + \mathcal{D} \cdot (\beta \vec{v}_s - \alpha \vec{v}_d) + \mathcal{T} \cdot \vec{J_e} = 0, \quad (4.59)$$

where we also include the spin transfer torque (STT) terms in addition to the spin orbit torque (SOT). If you want to take into account of the topological emergent fields as well, you can change α and β to $\tilde{\alpha}$ and $\tilde{\beta}$ following (4.53).

4.4.2.3 Spin Hall torque

We present the generalization of the Thiele equation in the context of controlling the skyrmions using the spin Hall current using the electric current flowing in an adjacent layer. One prominent example is putting together the ferromagnetic material with a heavy metal layer. By driving polarized currents along the heavy metal layer, one can pump the spin into the ferromagnetic material which has become standard experimental techniques.

From the LLG equation by including the spin Hall effect term in (4.40), we can derive the corresponding Thiele contribution following the procedure described in the section §4.4.1.

$$f_i^{SHT} = -\frac{1}{\gamma_0} \frac{g \mu_B \theta_{SH}}{2 e M_s^2 t_F} [\vec{M} \times (\hat{z} \times \vec{J}_{HM})]_j \left(-\frac{\partial M_j}{\partial x_i} \right), \quad (4.60)$$

which can be evaluated to give

$$f_i^{SHT} = \mathcal{B}_{ij} J_{HM,j},$$
$$\mathcal{B}_{ij} = -\frac{1}{\gamma_0} \frac{g\mu_B \theta_{SH}}{2eM_s^2 t_F} \epsilon_{zlj} \epsilon_{lmn} \frac{\partial M_m}{\partial x_i} M_n. \qquad (4.61)$$

Here $\mathcal{B}_{ij}$ is a second rank tensor. The corresponding Thiele equation comes as

$$\vec{G} \times \vec{v}_d + \alpha D \cdot \vec{v}_d + 4\pi \mathcal{B} \cdot \vec{J}_{HM} = 0. \qquad (4.62)$$

This form of Thiele equation can be also combined with the (4.55) that has the contribution of spin transfer torque terms due to the conduction electrons or the contributions due to the spin orbit torque given in (4.59). Those contributions compete each other and frequently one dominates the others depending on the physical situations.

4.4.3 Generalization with transverse velocity

Before ending the discussion of the Thiele equation, we want to mention that there is another type of generalizations of the Thiele equation. The translation motion of skyrmion center can also have the transverse motion, and thus we can generalize that with the following form

$$\begin{aligned} M_i &= M_i(x_j - X_j), \\ X_j &= (v_j^d + R_{jk} v_k^d) t, \end{aligned} \qquad (4.63)$$

where v_j^d describes a longitudinal motion, while $R_{jk} v_k^d = R\epsilon_{jk} v_k^d$ describes the transverse linear motion of the skyrmion center [85]. This is the subject of the modeling Hall viscosity in Chapter 7.

We note that this generalization is different from the rotation motion of skyrmions that can take the following ansatz:

$$\vec{M}'(\vec{x}, t) = R_{\hat{z}}(\phi) \cdot \vec{M}\left[R_{\hat{z}}^{-1}(\phi) \cdot (\vec{x} - \vec{v}_d t)\right], \qquad (4.64)$$

where $R_{\hat{z}}(\phi)$ describes the rotation of the magnetization vector $\vec{M}$ with respect to $\hat{z}$ direction. The rotation motion of the skyrmions and its interaction with the magnons are described in §6.3.3.

It is also possible to go beyond the rigid limit of the skyrmions. This can be treated approximately by including a term $\vec{m}_d(\partial_t \vec{X}(t))$ in the parametrization of the spin $\vec{m}(t,\vec{x}) = \vec{m}_s(\vec{x} - \vec{X}(t)) + \vec{m}_d(\partial_t \vec{X}(t)) + \vec{m}_f$, where $\vec{m}_s$ describes the motion of the skyrmion center with a coordinate $\vec{X}(t)$ and the fast mode $\vec{m}_f$ contains the magnon contributions [141]. Effectively this includes the mass term with $\mathcal{M} = \frac{1}{2} \int d^2x \vec{m}_s \cdot \left[\partial \vec{m}_d / \partial \dot{R}_i \times \partial \vec{m}_s / \partial \dot{x}_i \right]$.

4.5 Spin effects on transport

In this section we include the effects of spin in the theory of thermodynamics of irreversible process [86][87]. This is a direct extension of our discussion of irreversible thermodynamics by including spin chemical potential to the electric scalar potential presented in §2.6. In §2.6.1, we mentioned the chemical potential μ can be more general than the electric scalar potential. In [86], this idea was implemented to treat the electrons with up and down spins as two different charge carriers with chemical potentials $\mu_\uparrow$ and $\mu_\downarrow$, respectively, following the idea that the chemical potential and electric potential adds in Electrochemistry [88].

Specifically, we consider three different currents in 2 or 3 spatial dimensions. We use $\vec{Q}, \vec{J}_\uparrow, \vec{J}_\downarrow$ as vectors in the corresponding dimensions. $\vec{J}_\uparrow$ and $\vec{J}_\downarrow$ are the currents with corresponding electrochemical potentials $\mu_\uparrow = \mu_e + \mu_0 + \delta\mu$ and $\mu_\downarrow = \mu_e + \mu_0 - \delta\mu$, respectively. Here $\mu_e = e\phi$ and ϕ is electrostatic potential. The symmetric combination of these currents is the electric current $\vec{J}_e = \vec{J}_\uparrow + \vec{J}_\downarrow$, while the antisymmetric combination gives the so-called spin current $\vec{J}_s = \vec{J}_\uparrow - \vec{J}_\downarrow$ which is the flow of the spin angular momentum.

The entropy current (2.109) can be generalized to include the spin contribution.

$$\frac{ds}{dt} = \vec{\nabla}\left(\frac{1}{T}\right) \cdot \vec{J}_U - \vec{\nabla}\left(\frac{\mu_\uparrow}{T}\right) \cdot \vec{J}_\uparrow - \vec{\nabla}\left(\frac{\mu_\downarrow}{T}\right) \cdot \vec{J}_\downarrow . \qquad (4.65)$$

Spin Dynamics

This can be recast into more convenient form

$$\frac{ds}{dt} = -\frac{1}{T^2}\vec{Q}\cdot(\vec{\nabla}T) - \frac{1}{T}\vec{\nabla}\mu_\uparrow \cdot \vec{J}_\uparrow - \frac{1}{T}\vec{\nabla}\mu_\downarrow \cdot \vec{J}_\downarrow. \tag{4.66}$$

where $\vec{Q} = \vec{J}_U - \mu_\uparrow \vec{J}_\uparrow - \mu_\downarrow \vec{J}_\downarrow$. Now the dynamical equation connecting the currents and generalized forces are

$$\begin{pmatrix} \vec{Q} \\ \vec{J}_\uparrow \\ \vec{J}_\downarrow \end{pmatrix} = \begin{pmatrix} L_{qq} & L_{q\uparrow} & L_{q\downarrow} \\ L_{q\downarrow} & L_{\uparrow\uparrow} & L_{\uparrow\downarrow} \\ L_{q\uparrow} & L_{\downarrow\uparrow} & L_{\downarrow\downarrow} \end{pmatrix} \begin{pmatrix} -\vec{\nabla}T/T^2 \\ -\vec{\nabla}\mu_\uparrow/T \\ -\vec{\nabla}\mu_\downarrow/T \end{pmatrix}, \tag{4.67}$$

where $\vec{\nabla}\mu_{\uparrow,\downarrow} = \vec{\nabla}\mu_0 + e\vec{E} \pm \vec{\nabla}(\delta\mu) = \vec{\nabla}\bar{\mu} \pm \vec{\nabla}(\delta\mu)$ with a generalized potential $\bar{\mu}$. The dynamical coefficients L are $d \times d$ matrices for d spatial dimensions, while the vectors are d dimensional column vectors. Note that we already reduced the number of parameters using the Onsager relations along with the symmetry that a positive spin that is a majority spin in a field H is a minority spin in a field $-H$ and vice versa. Thus $L_{\uparrow q}(H) = L_{q\uparrow}(-H) = L_{q\downarrow}(H)$ and $L_{\downarrow q}(H) = L_{q\downarrow}(-H) = L_{q\uparrow}(H)$ [86]. It is interesting to realize that there is a spin mixing contributions $L_{\uparrow\downarrow}$ an $L_{\downarrow\uparrow}$ that are built in this formulation. In the spin mixing process, the charge carriers retains their momenta. It turns out that $L_{\uparrow\downarrow}$ and $L_{\downarrow\uparrow}$ are independent because $L_{\uparrow\downarrow}(H) = L_{\downarrow\uparrow}(-H) = L_{\uparrow\downarrow}(H)$.

It is useful to separate the electric charge transport and spin transport. To separate the spin potential $\delta\mu$ from the electrochemical potential $\mu_0 + \mu_e$, we add and subtract the second and third columns in the 3×3 matrix in (4.67). Then we add and subtract the second and third row of (4.67). Then

$$\begin{pmatrix} \vec{Q} \\ \vec{J}_e \\ \vec{J}_s \end{pmatrix} = \begin{pmatrix} L_{QeS} \end{pmatrix} \begin{pmatrix} -\vec{\nabla}T/T^2 \\ -\vec{\nabla}\bar{\mu}/T \\ -\vec{\nabla}\delta\mu/T \end{pmatrix}, \tag{4.68}$$

$$\begin{pmatrix} L_{QeS} \end{pmatrix} = \begin{pmatrix} L_{qq} & L_{q\uparrow} + L_{q\downarrow} & L_{q\uparrow} - L_{q\downarrow} \\ L_{q\downarrow} + L_{q\uparrow} & L_{\uparrow\uparrow} + L_{ss} + L_{\downarrow\downarrow} & L_{\uparrow\uparrow} - L_{aa} - L_{\downarrow\downarrow} \\ L_{q\downarrow} - L_{q\uparrow} & L_{\uparrow\uparrow} + L_{aa} - L_{\downarrow\downarrow} & L_{\uparrow\uparrow} - L_{ss} + L_{\downarrow\downarrow} \end{pmatrix},$$

where $\vec{J}_e = \vec{J}_\uparrow + \vec{J}_\downarrow, \vec{J}_s = \vec{J}_\uparrow - \vec{J}_\downarrow, L_{ss} = L_{\uparrow\downarrow} + L_{\downarrow\uparrow}$ and $L_{aa} = L_{\uparrow\downarrow} - L_{\downarrow\uparrow}$.

Let us connect the coefficients to the known transport coefficients. To do so, we assume that there is no spin mixing $L_{\uparrow\downarrow} = L_{\downarrow\uparrow} = 0$. Then the conductivities with condition $\vec{\nabla}T = 0$ gives

$$\sigma = \sigma_\uparrow + \sigma_\downarrow = -e\vec{J}_e/(\vec{\nabla}\bar{\mu}_e/e)$$
$$= e^2(L_{\uparrow\uparrow} + L_{\downarrow\downarrow})/T, \quad (4.69)$$

which determine the coefficients as $L_{\uparrow\uparrow} = \sigma_\uparrow T/e^2$ and $L_{\downarrow\downarrow} = \sigma_\downarrow T/e^2$. The absolute thermoelectric power can be obtained with $\vec{J}_e = 0$ as discussed in (2.115)

$$\epsilon_{tot} = \epsilon_\uparrow + \epsilon_\downarrow = -(\vec{\nabla}\bar{\mu}/e)/\vec{\nabla}T$$
$$= -(L_{q\downarrow}/L_{\uparrow\uparrow} + L_{q\uparrow}/L_{\downarrow\downarrow})/(eT), \quad (4.70)$$

which fixes $L_{q\downarrow} = -eTL_{\uparrow\uparrow}\epsilon_\uparrow = -T^2\sigma_\uparrow\epsilon_\uparrow/e$ and $L_{q\uparrow} = -eTL_{\downarrow\downarrow}\epsilon_\downarrow = -T^2\sigma_\downarrow\epsilon_\downarrow/e$. Similarly, the heat conductivity similar to (2.114) can be evaluated for $\vec{J}_e = 0$.

$$\kappa = -\frac{\vec{J}_Q}{\vec{\nabla}T} = \frac{1}{T^2}\left(L_{qq} - \frac{L_{q\uparrow}L_{q\downarrow}}{L_{\uparrow\uparrow}} - \frac{L_{q\downarrow}L_{q\uparrow}}{L_{\downarrow\downarrow}}\right). \quad (4.71)$$

This fix the L_{qq} as $L_{qq} = T^2\kappa + T^3\epsilon_\uparrow\epsilon_\downarrow(\sigma_\uparrow + \sigma_\downarrow)$.

Thus the coefficient matrix L_{QeS} in terms of κ, $\sigma_{\uparrow,\downarrow}$ and $\epsilon_{\uparrow,\downarrow}$ becomes

$$\begin{pmatrix} T^2\kappa + T^3\epsilon_\uparrow\epsilon_\downarrow(\sigma_\uparrow + \sigma_\downarrow) & -T^2(\sigma_\uparrow\epsilon_\uparrow + \sigma_\downarrow\epsilon_\downarrow)/e & -T^2(\sigma_\uparrow\epsilon_\uparrow - \sigma_\downarrow\epsilon_\downarrow)/e \\ -T^2(\sigma_\uparrow\epsilon_\uparrow + \sigma_\downarrow\epsilon_\downarrow)/e & (\sigma_\uparrow + \sigma_\downarrow)T/e^2 & (\sigma_\uparrow - \sigma_\downarrow)T/e^2 - L_{aa} \\ T^2(\sigma_\uparrow\epsilon_\uparrow - \sigma_\downarrow\epsilon_\downarrow)/e & (\sigma_\uparrow - \sigma_\downarrow)T/e^2 + L_{aa} & (\sigma_\uparrow + \sigma_\downarrow)T/e^2 \end{pmatrix}.$$
$$(4.72)$$

Here we assume that the conductivities due to the spin mixing is small compared to the direct conductivities. In that case, the spin mixing contribution can be captured by their difference $L_{\uparrow\downarrow} - L_{\downarrow\uparrow} = L_{aa}$.

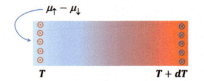 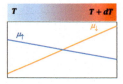

Figure 4.17 Spin Seebeck effect. Left: Electrons with spin-up (red arrow head) and spin-down (blue arrow tail) components have different Seebeck coefficients in a metallic magnet under a temperature gradient. Right: Illustration of the electrochemical potentials for spin-up $\mu_\uparrow$ and spin-down $\mu_\downarrow$ electrons.

4.5.1 Spin Seebeck effect

In this section we consider the newly developed spin Seebeck effect [89][90][91][92]. In ferromagnetic metal, all the three currents $\vec{Q}, \vec{J}_e, \vec{J}_S$ exist, while the charge current portion is suppressed in the magnetic insulator. We would like to mention the importance of magnon contribution to the spin Seebeck effect. Magnons can be treated as particle-like excitations similar to electrons. There are similarities between the electron contributions to the electronic transports in metal and the Magnonic contributions to the heat transports in insulator. Thus, the physical understanding developed in previous sections can be directly applied with minimal modifications.

Similar to the Seebeck effect that two different conducting materials have two different Seebeck coefficients, the electrons with two different spin polarizations have different Seebeck coefficients. When a metallic magnet is experiencing a temperature gradient, these two spin polarizations generate different amounts of flow, resulting in spin voltage $\mu_\uparrow - \mu_\downarrow$ that is proportional to the applied temperature gradient as in Fig. 4.17. Thus the magnet acts as a thermocouple in the spin sector.

Quantitatively, the electrochemical potential $\mu_\uparrow = \mu_0 + \delta\mu + \mu_e = \mu_\uparrow^c + e\phi$ has the spin dependent chemical potential $\mu_\uparrow^c$ that depends on the temperature and its density $n_\uparrow$. The gradient of $\mu_\uparrow$ has three different contributions. The electrochemical potential for the spin-down component $\mu_\downarrow = \mu_0 - \delta\mu + \mu_e = \mu_\downarrow^c + e\phi$ can be described similarly.

$$\vec{\nabla}\mu_{\uparrow,\downarrow} = \left(\frac{\partial \mu_{\uparrow,\downarrow}^c}{\partial T}\right) \vec{\nabla}T + \left(\frac{\partial \mu_{\uparrow,\downarrow}^c}{\partial n_{\uparrow,\downarrow}}\right) \vec{\nabla} n_{\uparrow,\downarrow} + e\vec{\nabla}\phi, \quad (4.73)$$

where the first term on the right-hand side measures the entropy contribution from the temperature gradient, while the second term comes from the density accumulation. Thus

$$\vec{\nabla}\delta\mu = -eS_s \vec{\nabla}T,$$
$$S_s = \epsilon_s = -\frac{1}{e}\left(\frac{\partial \mu_{\uparrow}^c}{\partial T} - \frac{\partial \mu_{\downarrow}^c}{\partial T}\right) = \epsilon_{\uparrow} - \epsilon_{\downarrow}, \quad (4.74)$$

where $\delta\mu = \mu_{\uparrow} - \mu_{\downarrow}$ and $S_s = \epsilon_s$ is the spin Seebeck coefficient. The definition of the Seebeck coefficient (thermoelectric power) is given in (2.115). Here we assume that the $\vec{\nabla}n_{\uparrow}$ and $\vec{\nabla}n_{\downarrow}$ decay within the spin diffusion length and do not contribute significantly, which were relevant for the study done in [89].

The measurement of the spin Seebeck effect utilize the inverse spin Hall effect that we discuss in §4.2.5 [89][92]. The setup is depicted in the left panel of Fig. 4.18. The temperature gradient ∇T and the in-plane magnetic field $\vec{H}_0$ or $\vec{H}$ are set along the $\hat{z}$ direction. The electron spins are polarized along the direction of the magnetic field (here the Nernst effects that play roles with transverse magnetic field are suppressed). Electrons with spin $(+\hat{z})$ polarized parallel to the magnetic field accumulate in the far end of the film's long side, while those with spins anti-parallel $(-\hat{z})$ in the near end.

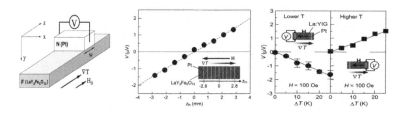

Figure 4.18 Measurement of the spin Seebeck effect using inverse spin Hall effect. Left: The $Ni_{81}Fe_{19}$ film has Pt wires attached to the ends of the film. Middle and Right: Demonstration of spin Seebeck effect by the electromotive forces in the Pt wires. Reproduced with permission from [92].

Focusing on this near end, the in-film electromotive force due to the spin Hall effect vanishes because the current $\vec{J}_S$ is parallel to the spin polarization. This is demonstrated by the measurement of the potential without the Pt wire (not displayed here). Once the Pt wire is placed on top of the film, the accumulated electrons move up to the wire, produce the electric current $\vec{J}_e \propto +\hat{x}$, and thus bring about the electromotive force $\vec{E}_{SHE} \propto \vec{J}_e \times \vec{\sigma} = \hat{x} \times (-\hat{z}) = -\hat{y}$, that confirms the middle and right panels of Fig. 4.18. The negatively charged electrons move against the direction of the electromotive force. As the temperature gradient is increased, there are more electrons accumulated and bigger current to produce more electric potential difference. The signs of the electric voltage at opposite ends are opposite as demonstrated in the right panel of Fig. 4.18. In the middle panel of Fig. 4.18, we can see that the electric potential coming from the inverse spin Hall effect is proportional to the displacement from the center along the direction of the temperature gradient. This is consistent with the electrochemical difference depicted in the right panel of Fig. 4.17.

While the explanation given by the electrochemical difference of conduction electron spins seems to be plausible, the conduction electrons alone cannot explain the spin Seebeck effect. This is because the conduction electron's short spin-flip diffusion length (that is about several nanometers in a NiFe alloy) fails to explain the propagation of spin information to a large length scale, several millimeters [93]. It was further demonstrated that a ferromagnetic insulator $Y_3Fe_5O_{12}$ (YIG) can carry spin currents by magnon excitations and a direct observation of the spin Seebeck effect in insulator $LaY_2Fe_5O_{12}$ [90]. These experiments indicates that, contrary to the conventional view that the spin current is carried by conduction electrons, the magnon can be a carrier for the spin Seebeck effect. There is a theoretical explanation for the spin Seebeck effect through a non-equilibrium between the magnon system in the ferromagnet and the conduction electron system in the nonmagnetic metal [93]. We come back to this when we discuss the skyrmion Seebeck effect in §6.3.2.

One can evaluate the spin current $\vec{J}_S = \vec{J}_\uparrow - \vec{J}_\downarrow$ using the transport coefficients described in (4.68). For this purpose, we consider a non-zero heat current $\vec{Q}$ without electric current $\vec{J}_e = 0$ along with no spin mixing $L_{\uparrow\downarrow} = L_{\downarrow\uparrow} = 0$. Then, (4.68), for general $\vec{Q}$, has the form

$$\begin{pmatrix} 0 \\ \vec{J}_s \end{pmatrix} = \begin{pmatrix} L_{q\downarrow} + L_{q\uparrow} & L_{\uparrow\uparrow} + L_{\downarrow\downarrow} & L_{\uparrow\uparrow} - L_{\downarrow\downarrow} \\ L_{q\downarrow} - L_{q\uparrow} & L_{\uparrow\uparrow} - L_{\downarrow\downarrow} & L_{\uparrow\uparrow} + L_{\downarrow\downarrow} \end{pmatrix} \begin{pmatrix} -\vec{\nabla}T/T^2 \\ -\vec{\nabla}\bar{\mu}/T \\ -\vec{\nabla}\delta\mu/T \end{pmatrix}. \quad (4.75)$$

There are two different cases that simplifies: (A) imposing a condition $\vec{\nabla}\delta\mu = 0$ or (B) imposing a condition $\vec{\nabla}\bar{\mu} = 0$.

First we consider the case (B) with $\vec{\nabla}\delta\mu = 0$. From the first equation of (4.75), $\vec{\nabla}\bar{\mu} = -(L_{q\downarrow} + L_{q\uparrow})/(L_{\uparrow\uparrow} + L_{\downarrow\downarrow})(\vec{\nabla}T/T)$. Plugging this into the second equation of (4.75), we get

$$\vec{J}_s^A = -2 \frac{L_{\downarrow\downarrow} L_{q\downarrow} - L_{\uparrow\uparrow} L_{q\uparrow}}{L_{\uparrow\uparrow} + L_{\downarrow\downarrow}} \frac{\vec{\nabla}T}{T^2}$$

$$= -\frac{2}{e} \frac{\sigma_\uparrow \sigma_\downarrow}{\sigma_\uparrow + \sigma_\downarrow} (\epsilon_\uparrow - \epsilon_\downarrow) \vec{\nabla}T \quad (4.76)$$

$$= -\frac{\sigma}{2e}(1 - P_\sigma^2) S_s \vec{\nabla}T.$$

Here we use $P_\sigma = (\sigma_\uparrow - \sigma_\downarrow)/(\sigma_\uparrow + \sigma_\downarrow)$ and $2\sigma_\uparrow \sigma_\downarrow/(\sigma_\uparrow + \sigma_\downarrow) = \sigma(1 - P_\sigma^2)/2$ with $\sigma = \sigma_\uparrow + \sigma_\downarrow$. This case has been considered in [94].

The case (B) with the condition $\vec{\nabla}\bar{\mu} = 0$ can be similarly evaluated as

$$\vec{J}_s^B = -2 \frac{L_{\downarrow\downarrow} L_{q\downarrow} + L_{\uparrow\uparrow} L_{q\uparrow}}{L_{\uparrow\uparrow} - L_{\downarrow\downarrow}} \frac{\vec{\nabla}T}{T^2}$$

$$= \frac{2}{e} \frac{\sigma_\uparrow \sigma_\downarrow}{\sigma_\uparrow - \sigma_\downarrow} (\epsilon_\uparrow + \epsilon_\downarrow) \vec{\nabla}T \quad (4.77)$$

$$= \frac{\sigma_s}{2e} \left(\frac{1}{P_\sigma^2} - 1 \right) S_{tot} \vec{\nabla}T,$$

where $\sigma_s = \sigma_\uparrow - \sigma_\downarrow$ and $S_{tot} = \epsilon_{tot} = \epsilon_\uparrow + \epsilon_\downarrow$. One can also consider the isothermal case $\vec{Q} = 0$ without restricting the electric current.

4.5.2 Spin Peltier effect

The Peltier coefficient describes the amount of heat that is carried by an electrical current when it passes through a material. In the context

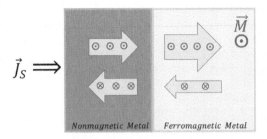

Figure 4.19 Illustration of the spin Peltier effect. A spin current is pushed through a nonmagnetic metal and ferromagnetic metal interface. The Peltier heat current vanishes in the nonmagnetic metal while it is non-zero in the ferromagnetic metal because the Peltier coefficients are different for the majority and minority spins.

of spin current, this can be made clear by using a pure spin current in a nonmagnetic material as illustrated in Fig. 4.19 [95][86]. The electric current as well as the heat current vanish because up-spin electrons move toward right while the down-spin electrons move toward the left.

On the other hand, the ferromagnetic metal serves as spin polarizer and provides better conductivity for the majority spin (up-spin depicted as a red arrow head). Thus the heat current for the majority spin is larger than that of the minority spin (down-spin depicted as a blue arrow tail), leading in a net heat current.

More quantitatively, we can derive the temperature gradient that develops in the ferromagnetic region when a spin current is accompanied by a charge current as in Fig. 4.19. For simplicity, we assume that no heat can enter or leave the ferromagnetic metal ($\vec{Q} = 0$) and disregard Joule heating. Then the first line of (4.68) along with (4.72) gives

$$\vec{\nabla}T = -\frac{[\sigma_\uparrow \epsilon_\uparrow + \sigma_\downarrow \epsilon_\downarrow]\vec{\nabla}\mu_c - [\sigma_\uparrow \epsilon_\uparrow - \sigma_\downarrow \epsilon_\downarrow]\vec{\nabla}\mu_s}{e[\kappa + T\epsilon_\uparrow \epsilon_\downarrow \sigma]} T, \qquad (4.78)$$

where $\sigma = \sigma_\uparrow + \sigma_\downarrow$, $\bar{\mu} = \mu_c = \mu_\uparrow + \mu_\downarrow$ is the charge part of the electrochemical potential, and $\delta\mu = \mu_s = \mu_\uparrow - \mu_\downarrow$ is the spin part of the electrochemical potential. With slightly different notation compared to the previous section, we set $\sigma\epsilon = \sigma_\uparrow\epsilon_\uparrow + \sigma_\downarrow\epsilon_\downarrow$ and thus

$\epsilon_{\uparrow,\downarrow} = \epsilon - (P_\sigma \mp 1)\epsilon_s/2$. Then the spin term in (4.78) gives the result

$$\vec{\nabla} T_s = \frac{T}{e[\kappa + T\epsilon_\uparrow \epsilon_\downarrow \sigma]} \left(\epsilon \sigma_s + \frac{\sigma}{2}(1 - P_\sigma^2)\epsilon_s \right) \vec{\nabla} \mu_s \, , \qquad (4.79)$$

where $\sigma_s = \sigma_\uparrow - \sigma_\downarrow$ and $\vec{\nabla} T_s$ is the temperature gradient due to the spin current. This result is apparently different from that of [95]. There are two equivalent, yet different looking, definitions we employed in the previous sections for the transport coefficients, for example (2.117) and (2.118). Here we use (2.117) which is different from those in [95].

Chapter 5

Ward Identity

One of the attractive aspects of the Ward identity approach is the systematic and consistent ways to incorporate all possible and universal physical quantities. The Ward identities utilize various underlying symmetries and the corresponding conservation laws and are much more general than the original account due to Ward. In the context of hydrodynamics that builds on symmetry principles, Ward identity provides a powerful tool to capture its transport coefficients such as conductivities and viscosities along with the thermodynamic quantities in terms of the so-called contact terms. It is even more powerful in the context of skyrmion physics, where its topological charges are also incorporated in the generalized Ward identities. While Ward identity itself does not compute a specific physical quantity except the special cases we encounter below, it helps a priori unknown physical quantity be connected to known ones. For example, the Hall viscosity, through Ward identities, can be related to either Hall conductivity or angular momentum depending on mutually commuting quantum mechanical observables. In an ideal environment, the Hall viscosity can be shown to be directly related to the topological charge of skyrmions.

Skyrmions and Hall Transport
Bom Soo Kim
Copyright © 2023 Jenny Stanford Publishing Pte. Ltd.
ISBN 978-981-4968-34-8 (Hardcover), 978-1-003-37253-0 (eBook)
www.jennystanford.com

5.1 Symmetries and Ward identity

5.1.1 Symmetry and equation of motion

In a simple term, the Ward identity greatly facilitates the consequences of existing symmetries in terms of conservation equations. Let us consider the continuity equation.

$$\frac{\partial \rho}{\partial t} + \vec{\nabla} \cdot \vec{J} = 0 ,\qquad(5.1)$$

where $\rho = J^0$ is a 'charge' density and J^i a 'current' density. Here charge and current refer to some general conserved quantity due to an existing symmetry. For example, electric charge and current is the consequence of a conserved $U(1)$ symmetry.

This continuity equation (5.1) is a simple example of Ward identity that connects the charge ρ and the current $\vec{J}$ in a specific way, which is a consequence of the local charge conservation. It is more convenient to use Fourier transform and put the equation in the momentum space.

$$\omega \rho + k_i J^i = 0 ,\qquad(5.2)$$

where ω and k_i are the frequency and the momentum along i direction. While the fully developed quantum field theoretic Ward identities are much more complicated with various different terms and higher derivatives, this simple equation (5.4) already reveals an essential property of Ward identity: There is a definite relation between the charge density and associated current.

The original Ward identity [96][97] has played important roles in checking consistency conditions for certain scattering amplitudes of quantum electrodynamics. In technical terms, the so-called quantum field theory ultra-violet divergences (infinities) may be removed by the 'renormalization' of electron mass and charge to all orders of perturbation theory. Essentially, various loop computations are required to have miraculous cancellations among themselves, revealing the underlying symmetry. For Ward, the symmetry concerned was $U(1)$ gauge symmetry of the electrodynamics. When a scattering amplitude $\mathcal{M}$ is coupled to a photon, it can be described as $\mathcal{M} = \epsilon_\mu \mathcal{M}^\mu$ with a (transverse) photon polarization vector ϵ_μ in a vector potential represented as $A_\mu(x^\mu) \propto \int d^4k\, \epsilon_\mu(k_\mu) e^{-ik_\mu x^\mu}$ with a longitudinal momentum k_μ. Under the guage transformation, the vector potential transforms as $A_\mu \to A_\mu + \partial_\mu \lambda$ with a parameter λ and the scattering amplitude as $\epsilon_\mu \mathcal{M}^\mu \to (\epsilon_\mu + \lambda k_\mu) \mathcal{M}^\mu$. Gauge invariance forces the condition to be satisfied as

$$k_\mu \mathcal{M}^\mu = 0, \qquad (5.3)$$

which is a distilled form of Ward identity. Thus, the unphysical 'longitudinal' polarization of the photon does not contribute to the scattering amplitudes. Interested readers can consult various quantum field theory textbooks.

In recent years, Ward identity has been generalized in many different ways in various sub-disciplines of science. We emphasize two distinct aspects. First, Ward identity can be used in the presence of any conserved current $\partial_\mu J^\mu = 0$, of course, with some caution that are associated with specific situations. One can interpret this Ward identity as the quantum counter part to the Noether's theorem in classical physics. In fact, the Ward identity is much more powerful in general and includes the so-called contact terms, which manifest themselves when two operators coincide at the same space-time position. In quantum mechanics language,

$$\partial_\mu \langle 0 | \mathcal{T} J^\mu(x) \cdots | 0 \rangle = \text{contact terms}, \qquad (5.4)$$

where $\mathcal{T}$ is time-ordered product, $\langle 0 | \cdots | 0 \rangle$ is the expectation value, and the contact terms arise when two operators meet at the same space and time point. While the contact terms did not play a role in the original form of Ward identity, we will see below that they play crucial roles in the generalized Ward identity. The formula (5.4) is nothing but the continuity equation discussed in (5.1) without the advertised contact terms.

Second, modern Ward identities have been generalized to incorporate a commutator with two different operators. For example, one can construct a commutator for the non-abelian global current satisfying $\partial_\mu J_a^\mu = K_a$. We will explain this in detail as a toy model in §5.2.1. More interesting case comes with the stress energy tensors. Space-time translation symmetry can be realized as the corresponding conservation equations, $\partial_\mu T^{\mu\nu} = 0$, with a stress energy tensor $T^{\mu\nu}$, similar to the continuity equation $\partial_\mu J^\mu = 0$. One can use the commutator of two momentum density $[T^{0i}(t,\vec{x}), T^{0j}(t',\vec{x}')]$, which is related to the conductivities as we discuss below in detail. When one uses the conservation equations $\partial_0 T^{0i} = \partial_k T^{ki}$ after taking time derivatives on the commutator, we get a different commutator with momentum flux $[T^{ik}(t,\vec{x}), T^{jl}(t',\vec{x}')]$, which is related to viscosities. Thus the Ward identity relates the conductivities to the viscosities, which sometimes provides highly non-trivial and interesting relations. These are used in the quantum Hall fluids and show interesting results presented in §5.4.

For an another level of non-trivial extensions, we observe that the equal-time momentum-momentum commutator density actually can accommodate the topological charge of skyrmions

$$[T^{0i}(t,\vec{x}), T^{0j}(t,\vec{x}')] = i\left(-\partial_i T^{0j} + \partial_j T^{0i} + i\epsilon^{ij}c\right)\delta^{(2)}(\vec{x}-\vec{x}'), \quad (5.5)$$

where c is the topological skyrmion charge density. The terms on the right-hand side survive only when the two momentum operators meets at the same spatial points due to the spatial delta function. These are the contact terms that we mentioned above. The corresponding Ward identity becomes simple in the presence of the translation symmetry, where the expectation value of the momentum generator is a constant. Then only the topological charge density survives, which provides a very simple and interesting relation. We will consider this in §5.5.

5.1.2 Geometric understanding of Ward identities

Fortunately, there is a simple geometric picture that clearly illustrates the physical origin of the Ward identities in the context of commutator of two stress energy tensors similar to (5.5), which is directly applicable to see the consequences in the presence of a mass gap.

We consider two spatial dimensions for a simple and clear picture. There are only two independent area-preserving shear transformations. They are either horizontal shear or a vertical shear. For example, the latter changes the vertical location by the amount proportional to the horizontal distance without changing the horizontal location. With the coordinate system (x,y), the corresponding linear transform can be written as $(x',y') = (x, y + \epsilon x)$ with a constant c. Then horizontal shear has the form $(x',y') = (x + \epsilon y, y)$. These shear transformation can be written as the following 2×2 matrices in the column vector basis $(x,y)^T$ with T being a transpose.

$$S_h = \begin{pmatrix} 1 & \epsilon \\ 0 & 1 \end{pmatrix}, \quad S_v = \begin{pmatrix} 1 & 0 \\ \epsilon & 1 \end{pmatrix}. \tag{5.6}$$

For our purpose, there are more convenient transformations that are equivalent to these. Let us take a square as an illustration. One shear transformation elongates one side and squeezes the other as illustrated in Fig. 5.1 (a). The other independent one stretches along the diagonal direction along 45 degree from the lower left corner as in Fig. 5.1 (b). These shear transformations can be written in matrix forms as

$$a = \begin{pmatrix} 1+\tilde{\epsilon} & 0 \\ 0 & 1-\tilde{\epsilon} \end{pmatrix}, \quad b = \begin{pmatrix} 1 & \tilde{\epsilon}' \\ \tilde{\epsilon}' & 1 \end{pmatrix}, \tag{5.7}$$

where $\tilde{\epsilon}$ and $\tilde{\epsilon}'$ are infinitesimal transformation parameters. These two matrices, a and b, turn out to be non-commutative with each other.

Interestingly, the operation $b^{-1}a^{-1}ba$ produces a net rotation.

$$b^{-1}a^{-1}ba = \begin{pmatrix} 1 & -2\tilde{\epsilon}\tilde{\epsilon}' \\ 2\tilde{\epsilon}\tilde{\epsilon}' & 1 \end{pmatrix} + \mathcal{O}(\tilde{\epsilon},\tilde{\epsilon}')^3. \tag{5.8}$$

Thus a certain combination of shear transformations can generate a rotation [48]. This is illustrated in Fig. 5.1 (c). A commutator of two stress tensors, generating shear transformations, can be related to angular momentum operator.

We note that the two matrices (without the diagonal components) given in (5.7) are intimately related to Pauli matrices σ_3 and σ_1.

$$\bar{\sigma}_1 = \sigma_x = \begin{pmatrix} 0 & 1 \\ 1 & 0 \end{pmatrix}, \tag{5.9}$$

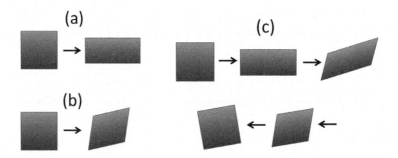

Figure 5.1 Two in-equivalent area preserving shear transformations in 2 spatial dimensions. (a): Elongation along the horizontal direction of a square. (b): Elongation along the diagonal direction of the square. (c): A combination of the two shear transformations produces a rotation.

$$\bar{\sigma}_2 = i\sigma_y = \begin{pmatrix} 0 & 1 \\ -1 & 0 \end{pmatrix}, \tag{5.10}$$

$$\bar{\sigma}_3 = \sigma_z = \begin{pmatrix} 1 & 0 \\ 0 & -1 \end{pmatrix}. \tag{5.11}$$

In 2+1 dimensions, we can construct three generators by contracting with the Pauli matrices with stress energy tensors as

$$Q^{ij} = \int d^2x \, x^i T^{0j},$$

$$Q_a = \frac{1}{2}(\bar{\sigma}_a)_{ij} Q^{ij} = \int d^2x \, S_a^0, \tag{5.12}$$

where $i, j = 1, 2$ and $a, b, c = 1, 2, 3$. The associated current density is

$$S_a^\mu = \frac{1}{2}(\bar{\sigma}_a)_{ij} x^i T^{\mu j}. \tag{5.13}$$

This generates the following algebra, which is $SL(2, \mathbb{R})$.

$$[Q_a, Q_b] = i f_{ab}^{\ c} Q_c. \tag{5.14}$$

The structure constants are antisymmetric on the first pair of indices and

$$f_{13}^2 = f_{12}^3 = f_{23}^1 = +1 \,. \tag{5.15}$$

In general the charges Q_1 and Q_3 are not conserved as one can check as follows:

$$\partial_\mu S_a^\mu = \frac{1}{2} (\bar{\sigma}_a)_{ij} T^{ij} \quad \longrightarrow \quad \partial_t Q_a = \frac{1}{2} \int d^2 x \, (\bar{\sigma}_a)_{ij} T^{ij} \,, \tag{5.16}$$

where we use the conservatin equation $\partial_\mu T^{\mu i} = 0$ and discard the surface terms. Q_2 is a symmetry when the system has rotational symmetry that implies $T^{ij} = T^{ji}$. It generates $SO(2)$ rotations in the plane. Note that all Q_a can be accidentally conserved in isotropic states $\langle T^{ij} \rangle \propto \delta^{ij}$. A representation of the generators acting on local operators is ($\vec{x} = (x,y)$)

$$S_1 = -\frac{i}{2}(x\partial_y + y\partial_x) \,,$$

$$S_2 = -\frac{i}{2}(x\partial_y - y\partial_x) \,, \tag{5.17}$$

$$S_3 = -\frac{i}{2}(x\partial_x - y\partial_y) \,.$$

The generators should satisfy the symmetry algebra. Therefore, the equal time commutator should satisfy the identity

$$\langle [S_a^0(t,\vec{x}), S_b^0(t,\vec{x}')] \rangle = i f_{ab}^{\ c} \langle S_c^0(t,\vec{x}) \rangle \delta^{(2)}(\vec{x} - \vec{x}') \,. \tag{5.18}$$

In particular

$$\langle i [S_1^0(t,\vec{x}), S_3^0(t,\vec{x}')] \rangle = - \langle S_2^0(t,\vec{x}) \rangle \delta^{(2)}(\vec{x} - \vec{x}') \,. \tag{5.19}$$

Note that

$$\int d^2 x \int d^2 x' \, \langle i [S_1^0(t,\vec{x}), S_3^0(t,\vec{x}')] \rangle = \langle i [Q_1, Q_3] \rangle = -\frac{\langle L_{xy} \rangle}{2} \,, \tag{5.20}$$

where $\langle L_{xy} \rangle = L$ is the total angular momentum.

Later in this chapter we are going to see that a commutator of two momentum generators, T^{0i}, followed by using the conservation equation $\partial_0 T^{0i} = -\partial_k T^{ki}$ similar to (5.1), produces a nontrivial result reported in the context of quantum Hall fluid. In one version, Hall viscosity is directly related to angular momentum of the gapped quantum Hall fluid [48][42], where the angular momentum manifest itself by the broken translation symmetry. In a different version, the same Hall viscosity is directly related to the Hall conductivity in the quantum Hall systems [43], where the presence of the translation symmetry ensures that there is no spontaneously generated angular momentum. These two independent relations seemingly contradict each other, yet they are consistent as these two cases utilize two mutually exclusive sets of measurable quantities that are allowed by quantum exclusion principles. This point has been discussed in §1.5 and in Chapter 3, the same Ward identity can produce two mutually exclusive and independent relations depending on the choices of the symmetries [98][99][22].

5.2 An example

Before considering the Ward identities involved with stress energy tensor and its generalization with topological charge, we present a conceptually simpler Ward identity with global currents. Here we attempt to develop mathematical machinery for the computations involved with the Ward identities without explaining the physical backgrounds related to current algebra in detail.

5.2.1 Ward identity for global currents

Here we introduce the Ward identity for a non-abelian global current J_a^μ with the group index a that satisfies the conservation equation in $d+1$ dimensions with index $\mu, \nu = 0, 1, \cdots d$ in the presence of the background electromagnetic fields A_μ^a and the corresponding field strength tensors $F_{\mu\nu}^a$. The current satisfies the equation

$$\partial_\mu J_a^\mu(t,\vec{x}) = K_a(t,\vec{x}) , \qquad (5.21)$$

along with the equal time current algebra

$$[J_a^0(t,\vec{x}), J_b^0(t,\vec{x}')] = iC_{ab}^{\ \ c} J_c^0(t,\vec{x}) \delta^{(d)}(\vec{x}-\vec{x}') + S_{ab}(t,\vec{x},\vec{x}') , \qquad (5.22)$$

where C_{abc} are the structure constant of the non-Abelian group $[T_a, T_b] = iC_{ab}^c T_c$ for the group generators T^a. The other factors K_a and S_{ab} (called a Schwinger term) are related to anomaly contributions [100]. For example,

$$K_a = -\frac{c}{8}\epsilon^{\mu\nu\rho\sigma} tr\left(T_a F_{\mu\nu} F_{\rho\sigma}\right) = -\frac{c}{4} d_{abc} \epsilon^{0ijk} F^b_{0i} F^c_{jk},$$
$$S_{ab}(t,\vec{x},\vec{x}') = -\frac{c}{8}\epsilon^{0ijk} d_{abc} \partial_i \left(F^c_{jk} \delta^{(3)}(\vec{x}-\vec{x}')\right),$$
(5.23)

where $F^a_{\mu\nu}$ is the field strength tensor for background electromagnetic fields, $d_{abc} = tr(\{T_a, T_b\} T_c)$ for the three spatial dimensions $d = 3$. Note that the first two indices in d_{abc} are symmetric, while they are antisymmetric in C_{abc}.

The current algebra satisfies

$$[Q_a(t), Q_b(t)] = iC_{ab}^c Q_c(t) + S_{ab}(t),$$
(5.24)

in terms of the charge $Q_a(t) = \int d^d x J_a^0(t,\vec{x})$. This is a little more complicated than the simple abelian conserved current $\partial_\mu J^\mu = 0$. Nevertheless, the complexity serves well for our discussions below for the full fledged Ward identities.

As retarded correlation functions can be directly related to physical observables such as conductivity or viscosity, we consider the following retarded correlation functions:

$$G^{\mu\nu}_{R\,ab}(t,\vec{x};t',\vec{x}') = i\theta(t-t')\langle[J_a^\mu(t,\vec{x}), J_b^\nu(t',\vec{x}')]\rangle,$$
(5.25)

where $\theta(t-t') = (1 + \text{sgn}(t-t'))/2$ is the Heaviside step function in terms of the sign function, sgn, and $\langle\cdots\rangle$ is the vacuum expectation value. Here we assume neither time translation invariance, nor space translation invariance to facilitate a general discussion.

Ward Identity

We start with two time derivatives on G_R^{00} as

$$\partial_t \partial_{t'} G_{R\,ab}^{00}(t,\vec{x};t',\vec{x}')$$
$$= i\theta(t-t')\partial_t \partial_{t'} \langle [J_a^0(t,\vec{x}), J_b^0(t',\vec{x}')] \rangle - i\delta'(t-t') \langle [J_a^0(t,\vec{x}), J_b^0(t,\vec{x}')] \rangle$$
$$+ i\delta(t-t')\partial_{t'} \langle [J_a^0(t,\vec{x}), J_b^0(t',\vec{x}')] \rangle - i\delta(t-t')\partial_t \langle [J_a^0(t,\vec{x}), J_b^0(t',\vec{x}')] \rangle$$
$$= \partial_i \partial_{j'} G_{R\,ab}^{ij}(t,\vec{x};t',\vec{x}') + i\delta'(t-t') \langle [J_a^0(t,\vec{x}), J_b^0(t,\vec{x}')] \rangle$$
$$- i(\partial_t - \partial_{t'})\{\delta(t-t') \langle [J_a^0(t,\vec{x}), J_b^0(t,\vec{x}')] \rangle\} \qquad (5.26)$$
$$= \partial_i \partial_{j'} G_{R\,ab}^{ij}(t,\vec{x};t',\vec{x}') + i\delta'(t-t') \langle [J_a^0(t,\vec{x}), J_b^0(t,\vec{x}')] \rangle$$
$$- i[(\partial_t - \partial_{t'})\delta(t-t')] \cdot \langle [J_a^0(t,\vec{x}), J_b^0(t,\vec{x}')] \rangle$$
$$- i\delta(t-t')\partial_t \langle [J_a^0(t,\vec{x}), J_b^0(t,\vec{x}')] \rangle + i\delta(t-t')\partial_{t'} \langle [J_a^0(t',\vec{x}), J_b^0(t',\vec{x}')] \rangle$$
$$= \partial_i \partial_{j'} G_{R\,ab}^{ij}(t,\vec{x};t',\vec{x}') - i\delta'(t-t') \langle [J_a^0(t,\vec{x}), J_b^0(t,\vec{x}')] \rangle \, ,$$

where we use the delta function in the argument of a derivative to change t' into t. Note that we use $'$ for both a set of coordinate x'^μ and also a derivative $\delta'(t-t') = \partial_{t-t'}(t-t')$ in this chapter. We also used the conservation equation (5.21) in the second expression.

$$i\theta(t-t')\partial_t \partial_{t'} \langle [J_a^0(t,\vec{x}), J_b^0(t',\vec{x}')] \rangle$$
$$= i\theta(t-t') \langle [\partial_i J_a^i(t,\vec{x}) + K_a, \partial_{j'} J_b^0(t',\vec{x}') + K_b] \rangle \qquad (5.27)$$
$$= \partial_i \partial_{j'} (i\theta(t-t') \langle [J_a^i(t,\vec{x}), J_b^0(t',\vec{x}')] \rangle)$$
$$= \partial_i \partial_{j'} G_{R\,ab}^{ij}(t,\vec{x};t',\vec{x}') \, .$$

Note that K_a comes from a background fields, and we take their commutators with the generators J_a^0 vanish.

Using the current algebra (5.22), which is input from the Ward identity point of view, we can get a Ward identity for global currents in coordinate space

$$\partial_t \partial_{t'} G_{R\,ab}^{00}(t,\vec{x};t',\vec{x}') = \partial_i \partial_{j'} G_{R\,ab}^{ij}(t,\vec{x};t',\vec{x}')$$
$$+ \delta'(t-t') C_{ab}^{\ c} \langle J_c^0(t,\vec{x}) \rangle \delta^{(d)}(\vec{x}-\vec{x}') \qquad (5.28)$$
$$- i\delta'(t-t') S_{ab}(t,\vec{x},\vec{x}') \, .$$

For a steady-state configuration, one point function $\langle J_c^0(t,\vec{x})\rangle$, identified as current density $\rho_c(t,\vec{x})$, is a constant, independent of time. This Ward identity captures the anomaly contribution through the last line. For even spatial dimensions, the Schwinger term vanishes and no anomaly terms are present.

5.2.2 Ward identity in momentum space

One might be interested in obtaining the Ward identity (5.28) in the momentum space, where it is expressed in terms of frequency and momentum. Due to the lack of translation invariance for the example considered in §5.2.1, we consider two sets of frequencies and momenta $(p_0,\vec{p}),(q_0,\vec{q})$ with the following form:

$$G_{R\,ab}^{\mu\nu}(t,\vec{x};t',\vec{x}') = \int \frac{dp_0 d^d\vec{p}\, dq_0 d^d\vec{q}}{(2\pi)^{2d+2}} e^{ip_0 t + i\vec{p}\cdot\vec{x}} e^{-iq_0 t' - i\vec{q}\cdot\vec{x}'} \tilde{G}_{R\,ab}^{\mu\nu}(p_0,\vec{p};q_0,\vec{q}),$$

$$J_a^\mu(t,\vec{x}) = \int \frac{dp_0 d^d\vec{p}}{(2\pi)^{d+1}} e^{ip_0 t + i\vec{p}\cdot\vec{x}} \tilde{J}_a^\mu(p_0,\vec{p}), \qquad (5.29)$$

$$F_{jk}^c(t,\vec{x}) = \int \frac{dp_0 d^d\vec{p}}{(2\pi)^{d+1}} e^{ip_0 t + i\vec{p}\cdot\vec{x}} \tilde{F}_{jk}^c(p_0,\vec{p}),$$

where d is the number of the spatial dimensions. We want to compute the following regularized expression to deal with the case without translation symmetry.

$$I_{ab}(\vec{\epsilon}) = \partial_t \partial_{t'} \int d^d x \int d^d x'\, e^{-\vec{\epsilon}\cdot\vec{x} - \vec{\epsilon}\cdot\vec{x}'} G_{R\,ab}^{00}(t,\vec{x};t',\vec{x}'), \qquad (5.30)$$

where the constant vector $\vec{\epsilon}$ is introduced to regulate the integrals. One can think of this calculation as taking the Fourier transform of the correlator and taking the zero momentum limit in a symmetric way. Eventually we will take $\vec{\epsilon}\to 0$ limit. It is useful as each different power of $\vec{\epsilon}$ turns out to give an independent set of Ward identities at zero momentum limit. Similar techniques were used in [99] to derive quantum field theory Ward identities.

In terms of the parameter $\vec{\epsilon}$, the momentum space expression of $\tilde{I}(\omega, s, \vec{\epsilon})$ for the left side of the equation (5.28) becomes

$$\tilde{I}_{ab}(\omega, s, \vec{\epsilon}) = \int dt dt' \, e^{-i\omega t + is t'} I_{ab}(\vec{\epsilon})$$
$$= \omega s \, \tilde{G}^{00}_{R\,ab}(\omega, \vec{p}; s, \vec{q})\big|_{\vec{p}=\vec{\epsilon}, \vec{q}=-\vec{\epsilon}} \,, \tag{5.31}$$

where we use (5.29) to evaluate (5.30). While this looks complicated, the computation is straightforward as the time and space integrals of $t, t', \vec{x}, \vec{x}'$ produce delta functions $\delta(\omega - p_0), \delta(s - q_0), \delta(\vec{p} - \vec{\epsilon}), \delta(\vec{q} + \vec{\epsilon})$, and the corresponding frequency and momentum integrals replace $p_0, q_0, \vec{p}, \vec{q}$ to $\omega, s, \vec{\epsilon}, -\vec{\epsilon}$.

The right side of the equation (5.28) can be evaluated

$$\tilde{I}_{ab}(\omega, s, \vec{\epsilon})$$
$$= \int dt d^d\vec{x} dt' d^d\vec{x}' \, e^{-i\omega t + is t'} e^{-i\vec{\epsilon}\cdot\vec{x} - i\vec{\epsilon}\cdot\vec{x}'} \left[\partial_i \partial_{j'} G^{ij}_{R\,ab}(t, \vec{x}; t', \vec{x}') \right.$$
$$\left. - i\delta'(t - t')\{iC^c_{ab}\langle J^0_c(t,\vec{x})\rangle \delta^{(d)}(\vec{x} - \vec{x}') + S_{ab}(t, \vec{x}, \vec{x}')\} \right] \tag{5.32}$$
$$= p_i q_j \tilde{G}^{ij}_{R\,ab}(\omega, \vec{p}; s, \vec{q})\big|_{\vec{p}=-\vec{q}=\vec{\epsilon}} + isC^c_{ab}\langle \tilde{J}^0_c(\omega - s, 2\vec{\epsilon})\rangle + S_{ab}(\omega, s, \vec{\epsilon}) \,.$$

We compute the last two terms below. Putting them together we get the general Ward identity in momentum space

$$\omega s \, \tilde{G}^{00}_{R\,ab}(\omega, \vec{p}; s, \vec{q})\big|_{\vec{p}=\vec{\epsilon}, \vec{q}=-\vec{\epsilon}} = p_i q_j \tilde{G}^{ij}_{R\,ab}(\omega, \vec{p}; s, \vec{q})\big|_{\vec{p}=\vec{\epsilon}, \vec{q}=-\vec{\epsilon}}$$
$$+ isC^c_{ab}\langle \tilde{J}^0_c(\omega - s, 2\vec{\epsilon})\rangle + S_{ab}(\omega, s, \vec{\epsilon}) \,. \tag{5.33}$$

The term $p_i q_j \tilde{G}^{ij}_{R\,ab}(\omega, \vec{p}; s, \vec{q})\big|_{\vec{p}=\vec{\epsilon}, \vec{q}=-\vec{\epsilon}}$ will vanish when there is no momentum pole for $\vec{\epsilon} \to 0$. Even in this case, the term $\omega s \, \tilde{G}^{00}_{R\,ab}(\omega, \vec{p}; s, \vec{q})\big|_{\vec{p}=\vec{\epsilon}, \vec{q}=-\vec{\epsilon}}$ will not vanish in the presence of the other non-zero contributions on the right hand side of (5.33).

Here we explicitly derive the last two terms in (5.33).

$$\int dtdt' d^d\vec{x}d^d\vec{x}' e^{-i\omega t+ist'-i\vec{\epsilon}\cdot\vec{x}-i\vec{\epsilon}\cdot\vec{x}'} \delta'(t-t')C_{ab}^c \langle J_c^0(t,\vec{x})\rangle \delta^{(d)}(\vec{x}-\vec{x}')$$

$$= \int dtdt' d^d\vec{x} \int \frac{dp_0 d^d\vec{p}}{(2\pi)^{d+1}} e^{i(p_0-\omega)t+ist'+i(\vec{p}-2\vec{\epsilon})\cdot\vec{x}} \delta'(t-t')C_{ab}^c \langle \tilde{J}_c^0(p_0,\vec{p})\rangle$$

$$= \int dtdt' \frac{dp_0}{2\pi} e^{-i(\omega-p_0)t+ist'} \left[\frac{1}{2}(\partial_t - \partial_{t'})\delta(t-t')\right] C_{ab}^c \langle \tilde{J}_c^0(p_0, 2\vec{\epsilon})\rangle \quad (5.34)$$

$$= \int dtdt' \frac{dp_0}{2\pi} e^{-i(\omega-p_0)t+ist'} \left[\frac{i}{2}(\omega-p_0+s)\right] \delta(t-t') C_{ab}^c \langle \tilde{J}_c^0(p_0, 2\vec{\epsilon})\rangle$$

$$= is C_{ab}^c \langle \tilde{J}_c^0(\omega-s, 2\vec{\epsilon})\rangle = i\omega C_{ab}^c \langle \tilde{J}_c^0(s-\omega, -2\vec{\epsilon})\rangle,$$

where, to derive the second line, we use the expression for the Fourier transform $\langle J_c^0(t,\vec{x})\rangle = \int (dp_0 d^d\vec{p})/((2\pi)^{d+1}) e^{ip_0 t+i\vec{p}\cdot\vec{x}} \langle \tilde{J}_c^0(p_0,\vec{p})\rangle$, followed by the integral over $\vec{x}'$ with the delta function $\delta^{(d)}(\vec{x}-\vec{x}')$. One can check that the last identity comes when the delta function identity $\delta'(t-t')C_{ab}^c \langle J_c^0(t,\vec{x})\rangle \delta^{(d)}(\vec{x}-\vec{x}') = \delta'(t-t')C_{ab}^c \langle J_c^0(t',\vec{x}')\rangle \delta^{(d)}(\vec{x}-\vec{x}')$ is used. By combining the last two terms, one can write them in a symmetric form as $i(s/2)C_{ab}^c \langle \tilde{J}_c^0(\omega-s, 2\vec{\epsilon})\rangle + i(\omega/2)C_{ab}^c \langle \tilde{J}_c^0(s-\omega, -2\vec{\epsilon})\rangle$. It is also straightforward to derive the term with $\langle \tilde{J}_c^0(\omega-s, 2\vec{\epsilon})\rangle$.

The Schwinger term can be written

$$S_{ab}(\omega, s, \vec{\epsilon})$$

$$= -i \int dtdt' d^d\vec{x}d^d\vec{x}' e^{-i\omega t+ist'-i\vec{\epsilon}\cdot\vec{x}-i\vec{\epsilon}\cdot\vec{x}'} \delta'(t-t') S_{ab}(t,\vec{x},\vec{x}')$$

$$= i\frac{c}{8} \int dtdt' d^d\vec{x}d^d\vec{x}' e^{-i\omega t+ist'-i\vec{\epsilon}\cdot\vec{x}-i\vec{\epsilon}\cdot\vec{x}'} \delta'(t-t') \epsilon^{0ijk} d_{abc} \partial_i \left(F_{jk}^c \delta^{(3)}(\vec{x}-\vec{x}')\right)$$

$$= \frac{ic}{8} \epsilon^{0ijk} d_{abc} \int dtdt' \, e^{-i\omega t+ist'} \delta'(t-t')$$

$$\times \int d^d\vec{x}d^d\vec{x}' \, e^{-i\vec{\epsilon}\cdot\vec{x}-i\vec{\epsilon}\cdot\vec{x}'} \partial_i \left(\int \frac{dp_0 d^d\vec{p}}{(2\pi)^{d+1}} e^{ip_0 t+i\vec{p}\cdot\vec{x}} \tilde{F}_{jk}^c(p_0,\vec{p}) \delta^{(3)}(\vec{x}-\vec{x}')\right)$$

$$= is\frac{ic}{8} \epsilon^{0ijk} d_{abc} \int d^d\vec{x}d^d\vec{x}' \int \frac{d^d\vec{p}}{(2\pi)^d} e^{-i\vec{\epsilon}\cdot\vec{x}-i\vec{\epsilon}\cdot\vec{x}'} e^{i\vec{p}\cdot\vec{x}}$$

$$\times \left[ip_i \tilde{F}_{jk}^c(\omega-s,\vec{p})\delta^{(3)}(\vec{x}-\vec{x}') + \tilde{F}_{jk}^c(\omega-s,\vec{p})\partial_i\left(\int \frac{d^d\vec{q}}{(2\pi)^d} e^{i\vec{q}\cdot(\vec{x}-\vec{x}')}\right)\right]$$

$$= -is\frac{c}{8} \epsilon^{0ijk} d_{abc}(p_i+q_i)\tilde{F}_{jk}^c(\omega-s, 2\vec{\epsilon})\Big|_{\vec{p}=2\vec{\epsilon},\vec{q}=-\vec{\epsilon}}$$

$$= i\omega \frac{c}{8} \epsilon^{0ijk} d_{abc}(q_i+p_i)\tilde{F}_{jk}^c(s-\omega, -2\vec{\epsilon})\Big|_{\vec{p}=\vec{\epsilon},\vec{q}=-2\vec{\epsilon}},$$

$$(5.35)$$

where we use a specific form of the Schwinger term for $d = 3$ given in (5.23) in the third line. The Fourier transform for F has the momentum $\vec{p}$, while the delta function representation $\delta^{(3)}(\vec{x} - \vec{x}') = \int (d^d \vec{q}/(2\pi)^d) e^{i\vec{q} \cdot (\vec{x} - \vec{x}')}$ has momentum $\vec{q}$. The last equality comes when we exchange the Fourier modes $(p_0, \vec{p})$ and $(q_0, \vec{q})$, where we note that $p_i + q_i$ in last expression has a relative sign compared to the previous one.

Combining all of them together, we get

$$\omega s \, \tilde{G}^{00}_{R\,ab}(\omega, \vec{p}; s, \vec{q})\big|_{\vec{p}=\vec{\epsilon}, \vec{q}=-\vec{\epsilon}} = p_i q_j \tilde{G}^{ij}_{R\,ab}(\omega, \vec{p}; s, \vec{q})\big|_{\vec{p}=\vec{\epsilon}, \vec{q}=-\vec{\epsilon}}$$
$$+ \frac{s}{2} i C^c_{ab} \langle \tilde{J}^0_c(\omega - s, 2\vec{\epsilon}) \rangle + \frac{w}{2} i C^c_{ab} \langle \tilde{J}^0_c(s - \omega, -2\vec{\epsilon}) \rangle$$
$$- is \frac{c}{8} \epsilon^{0ijk} d_{abc} (p_i + q_i) \tilde{F}^c_{jk}(\omega - s, 2\vec{\epsilon})\big|_{\vec{p}=2\vec{\epsilon}, \vec{q}=-\vec{\epsilon}}$$
$$+ i\omega \frac{c}{8} \epsilon^{0ijk} d_{abc} (q_i + p_i) \tilde{F}^c_{jk}(s - \omega, -2\vec{\epsilon})\big|_{\vec{p}=\vec{\epsilon}, \vec{q}=-2\vec{\epsilon}}.$$
(5.36)

This is a Ward identity associated with a global current with an anomaly contribution in terms of Schwinger term. Here we see the connections between the retarded Green's functions and one point functions, which can be expressed as physical observables using Kubo formula and thermodynamic properties depending on the symmetries present in a system. In particular, we see that rotational symmetry is powerful to organize the Green's functions. A set of Ward identity can be obtained by setting $\vec{\epsilon} \to 0$. Another set can be obtained for a linear order in the expansion of $\vec{\epsilon}$. More generally, different order of $\vec{\epsilon}$ expansion gives an independent Ward identity as different derivative orders provides different transport coefficients. We will see similar features for more interesting Ward identities involved with a topological charge.

5.3 Quantum field theory Ward identities

Here we construct Ward identities involved with stress energy tensors $T^{\mu\nu}$, which carries the information of fundamental physical quantities such as energy, momentum, pressure, and shear stress. The Ward identity with the stress energy tensors can be useful for identifying useful relations among the physical quantities of the insulating systems. And then, we add a certain conserved current J^μ to consider the charge dynamics. This description is useful for the conductors. In this section we focus on technical development of the Ward identities.

5.3.1 Ward identities for stress energy tensor

We present a simple quantum field theory Ward identities, which reflect the consequences of symmetries and the corresponding conservation equations.

$$\partial_\mu T^{\mu\nu} = 0. \tag{5.37}$$

First we construct simpler Ward identities without skyrmion topological charges.

Similar to §5.2.1, we consider a commutator of two energy currents, $T^{0j}(x^\mu)$ and $T^{0l}(x'^\mu)$ (they are the momentum densities in Lorentz invariant theories), multiplied by a Heaviside step function $\theta(x^0 - x'^0)$. The resulting quantity is called the retarded Green's function introduced in §2.5. Without the subscript R,

$$G^{0j0l}(x^\mu;x'^\mu) = i\theta(x^0 - x'^0)\langle[T^{0j}(x^\mu), T^{0l}(x'^\mu)]\rangle . \tag{5.38}$$

Let us examine a simple Ward identity by taking two time derivatives on this equation (5.38). Then there are 4 terms, one term with both derivatives acting on the two T^{0i}s, two terms with one derivative on T^{0i} and another on the step function, and one term with both derivatives on the step function.

Following a similar computation done in (5.26), these terms can be organized as

$$\begin{aligned}\partial'_0\partial_0 G^{0j0l}(x^\mu;x'^\mu) = &\ \partial_n\partial'_m G^{njml}(x^\mu;x'^\mu) \\ &- \delta'(x^0 - x'^0)C^{0j0l}(x^\mu;x'^\mu) \\ &- \delta(x^0 - x'^0)(\partial_0 - \partial'_0)C^{0j0l}(x^\mu;x'^\mu),\end{aligned} \tag{5.39}$$

where ∂' is a derivative acting on x'^μ. We use the conservation equation $\partial_0 T^{0j} = -\partial_m T^{mj}$ to rewrite the first term so that the term with both derivatives acting on the two T^{0i}s turns into the Green's function as $\partial_n \partial'_m G^{njml}(x^\mu; x'^\mu)$, where

$$G^{njml}(x^\mu; x'^\mu) = i\theta(x^0 - x'^0)\langle [T^{nj}(x^\mu), T^{ml}(x'^\mu)]\rangle . \tag{5.40}$$

The other three terms in (5.39) involve with delta function $\delta(x^0 - x'^0)$, which turns the Green's functions into equal time correlator that is also called contact terms.

$$C^{0j0l}(x^0, \vec{x}, \vec{x}') = i\langle [T^{0j}(x^0, \vec{x}), T^{0l}(x^0, \vec{x}')]\rangle . \tag{5.41}$$

Here we would like to rewrite the contact terms, the equal time commutators, in terms of one point functions. There is a long history of developing the equal time correlation functions for the stress energy tensor in the Lorentz invariant quantum systems [101][102][103]. The last reference contains a full list of equal time commutation relations for the stress energy tensors including Schwinger terms associated with the stress energy tensor similar to (5.22). We are interested in the commutator with two momentum densities given by

$$[T^{0i}(t, \vec{x}), T^{0j}(t, \vec{x}')] = i\left(-\partial_i T^{0j} + \partial_j T^{0i}\right) \delta^{(2)}(\vec{x} - \vec{x}') , \tag{5.42}$$

where the derivatives act on the coordinate $\vec{x}$, not $\vec{x}'$. Schwinger terms do not contribute on the right-hand side [103]. In this section, we do not consider the topological charge contribution presented in (5.5). By plugging the relation (5.42) into (5.39), we obtain the coordinate space Ward identities.

If we assume time translation symmetry, we perform a Fourier transform in the time direction by multiplying $\int d(x^0 - x'^0) e^{i\omega(x^0 - x'^0)}$ on both sides of (5.39) to arrive at our general result.

$$\begin{aligned}\omega^2 G^{0j0l}(\omega, \vec{x}, \vec{x}') = {} & \partial_n \partial'_m G^{njml}(\omega, \vec{x}, \vec{x}') \\ & + i\omega \epsilon^{jl} \epsilon^n_m \partial_n \langle T^{0m}(\vec{x})\rangle \delta(\vec{x} - \vec{x}') \\ & - \frac{1}{2}[\partial_l \partial_n \langle T^{nj}(\vec{x})\rangle + \partial_j \partial_m \langle T^{ml}(\vec{x})\rangle]\delta(\vec{x} - \vec{x}') ,\end{aligned} \tag{5.43}$$

where we use the conservation equation again when evaluating the last line. This is somewhat abstract, yet general, Ward identities, where all the relevant ingredients are expressed in terms of stress energy tensors. Here $G^{0j0l}(\omega,\vec{x},\vec{x}')$ and $G^{njml}(\omega,\vec{x},\vec{x}')$ are two spatial and four spatial index Green's functions, while $\langle T^{0m}(\vec{x})\rangle$, $\langle T^{nj}(\vec{x})\rangle$, and $\langle T^{ml}(\vec{x})\rangle$ are one point functions. These are identified in terms of physically relevant quantities in momentum space below, which depend on symmetries and specific physical systems.

5.3.2 Ward identities based on symmetries

Here we go one step further by employing some relevant symmetries depending on physical situations. As discussed in §1.5, not all quantum mechanical observables commute with each other. This means that certain physical quantities cannot be precisely determined at the same time. We keep angular momentum as it is convenient in organizing physical quantities in a nice way. And there is a tension in keeping both the spatial translation symmetry and the angular momentum. We consider these two cases one by one. Here we present them starting from a simple setup following [22][99].

5.3.2.1 Rotation and translation symmetries

Here we consider a physical system with the rotation and translation symmetries. Translation symmetry does not allow any non-trivial spatial dependence for the one point functions. Thus all the contact terms in the second and third lines in (5.43) vanish. This makes the Ward identity particularly simple. Then,

$$\omega^2 G^{0j0l}(\omega,\vec{x},\vec{x}') = \partial_n \partial'_m G^{njml}(\omega,\vec{x},\vec{x}') . \tag{5.44}$$

The presence of spatial translation symmetry allows us to introduce a Fourier transform for the coordinates. Focusing on 2 spatial dimensions,

$$G^{\mu\nu\alpha\beta}(\omega,\vec{x}-\vec{x}') = \frac{1}{(2\pi)^2} \int d^2\vec{q}\, e^{i\vec{q}\cdot(\vec{x}-\vec{x}')}\, \tilde{G}^{\mu\nu\alpha\beta}(\omega,\vec{q}) , \tag{5.45}$$

where μ,ν,α,β are the space time indices $0,1,2$, while i,j,k,l,m,n spatial indices $1,2$.

The tensor structures for the retarded Green's functions in the presence of the rotation symmetry is tightly constrained, which is quite useful in organizing the allowed physical quantities. In the momentum space, the two spatial index Green's function can be written as

$$\tilde{G}^{0i0k} = -i\omega\left[\delta^{ik}\kappa_\delta + q^i q^k \kappa_q\right] \\ - i\omega\left[\epsilon^{ik}\kappa_\epsilon + (\epsilon^{in} q_n q^k + \epsilon^{kn} q_n q^i)\kappa_{q\epsilon}\right], \quad (5.46)$$

where κ_δ and κ_q are the symmetric thermal conductivities, which can have momentum dependence as a function of q^2. Their more appropriate names might be momentum conductivities which are related to the response of the momentum density and current, T^{0i} and T^{ij}. They exist in the system with parity symmetry. Note that terms proportional to a single momentum q^i are not allowed by the rotational invariance. The other two, κ_ϵ and $\kappa_{q\epsilon}$, are the antisymmetric thermal conductivities that contribute in the absence of the parity symmetry. In the presence of electric charges and electromagnetic interactions, there are additional contributions with the electric conductivities. We extend the Ward identity for this case in §5.3.3.

The four spatial index Green's function has familiar tensor structures as we already reviewed in §2.3 in the presence of the rotation symmetry. They are related to the viscosities.

$$\tilde{G}^{njml} = -i\omega\left[\eta(\delta^{nm}\delta^{jl} + \delta^{nl}\delta^{mj} - \delta^{nj}\delta^{ml}) + \zeta\delta^{nj}\delta^{ml}\right] \\ - i\omega\left[\frac{\eta_H}{2}(\epsilon^{nm}\delta^{jl} + \epsilon^{nl}\delta^{jm} + \epsilon^{jm}\delta^{nl} + \epsilon^{jl}\delta^{nm})\right]. \quad (5.47)$$

Here η and ζ are the shear and bulk viscosities that are related to the symmetric part of the tensor. In the absence of parity symmetry, one can use the antisymmetric epsilon tensor. Thus we have also Hall viscosity η_H.

After a little algebra for (5.44) with the Fourier transform, adapting the tensor structures (5.46) and (5.47), we obtain

$$\omega^2\left[\delta^{jl}\kappa_\delta + \epsilon^{jl}\kappa_\epsilon + q^j q^l \kappa_q + (\epsilon^{jn} q_n q^l + \epsilon^{ln} q_n q^j)\kappa_{q\epsilon}\right] \\ = \delta^{jl} q^2 \eta + \epsilon^{jl} q^2 \eta_H + q^j q^l \zeta. \quad (5.48)$$

This equation seems to give a single Ward identity. This is not the case. Note that there are 4 independent tensor structures, and thus there are actually 4 independent Ward identities depending on these tensor structures.

$$\omega^2 \kappa_\delta = q^2 \eta, \qquad \omega^2 \kappa_\epsilon = q^2 \eta_H, \\ \omega^2 \kappa_q = \zeta, \qquad \omega^2 \kappa_{q\epsilon} = 0. \tag{5.49}$$

Thus thermal conductivity components are directly related to the viscosity tensors. In particular, the shear viscosity is directly related to the thermal conductivity κ_δ, while the Hall viscosity to another thermal conductivity κ_ϵ. We are going to see similar relations below with concrete models. These are the Ward identities for the physical systems with translation and rotation symmetry in momentum space. In the presence of the topological charge, the Ward identity has an interesting extension.

5.3.2.2 Without translation symmetry

Now we move on to physical systems without spatial translation symmetry. We still assume the time translation symmetry. Here we consider a slightly different approach with a coordinate space representation. One could take a similar approach done in §5.2.2 using the momentum space representation, which is also considered in the following section.

To develop the Ward identity in the coordinate space, we observe that there are two spatial derivatives in (5.43). To obtain the expressions without these spatial derivatives, we multiply two coordinates on the expression and integrate over the spaces.

$$\omega^2 \int d^2x d^2x' x^i x'^k G^{0j0l}(\omega, \vec{x}, \vec{x}') = \int d^2x d^2x' \, x^i x'^k \partial_n \partial'_m G^{njml}(\omega, \vec{x}, \vec{x}') \\ + \int d^2x d^2x' \, x^i x'^k \left[i\omega e^{jl} \epsilon^n_{\;m} \partial_n \langle T^{0m}(\vec{x}) \rangle \delta(\vec{x}-\vec{x}') \right. \tag{5.50} \\ \left. - \frac{1}{2}\left(\partial_l \partial_n \langle T^{nj}(\vec{x}) \rangle + \partial_j \partial_m \langle T^{ml}(\vec{x}) \rangle \right) \delta(\vec{x}-\vec{x}') \right].$$

We need more information to go further.

Here we would like to consider a vanishing angular frequency ω limit for this Ward identity, as it is simpler and more interesting. With rotational invariance, we can organize the two-spatial index Green's function as

$$G^{0j0l}(\omega,\vec{x},\vec{x}') = -i\omega\bar{\kappa}^{jl}(\vec{x},\vec{x}')\delta^{(2)}(\vec{x}-\vec{x}') + \mathcal{O}(\omega^2), \tag{5.51}$$

where $\bar{\kappa}^{jl}(\vec{x},\vec{x}')$ is the thermal conductivity similar to (5.46) with ¯ indicating the coordinate space. In this form, we note that the term, $\omega^2 G^{0j0l}$, on the left-hand side of (5.50) is already $\mathcal{O}(\omega^3)$. In fact, it is possible to argue that this is $O(\omega^2)$ or higher in frequency if there is an energy gap and no magnetic field [98][99], or if the only gapless degrees of freedom are ordinary Goldstone bosons. Otherwise there can be additional contributions at low frequencies. Keeping this in mind, we proceed to see what relations can be derived from the Ward identity.

Using the four index Green's function

$$G^{njml}(\omega,\vec{x},\vec{x}') = -i\omega\bar{\eta}^{njml}(\vec{x},\vec{x}')\delta^{(2)}(\vec{x}-\vec{x}') + \mathcal{O}(\omega^2), \tag{5.52}$$

similar to (5.47) in the presence of rotational invariance, the first term on the right-hand side of (5.50) can be organized as

$$-i\omega \int d^2x d^2x' \, x^i x'^k \partial_n \partial'_m \left(\bar{\eta}^{njml}\delta^{(2)}(\vec{x}-\vec{x}')\right) + \mathcal{O}(\omega^2)$$

$$= -i\omega \int d^2x d^2x' \, x^i x'^k \left[\delta^{(2)}(\vec{x}-\vec{x}')\partial_n \partial'_m \bar{\eta}^{njml} + \partial_n \bar{\eta}^{njml}\partial'_m \delta^{(2)}(\vec{x}-\vec{x}')\right.$$

$$\left. + \partial'_m \bar{\eta}^{njml}\partial_n \delta^{(2)}(\vec{x}-\vec{x}') + \bar{\eta}^{njml}\partial_n \partial'_m \delta^{(2)}(\vec{x}-\vec{x}')\right] + \mathcal{O}(\omega^2)$$

$$= -i\omega \int d^2x \partial_n \partial_m (x^i x^k)\bar{\eta}^{njml} + \mathcal{O}(\omega^2)$$

$$= -i\omega \int d^2x (\bar{\eta}^{ijkl} + \bar{\eta}^{kjil}) + \mathcal{O}(\omega^2), \tag{5.53}$$

where we integrate by parts the derivative of the delta function and do the integrals over the delta functions. Using the spatial viscosity tensor in the presence of the rotation symmetry

$$\begin{aligned}\bar{\eta}^{ijkl} = \bar{\eta}\left(\delta^{ik}\delta^{jl} + \delta^{il}\delta^{jk} - \delta^{ij}\delta^{kl}\right) + \bar{\zeta}\delta^{ij}\delta^{kl} \\ + \frac{\bar{\eta}_H}{2}(\epsilon^{ik}\delta^{jl} + \epsilon^{il}\delta^{jk} + \epsilon^{jk}\delta^{il} + \epsilon^{jl}\delta^{ik}),\end{aligned} \tag{5.54}$$

we obtain

$$-i\omega \int d^2x \, (\bar{\eta}^{ijkl} + \bar{\eta}^{kjil}) + \mathcal{O}(\omega^2) \tag{5.55}$$

$$= -i\omega \int d^2x \, \left(2\bar{\eta}\delta^{ik}\delta^{jl} + \bar{\zeta}(\delta^{ij}\delta^{kl} + \delta^{il}\delta^{jk}) + 2\bar{\eta}_H \epsilon^{jl}\delta^{ik}\right) + \mathcal{O}(\omega^2).$$

Note that these three viscosities $\bar{\eta}, \bar{\zeta}, \bar{\eta}_H$ are associated with different tensor structures.

As there is no translation invariance, the one point functions in the second and third lines in (5.50) are, in general, not constant. These are nothing but the thermodynamic quantities that play roles depending on the situations, such as pressure, angular momentum, magnetization, and so on. In the presence of rotational symmetry, the stress energy tensor can be put into a diagonal form

$$\langle T^{ij} \rangle = p\delta^{ij}, \tag{5.56}$$

where p is the pressure. Furthermore, if the angular momentum is non-zero and there is rotational invariance, the expectation value of the momentum density takes the following form

$$\langle T^{0i} \rangle = \frac{1}{2}\epsilon^{ik}\partial_k \ell_T, \tag{5.57}$$

where ℓ_T is nothing but the angular momentum density. For a homogeneous system with a boundary, $\langle T^{0i} \rangle$ becomes a momentum density at the boundary. The expression (5.57) can be motivated by a classical definition of angular momentum $\vec{L} = \vec{r} \times \vec{p}$ as written in an index notation, $L_i = \epsilon_{ijk} r_j p_k$. Its quantum version is presented below. By taking a derivative ∂_m followed by multiplying ϵ_{nmi}, we get $\epsilon_{nmi}\partial_m L_i = \epsilon_{nmi}\partial_m(\epsilon_{ijk} r_j p_k) = \epsilon_{nmi}\epsilon_{ijk}\delta_{mj} p_k = -2p_n$ for a constant momentum density $\vec{p} = \text{const.}$ and for $i, m = 1, 2$. By setting $i = 3$, we get $p_n = -(1/2)\epsilon_{nm}\partial_m L_z$. In (5.57), the expectation value along i direction signifies the broken translation symmetry as discussed in §1.5.

After performing the delta function integral for the terms of (5.50),

$$\int d^2x\, x^i x^k \left[i\omega e^{jl} \epsilon^n_{\ m} \partial_n \langle T^{0m}(\vec{x}) \rangle - \frac{1}{2}\left(\partial_l \partial_n \langle T^{nj}(\vec{x}) \rangle + \partial_j \partial_m \langle T^{ml}(\vec{x}) \rangle \right) \right]$$
$$= -\frac{1}{2}\int d^2x \left[i2\omega \delta^{ik} e^{jl} \ell_T + \delta^{il}\langle T^{kj} \rangle + \delta^{kl}\langle T^{ij} \rangle + \delta^{ij}\langle T^{kl} \rangle + \delta^{kj}\langle T^{il} \rangle \right]$$
$$= -\int d^2x \left[i\omega \delta^{ik} e^{jl} \ell_T + (\delta^{ij}\delta^{kl} + \delta^{il}\delta^{jk})p \right]. \tag{5.58}$$

Putting together (5.51), (5.55), and (5.58), the Ward identity (5.50) becomes

$$0 = -i\omega e^{jl} \delta^{ik} \int d^2x\, \ell_T - (\delta^{ij}\delta^{kl} + \delta^{il}\delta^{jk}) \int d^2x\, p$$
$$-i\omega \int d^2x \left(2\bar{\eta} \delta^{ik}\delta^{jl} + \bar{\zeta}(\delta^{ij}\delta^{kl} + \delta^{il}\delta^{jk}) + 2\bar{\eta}_H e^{jl}\delta^{ik} \right) + \mathcal{O}(\omega^2). \tag{5.59}$$

We find the conditions at zero frequency from the Ward identities

$$\int d^2x\, \bar{\eta}_H(\omega \to 0) = -\int d^2x\, \frac{\ell_T}{2},$$
$$\lim_{\omega \to 0} \int d^2x\, i\omega \bar{\eta} = 0, \tag{5.60}$$
$$\int d^2x\, \bar{\zeta}(\omega \to 0) = \frac{i}{\omega} \int d^2x\, p.$$

The first equation is the relation between the Hall viscosity and the angular momentum density that has been advertised in [42]. We explain this relation in more detail in §5.3.2.3. The second equation reveals that the shear viscosity vanishes when there is a mass gap in the system at the zero frequency limit. This contrast the situation considered in (5.49) in §5.3.2.1.

The last equation in (5.60) indicates that bulk viscosity of a system is related to pressure. It is a static response to an external strain when the pressure is non-zero. We can understand its origin using the generating functional. A partition function introduced in §2.5 is useful way to systematically investigate the responses of a physical system under the influence of external sources. Here we focus on a system's responses under the metric variations that are related to stress energy tensors. For the partition function,

$$\mathcal{Z}[g_{ij}] = \int \mathcal{D}\phi e^{iS[\phi, g_{ij}]}, \tag{5.61}$$

where ϕ denote the fields in a theory. If we change the metric $g_{ij} \to g_{ij} + \delta g_{ij}$, the first order variation of the generating function, logarithm of the partition function, is given by

$$\delta^{(1)} \ln \mathcal{Z} = -\frac{1}{2} \int dt d^2x \sqrt{-g} \langle T^{ij} \rangle \delta g_{ij}, \tag{5.62}$$

where $\sqrt{-g}$ is the determinant of the metric, whose variation is $\delta\sqrt{-g} = -1/(2\sqrt{-g}) \cdot \delta g = (\sqrt{-g}/2) g^{\mu\nu} \delta g_{\mu\nu}$ with $\delta g = \delta(\det g_{\mu\nu}) = g g^{\mu\nu} \delta g_{\mu\nu}$ from the differentiation of matrix determinant. The second order variation includes two kind of terms: those that come from the variation of the one-point function and those that come from the variation of the square root of the determinant. The latter corresponds to changes of the volume.

$$\delta^{(2)} \ln \mathcal{Z} = -\frac{1}{4} \int dt d^2x \sqrt{-g} \bigg[\frac{1}{2} g^{kl} \langle T^{ij} \rangle + \frac{1}{2} g^{ij} \langle T^{kl} \rangle \\ + \frac{\delta \langle T^{ij} \rangle}{\delta g_{kl}} + \frac{\delta \langle T^{kl} \rangle}{\delta g_{ij}} \bigg] \delta g_{ij} \delta g_{kl}, \tag{5.63}$$

where we symmetrize the result as the two variations δg_{ij} and δg_{kl} are symmetric. For $\langle T^{ij} \rangle = p g^{ij}$, the contact terms in the first line introduce the bulk viscosity, while those in the second line does pressure.

5.3.2.3 Momentum space Ward identity without translation symmetry

After obtaining the Ward identities (5.50), (5.59), and (5.60) in the coordinate space, we recast the same expressions in the momentum space. We take a slightly different approach by looking into various 'moments' of some physical quantities by multiplying additional coordinates. These were useful in defining the momentum (1.123) and the angular momentum (1.139). Another independent motivation comes from the ways to extract the Hall viscosity and the orbital angular momentum using the moments and to figure out their non-trivial connection [42]. (Interested readers can also consult [49][50][48].)

In quantum theory, the angular momentum operator can be written with a stress energy tensor.

$$L = \int d^2x \, \epsilon_{ij} x^i T^{0j} , \tag{5.64}$$

where the direction of the angular momentum is perpendicular to the 2 spatial plane we consider. The momentum density T^{0j} generates a translation,

$$[T^{0i}(\vec{x}), \mathcal{O}(\vec{x}')] = -\partial_i \mathcal{O}(\vec{x}) \delta^{(2)}(\vec{x} - \vec{x}') , \tag{5.65}$$

on an operator $\mathcal{O}$ at the same spatial point.

Motivated by the form with an operator multiplied by a coordinate, we consider the linear transformations generated by the following generators.

$$Q^{ij} = x^i T^{0j} . \tag{5.66}$$

The commutator associated with the generators are

$$i[Q^{ij}, Q^{kl}] = \delta^{jk} Q^{il} - \delta^{il} Q^{kj} + ix^i x^k [T^{0j}, T^{0l}] . \tag{5.67}$$

The last term vanishes for a simple quantum mechanics operator $T^{0i} = -i\partial_i$. In general the term yields non-vanishing contributions, for example in the presence of magnetic field that modifies the momentum density.

One can then understand the relation between the Hall viscosity and the angular momentum through a commutator of two shear transformations, which turns out to be proportional to the angular momentum density as explained in §5.1.2. More specifically,

$$[\sigma^1_{ij} \int d^2x Q^{ij}_B(\vec{x}), \sigma^3_{kl} \int d^2x' Q^{kl}_B(\vec{x}')] \propto \int d^2x \epsilon_{ij} Q^{ij}_B(\vec{x}) = L , \tag{5.68}$$

where σ^1 and σ^3 are Pauli matrices. This derivation using the geometric interpretation fits well with the Berry phase calculations

[42, 49, 50, 48]. However, there are some subtleties in the derivation. First, the angular momentum density defined in (5.64) is non-zero only if translation invariance is broken. Second, the integrals over space need to be regulated. These subtleties can be related to each other, for instance if the physical system is in a finite volume or confined in some way by an external potential. In order to establish a relation for the Hall viscosity of a homogeneous and infinite system, one needs to take a limit where translation invariance is recovered. This limit does not commute with the zero momentum limit. The latter has to be taken first in order to fix the value of the Hall viscosity.

Now, we consider the momentum space representation of the Ward identity without translation symmetry. While one might think the quantities on the left hand sides can be compared directly to the right hand ones in (5.60), they need to be regulated in general following §5.2.2.

If the relation $[T^{0j}(\vec{x}), T^{0l}(\vec{x})] = 0$ holds, (5.67) takes the form

$$\langle i[Q^{ij}(t,\vec{x}), Q^{kl}(t,\vec{x}')]\rangle = \left(\delta^{jk}Q^{il}(t,\vec{x}) - \delta^{li}Q^{kj}(t,\vec{x}')\right)\delta^{(2)}(\vec{x}-\vec{x}'). \quad (5.69)$$

When there is the time translation symmetry (without space translation symmetry), the commutator only depends on $t - t'$. Thus the corresponding retarded correlator has the form

$$Q^{ijkl}(t-t';\vec{x},\vec{x}') = i\theta(t-t')\langle [Q^{ij}(t,\vec{x}), Q^{kl}(t',\vec{x}')]\rangle, \quad (5.70)$$

where $Q^{ijkl}(t-t';\vec{x},\vec{x}') = x^i x'^k G^{0j,0l}(t-t';\vec{x},\vec{x}')$. The retarded correlators for the stress energy tensors are given in (5.38) and (5.40).

Now we compute the following quantity

$$I^{ijkl}(\vec{e}) \equiv \partial_t \partial_{t'} \int d^2x d^2x' \, e^{-i\vec{e}\cdot\vec{x} - i\vec{e}\cdot\vec{x}'} Q^{ijkl}(t-t';\vec{x},\vec{x}'). \quad (5.71)$$

After Fourier transform with respect to time,

$$\widetilde{I}^{ijkl}(\vec{e}) = \int d(t-t') e^{-i\omega(t-t')} I^{ijkl}(\vec{e}), \quad (5.72)$$

and use the Fourier transform of the retarded correlators

$$G^{\mu\nu\alpha\beta}(t-t';\vec{x},\vec{x}') = \int \frac{dp_0 d^2p d^2q}{(2\pi)^5} e^{ip_0(t-t')+i\vec{p}\cdot\vec{x}-i\vec{q}\cdot\vec{x}'} \tilde{G}^{\mu\nu\alpha\beta}(\omega,\vec{p},\vec{q}). \tag{5.73}$$

After simple manipulations using (5.73) and (5.71), the Fourier transform (5.72) can be written in the following form

$$\tilde{I}^{ijkl}(\vec{e}) = \omega^2 \frac{\partial}{\partial p_i} \frac{\partial}{\partial q_k} \tilde{G}^{0j0l}(\omega,\vec{p},\vec{q}) \Big|_{\vec{p}=\vec{e},\vec{q}=-\vec{e}}. \tag{5.74}$$

This expression is one side of the relation we are going to compare below.

Starting with (5.71) and taking the time derivatives explicitly we can get four different terms that can be computed similarly as in (5.39). We get

$$I^{ijkl}(\vec{e}) = \int d^2x d^2x' \, e^{-i\vec{e}\cdot\vec{x}-i\vec{e}\cdot\vec{x}'} \Big[x^i x'^k \, \partial_n \partial'_m G_R^{njml}(t-t';\vec{x},\vec{x}')$$
$$-\delta'(t-t')\{\delta^{ik}\langle Q^{il}(t,\vec{x})\rangle - \delta^{li}\langle Q^{kj}(t,\vec{x})\rangle\}\delta^{(2)}(\vec{x}-\vec{x}') \tag{5.75}$$
$$-\delta(t-t')\partial_{t-t'}\{\delta^{ik}\langle Q^{il}(t-t',\vec{x})\rangle - \delta^{li}\langle Q^{kj}(t-t',\vec{x})\rangle\}\delta^{(2)}(\vec{x}-\vec{x}') \Big],$$

where we use the equation of motion in the first line, the equal time correlator (5.69) in the second line. To derive the last line we use the time translation invariance. For the term with a ∂_t derivative on step function and a $\partial_{t'}$ derivative on Q, we have

$$-i\delta(t-t')\partial_t \langle [Q^{ij}(t,\vec{x}), Q^{kl}(t',\vec{x}')] \rangle$$
$$= -i\delta(t-t')\partial_{t-t'} \langle [Q^{ij}(t-t',\vec{x}), Q^{kl}(0,\vec{x}')] \rangle, \tag{5.76}$$

where we shift $t' \to 0, t \to t - t'$ in the second line.

The other term can be manipulated as

$$i\delta(t-t')\partial_{t'} \langle [Q^{ij}(t,\vec{x}), Q^{kl}(t',\vec{x}')] \rangle$$
$$= i\delta(t-t')\partial_{t'-t} \langle [Q^{ij}(0,\vec{x}), Q^{kl}(t'-t,\vec{x}')] \rangle \tag{5.77}$$
$$= -i\delta(t-t')\partial_{t-t'} \langle [Q^{ij}(0,\vec{x}), Q^{kl}(t-t',\vec{x}')] \rangle,$$

where we shift $t \to 0$, $t' \to t' - t$ in the second equation and assume that Q^{ij} is invariant under the time reversal in the third line. By evaluating the time derivative on both (5.76) and (5.77) and by changing $\delta(t-t')Q^{kl}(0,\vec{x}') \to \delta(t-t')Q^{kl}(t-t',\vec{x}')$ using the delta function, we combine them as

$$-i\delta(t-t')\partial_{t-t'}\left\langle [Q^{ij}(t-t',\vec{x}), Q^{kl}(t-t',\vec{x}')]\right\rangle$$
$$= -\delta(t-t')\partial_{t-t'}\left\{\delta^{jk}\langle Q^{il}(t-t',\vec{x})\rangle \right. \quad (5.78)$$
$$\left. -\delta^{li}\langle Q^{kj}(t-t',\vec{x})\rangle\right\}\delta^{(2)}(\vec{x}-\vec{x}')\,.$$

This is the last term in (5.75).

Now we can convert the linear operators Q^{ij} into $x^i T^{0j}$ of (5.75). Then we get

$$I^{ijkl}(\vec{\epsilon}) = \int d^2x d^2x'\, e^{-i\vec{\epsilon}\cdot\vec{x} - i\vec{\epsilon}\cdot\vec{x}'}\left[x^i x'^k\, \partial_n \partial'_m G_R^{njml}(t-t';\vec{x},\vec{x}')\right.$$
$$-\delta'(t-t')\{\delta^{jk}x^i\langle T^{0l}(t,\vec{x})\rangle - \delta^{li}x^k\langle T^{0j}(t,\vec{x})\rangle\}\delta^{(2)}(\vec{x}-\vec{x}') \quad (5.79)$$
$$\left. -\delta(t-t')\partial_{t-t'}\{\delta^{jk}x^i\langle T^{0l}(t-t',\vec{x})\rangle - \delta^{li}x^k\langle T^{0j}(t-t',\vec{x})\rangle\}\delta^{(2)}(\vec{x}-\vec{x}')\right].$$

After taking Fourier transform as in §5.2.2, we arrive

$$\tilde{I}^{ijkl}(\vec{\epsilon}) = \frac{\partial}{\partial p_i}\frac{\partial}{\partial q_k}\left[p_n q_m \tilde{G}_R^{njml}(\omega,\vec{p},\vec{q})\right]_{\vec{p}=-\vec{q}=\vec{\epsilon}}$$
$$-\omega\left[\delta^{jk}\frac{\partial}{\partial p_i}\langle \tilde{T}^{0l}(\omega,\vec{p})\rangle - \delta^{li}\frac{\partial}{\partial p_k}\langle \tilde{T}^{0j}(\omega,\vec{p})\rangle\right]_{\vec{p}=2\vec{\epsilon}}$$
$$+\frac{\partial}{\partial p_i}\left[\delta^{jk}p_n\langle \tilde{T}^{nl}(\omega,\vec{p})\rangle\right]_{\vec{p}=2\vec{\epsilon}} \quad (5.80)$$
$$-\frac{\partial}{\partial q_k}\left[\delta^{li}q_m\langle \tilde{T}^{mj}(\omega,\vec{q})\rangle\right]_{\vec{q}=2\vec{\epsilon}}.$$

Putting the left-hand side (5.74) and the right-hand side

$$\begin{aligned}0 =& \frac{\partial}{\partial p_i} \frac{\partial}{\partial q_k} \Big[-\omega^2 \tilde{G}_R^{0j0l}(\omega, \vec{p}, \vec{q}) + p_n q_m \tilde{G}_R^{njml}(\omega, \vec{p}, \vec{q}) \Big]_{\vec{p}=-\vec{q}=\vec{e}} \\ & - \omega \Big[\delta^{jk} \frac{\partial}{\partial p_i} \langle \tilde{T}^{0l}(\omega, \vec{p}) \rangle - \delta^{li} \frac{\partial}{\partial p_k} \langle \tilde{T}^{0j}(\omega, \vec{p}) \rangle \Big]_{\vec{p}=2\vec{e}} \\ & + \frac{\partial}{\partial p_i} \Big[\delta^{jk} p_n \langle \tilde{T}^{nl}(\omega, \vec{p}) \rangle \Big]_{\vec{p}=2\vec{e}} \\ & - \frac{\partial}{\partial q_k} \Big[\delta^{li} q_m \langle \tilde{T}^{mj}(\omega, \vec{q}) \rangle \Big]_{\vec{q}=2\vec{e}}.\end{aligned} \quad (5.81)$$

This equation reduces to our previous result (5.59) and (5.60) with angular momentum by 1) contracting with the combination of Pauli matrices, $(1/4)(\sigma_1)_{ij}(\sigma_3)_{kl}$, 2) further manipulating the second and third lines by changing the momentum derivative to coordinate, 3) taking $\omega \to 0$ limit, and 4) by ignoring the ω^2 term in the first line. Later we take an explicit example to demonstrate the last ingredient. This expression is also suitable for evaluating the Ward identity for small momentum expansion in terms of $\vec{e}$.

5.3.3 Ward identity with conserved current

In this section we have an important generalization of the Ward identity by including the conserved current that is described by $\partial_\mu J^\mu = 0$. While we can include any conserved current, it is particularly interesting to consider the electric current in the presence of an external electromagnetic field, especially magnetic field B. When the magnetic field is non-zero, there are some subtleties that have to be taken into account. For example, the stress energy tensor is no longer conserved in the presence of a constant magnetic field. The modified conservation equation has been introduced in (2.40).

$$\partial_\mu T^{\mu\nu} = F^{\nu\sigma} J_\sigma, \quad (5.82)$$

where we denote the antisymmetric field strength tensor as $F^{\nu\sigma} = \partial^\nu A^\sigma - \partial^\sigma A^\nu$ with the electromagnetic gauge potential A^μ. We can generalize the stress energy tensor as

$$\tilde{T}^{\mu\nu} = T^{\mu\nu} - F^{\nu\sigma}x_\sigma J^\mu . \tag{5.83}$$

Then the conservation equation for the still holds for this new tensor as $\partial_\mu \tilde{T}^{\mu\nu} = \partial_\mu T^{\mu\nu} - F^{\nu\sigma}(\partial_\mu x_\sigma)J^\mu - F^{\nu\sigma}x_\sigma \partial_\mu J^\mu = 0$ with $\partial_\mu J^\mu = 0$.

For the magnetic field perpendicular to the 2 dimensional plane, the magnetic field can be written as $B = (1/2)\epsilon_{ij}F^{ij}$. The conservation equation becomes

$$\partial_\mu T^{\mu i} = B\epsilon^i{}_n J^n . \tag{5.84}$$

The generalized momentum generator has the form

$$T_B^{0i} = T^{0i} - \frac{B}{2}\epsilon^i{}_n x^n J^0 \tag{5.85}$$

for describing the generators of spatial translations in the presence of a magnetic field.

Furthermore, the angular momentum we previously considered in (5.64) is not conserved.

$$\partial_0 L = \int d^2x \, \epsilon_{ij} x^i \partial_0 T^{0j} \neq 0 . \tag{5.86}$$

For a constant magnetic field, it is possible to define a generalized angular momentum operator that is conserved

$$L_B = \int d^2x \, \epsilon_{ij} x^i \left(T^{0j} - \frac{B}{2}\epsilon^j{}_n x^n J^0 \right) . \tag{5.87}$$

The density J^0 is the generator of global $U(1)$ transformations, so the conserved angular momentum generates a combination of geometric rotations and 'gauge' transformations on the physical states we consider.

We are ready to develop the Ward identity in the presence of conserved current along with the conserved (generalized) stress energy tensor. Similar to the previous section §5.3.1, we consider a retarded Green's function of two original energy currents, $T^{0j}(x^\mu)$ and

$T^{0l}(x'^\mu)$, as

$$G^{0j0l}(x^\mu;x'^\mu) = i\theta(x^0 - x'^0)\langle[T^{0j}(x^\mu), T^{0l}(x'^\mu)]\rangle, \tag{5.88}$$

which is the same as (5.38). We follow the same steps, taking two time derivatives on this equation and computing 4 different terms. The last two lines that involve with equal time commutator, with at least one time derivative on step function, in the expression (5.39) are not modified.

The first term in (5.39), $i\theta(x^0 - x'^0)\langle[\partial_0 T^{0j}(x^\mu), \partial_0' T^{0l}(x'^\mu)]\rangle$ has extra contributions as we have a generalized conservation equation, $\partial_0 T^{0i} = -\partial_j T^{ji} + B\epsilon^i{}_n J^n$. We get four different types of Green's functions with different combinations of T and J.

$$\begin{aligned}G^{\mu\nu\alpha\beta}(x;x') &= i\theta(x^0 - x'^0)\langle[T^{\mu\nu}(x), T^{\alpha\beta}(x')]\rangle, \\ G^{\mu\nu,\alpha}(x;x') &= i\theta(x^0 - x'^0)\langle[T^{\mu\nu}(x), J^\alpha(x')]\rangle, \\ G^{\mu,\alpha\beta}(x;x') &= i\theta(x^0 - x'^0)\langle[J^\mu(x), T^{\alpha\beta}(x')]\rangle, \\ G^{\mu\alpha}(x;x') &= i\theta(x^0 - x'^0)\langle[J^\mu(x), J^\alpha(x')]\rangle,\end{aligned} \tag{5.89}$$

along with the corresponding equal time commutators

$$\begin{aligned}C^{\mu\nu\alpha\beta}(x;x') &= i\langle[T^{\mu\nu}(x), T^{\alpha\beta}(x')]\rangle, \\ C^{\mu\nu,\alpha}(x;x') &= i\langle[T^{\mu\nu}(x), J^\alpha(x')]\rangle, \\ C^{\mu,\alpha\beta}(x;x') &= i\langle[J^\mu(x), T^{\alpha\beta}(x')]\rangle, \\ C^{\mu\alpha}(x;x') &= i\langle[J^\mu(\vec{x}), J^\alpha(x')]\rangle,\end{aligned} \tag{5.90}$$

where $x = (x^0;\vec{x})$ and $x' = (x'^0;\vec{x}')$.

After a straightforward computation following (5.26) and (5.39), these terms can be organized as

$$\begin{aligned}\partial_0'\partial_0 G^{0j0l}(x^\mu;x'^\mu) = &-\delta'(x^0 - x'^0)C^{0j0l}(x^\mu;x'^\mu) \\ &- \delta(x^0 - x'^0)(\partial_0 - \partial_0')C^{0j0l}(x^\mu;x'^\mu) \\ &+ \partial_n \partial_m' G^{njml}(x^\mu;x'^\mu) - B\epsilon^j{}_n \partial_m' G^{n,ml}(x^\mu;x'^\mu) \\ &- B\epsilon^l{}_m \partial_n G^{nj,m}(x^\mu;x'^\mu) + B^2 \epsilon^j{}_n \epsilon^l{}_m G^{nm}(x^\mu;x'^\mu),\end{aligned} \tag{5.91}$$

where ∂' is a derivative acting on x'^μ. This is the form of Ward identity with a conserved current in terms of Green's functions and equal time commutators.

5.3.3.1 Evaluating contact terms

To evaluate the contact terms associated with the C^{0j0l} in (5.91) in the presence of the magnetic field, we use (5.84). The first two lines can be recast into more convenient forms as follows:

$$-\delta'(x^0-x'^0)C^{0j0l}(x;x') - \delta(x^0-x'^0)(\partial_0 - \partial'_0)C^{0j0l}(x;x')$$
$$= -\frac{1}{2}(\partial_0 - \partial'_0)\big(\delta(x^0-x'^0)C^{0j0l}(x;x')\big) \quad (5.92)$$
$$+ \frac{1}{2}\delta(x^0-x'^0)\big(\partial_n C^{nj0l} - \partial'_m C^{0jml} - Be^j{}_n C^{n,0l} + Be^l{}_m C^{0j,m}\big).$$

We also reorganize the terms with magnetic field B as follows:

$$-Be^j{}_n \partial'_m G^{n,ml}(x;x') - Be^l{}_m \partial_n G^{nj,m}(x;x') + B^2 e^j{}_n e^l{}_m G^{nm}(x;x')$$
$$= Be^j{}_n \partial'_t G^{n,0l}(x;x') + Be^l{}_m \partial_t G^{0j,m}(x;x') - B^2 e^j{}_n e^l{}_m G^{nm}(x;x') \quad (5.93)$$
$$+ \delta(x^0-x'^0)\big(Be^j{}_n C^{n,0l}(x;x') - Be^l{}_m C^{0j,m}(x;x')\big),$$

where we use $\partial'_m G^{n,ml}(x;x') = -\partial'_t G^{n,0l}(x;x') - C^{n,0l}(x;x') + Be^l{}_m G^{nm}(x;x')$ with $\partial'_0 \theta(x^0-x'^0) = -\delta(x^0-x'^0)$.

Putting them together and performing the Fourier transform after multiplying $\int d(x^0-x'^0)e^{i\omega(x^0-x'^0)}$ on both sides, (5.91) turns into

$$\omega^2 G^{0j0l} - i\omega Be^j{}_n G^{n,0l} + i\omega Be^l{}_m G^{0j,m} + B^2 e^j{}_n e^l{}_m G^{nm}$$
$$= \partial_n \partial'_m G^{njml} + i\omega C^{0j0l} + \frac{1}{2}\big(\partial_n C^{nj0l} - \partial'_m C^{0jml}\big) \quad (5.94)$$
$$+ \frac{B}{2}e^j{}_n C^{n,0l} - \frac{B}{2}e^l{}_m C^{0j,m}.$$

We organize the equation by keeping the Green's funtions with two spatial dimensions in the left hand side, while putting the contact terms to the right hand side.

To go further, we make use of the fact that T_B is the translation generator in the presence of the magnetic field that satisfies

$$i[T_B^{0i}(\vec{x}), \mathcal{O}(\vec{x}')] = \partial_i \mathcal{O}(\vec{x})\delta^{(2)}(\vec{x}-\vec{x}'), \tag{5.95}$$

where $\mathcal{O}$ is an operator. With this identity we can simplify the contact terms.

$$\begin{aligned} i\omega C^{0j0l} &= i\omega C_B^{0j0l} + \frac{i\omega B}{2}\left[\epsilon^j{}_n x^n C_B^{0,0l} + \epsilon^l{}_m x'^m C_B^{0j,0}\right] \\ &= i\omega\left[\partial_j\langle T_B^{0l}\rangle - \partial_l\langle T_B^{0j}\rangle\right]\delta^{(2)}(\vec{x}-\vec{x}') \\ &\quad - \frac{i\omega B}{2}\left[\epsilon^j{}_n x^n \partial_l\langle J^0\rangle - \epsilon^l{}_m x'^m \partial_j\langle J^0\rangle\right]\delta^{(2)}(\vec{x}-\vec{x}'), \end{aligned} \tag{5.96}$$

where the subscript B indicates that the stress energy tensor is T_B given in (5.85).

$$\begin{aligned} \partial_n C^{nj0l} &= \partial_n C_B^{nj0l} + \frac{B}{2}\epsilon^l{}_m x'^m \partial_n C^{nj,0} \\ &= \partial_n C_B^{nj0l} = -\partial_l\partial_n\langle T^{nj}\rangle\delta^{(2)}(\vec{x}-\vec{x}'), \end{aligned} \tag{5.97}$$

where we used the fact that the equal time commutators with the charge density vanish, $C^{nj,0} = [T^{nj}(\vec{x}), J^0(\vec{x})] = 0$. Similarly,

$$\partial'_m C^{0jml} = \partial'_m C_B^{0jml} = \partial_j \partial_m\langle T^{ml}\rangle\delta^{(2)}(\vec{x}-\vec{x}'). \tag{5.98}$$

Finally, we also compute

$$\begin{aligned} \frac{B}{2}\left[\epsilon^j{}_n C^{n,0l} - \epsilon^l{}_m C^{0j,m}\right] &= \frac{B}{2}\left[\epsilon^j{}_n C_B^{n,0l} - \epsilon^l{}_m C_B^{0j,m}\right] \\ &= -\frac{B}{2}\left[\epsilon^j{}_n \partial_l\langle J^n\rangle + \epsilon^l{}_m \partial_j\langle J^m\rangle\right]\delta^{(2)}(\vec{x}-\vec{x}'). \end{aligned} \tag{5.99}$$

We also use $C^{n,m} = i[J^n(\vec{x}), J^m(\vec{x})] = 0$.

We collect all the contact terms in (5.94).

$$i\omega\delta^{(2)}(\vec{x}-\vec{x}')\left[\partial_j\langle T_B^{0l}\rangle - \partial_l\langle T_B^{0j}\rangle + \frac{B}{2}\{\epsilon^l_{\ m}x^m\partial_j\langle J^0\rangle - \epsilon^j_{\ n}x^n\partial_l\langle J^0\rangle\}\right]$$
$$-\frac{1}{2}\delta^{(2)}(\vec{x}-\vec{x}')\left[B\epsilon^j_{\ n}\partial_l\langle J^n\rangle + B\epsilon^l_{\ m}\partial_j\langle J^m\rangle + \partial_n\partial_l\langle T^{nj}\rangle + \partial_j\partial_m\langle T^{ml}\rangle\right]$$
$$= i\omega\epsilon^{jl}\left[\epsilon^n_{\ m}\partial_n\langle T^{0m}\rangle + B\langle J^0\rangle\right]\delta^{(2)}(\vec{x}-\vec{x}')$$
$$-\frac{1}{2}\left[\partial_l(\partial_n\langle T^{nj}\rangle + B\epsilon^j_{\ n}\langle J^n\rangle) + \partial_j(\partial_m\langle T^{ml}\rangle + B\epsilon^l_{\ m}\langle J^m\rangle)\right]\delta^{(2)}(\vec{x}-\vec{x}'). \tag{5.100}$$

In the first line, we simplify the expression by introducing the explicit form of T_B in (5.85), group terms, and use $\epsilon_{ij}\epsilon_{kl} = \delta_{ik}\delta_{jl} - \delta_{il}\delta_{jk}$.

If the angular momentum is non-zero and there is rotational invariance, the expectation value of the momentum density takes the form as before $\langle T^{0i}\rangle = (1/2)\epsilon^{ik}\partial_k\ell_T$, which is given in (5.64). For a homogeneous system with a boundary, this becomes a momentum density at the edge. In addition, there can be a non-zero magnetization

$$\langle J^i\rangle = \epsilon^{ik}\partial_k M. \tag{5.101}$$

This expression is familiar in the context of the bound current due to magnetization from electrodynamics $\vec{J}_m = \vec{\nabla}\times\vec{M}$. It takes the form $J_m^i = \epsilon^{ijk}\partial_j M_k = \epsilon^{ij}\partial_j M$ when the magnetization is along $M_z = M$ direction and we write xy coordinates only.

Then the Ward identity becomes

$$\omega^2 G^{0j0l} - i\omega B\epsilon^j_{\ n}G^{n,0l} + i\omega B\epsilon^l_{\ m}G^{0j,m} + B^2\epsilon^j_{\ n}\epsilon^l_{\ m}G^{nm}$$
$$= \partial_n\partial'_m G^{njml} + i\omega\epsilon^{jl}\left[B\langle J^0\rangle - \frac{1}{2}\partial^2\ell_T\right]\delta^{(2)}(\vec{x}-\vec{x}') \tag{5.102}$$
$$-\frac{1}{2}\left[\partial_l\{\partial_n\langle T^{nj}\rangle - B\partial_j M\} + \partial_j\{\partial_m\langle T^{ml}\rangle - B\partial_l M\}\right]\delta^{(2)}(\vec{x}-\vec{x}').$$

Here the Green's functions $G^{\cdots}(\vec{x},\vec{x}')$ have the arguments on both the coordinates $\vec{x}$ and $\vec{x}'$. In addition to the terms that appear due to the conservation equation of the stress energy tensor, there are several contributions from contact terms that affect the response of the system to external sources. The terms in the last line do not contain factors of ω, thus they are associated with the response to static changes in the spatial metric, *i.e.* to the elasticity tensor. If there is rotational invariance, the expectation value of the stress tensor has the form $\langle T^{ij} \rangle = p \delta^{ij}$.

The Ward identity (5.102) contains the contact terms which play an important role when the system is not translation invariant. One can go further to apply this to interesting physical systems. We come back to this below with some additional ingredients after discussing the translation invariant cases first.

5.3.3.2 Ward identity with translation invariance

We proceed to consider more useful Ward identity specialized on various symmetries. First, we consider the simplest case where the system is in a homogeneous state. In this case the expectation values of operators are independent of the position, and the terms with derivatives in the contact terms vanish.

The Ward identity (5.102) reduces to

$$\omega^2 G^{0j0l}(\omega, \vec{x} - \vec{x}') - i\omega B e^j{}_n G^{n,0l}(\omega, \vec{x} - \vec{x}')$$
$$+ i\omega B e^l{}_m G^{0j,m}(\omega, \vec{x} - \vec{x}') + B^2 e^j{}_n e^l{}_m G^{nm}(\omega, \vec{x} - \vec{x}') \quad (5.103)$$
$$= i\omega e^{jl} B \rho \, \delta^{(2)}(\vec{x} - \vec{x}') + \partial_n \partial'_m G^{njml}(\omega, \vec{x} - \vec{x}'),$$

where the Green's functions $G^{\cdots}(\omega, \vec{x} - \vec{x}')$ depend only on the difference, $\vec{x} - \vec{x}'$, due to the translation invariance. $\rho = \langle J^0 \rangle$ is the charge density.

After taking Fourier transform with respect to $\vec{x} - \vec{x}'$ using (5.45), we get

$$\omega^2 G^{0j0l}(\omega, \vec{q}) - i\omega B e^j{}_n G^{n,0l}(\omega, \vec{q})$$
$$+ i\omega B e^l{}_m G^{0j,m}(\omega, \vec{q}) + B^2 e^j{}_n e^l{}_m G^{nm}(\omega, \vec{q}) \quad (5.104)$$
$$= i\omega e^{jl} B \rho + q_n q_m G^{njml}(\omega, \vec{q}).$$

This is the momentum space Ward identity in terms of Green's functions. As we already mentioned above, rotational invariance is useful for organizing the Green's functions. The four spatial index Green's function $G^{njml}(\omega, \vec{q})$ is related to shear, bulk, and Hall viscosity as in (5.47).

It is illuminating to compare the Ward identity to other calculations in the literature. We are not aware of an example where both the viscosities and the finite momentum current correlators were computed in a relativistic system. Nevertheless, it is possible to compare our result with the results in [104, 105, 106] at zero momentum $q^2 = 0$, which use the anti-de Sitter space conformal field theory (AdS/CFT) correspondence to compute field theory correlators. These references explicitly compute the current correlators for the holographic dual of a magnetically charged black hole in asymptotically AdS_4 spacetime. In the 2 + 1 dimensional field theory, the magnetically charged black hole implies that there is a non-zero background magnetic field turned on. Explicit comparison is done in [22]. Interested readers are encouraged to check out recent developments of string theory through AdS/CFT correspondence, especially toward the applications of the interesting systems in condensed matter.

5.3.3.3 Organizing two spatial index Green's functions

Due to the current operator J^μ in addition to stress energy tensor $T^{\mu\nu}$, there are four different types of Green's functions with two spatial indices in (5.102) and (5.104). These are related by the Fourier transform in the presence of the space translation symmetry. $G^{a,b}(\omega; \vec{x} - \vec{x}') = \int (d^2q/(2\pi)^2) e^{i\vec{q}\cdot(\vec{x}-\vec{x}')} \widetilde{G}^{a,b}(\omega, \vec{q})$, where $a, b = i, j, \cdots, 0i, 0j, \cdots$.

In the presence of the rotational invariance, these Green's functions have the same tensor structures. We focus on the momentum space representation, which can be written in the following form:

$$\widetilde{G}^{a,b} = \Pi_\delta \delta^{ab} + \Pi_\epsilon \epsilon^{ab} + q^a q^b \Pi_q + (\epsilon^{bo} q^a + \epsilon^{ao} q^b) q_o \Pi_{q\epsilon}. \quad (5.105)$$

There are four different types, $\Pi^{TT}, \Pi^{TJ}, \Pi^{JT}$, and Π^{JJ}. We have already introduced the first type as the thermal conductivity tensors in (5.46). Plugging them into (5.104) and doing some algebra, we arrive at

$$\delta^{jl}\left[\omega^2\Pi_\delta^{TT} + i\omega B(\Pi_\epsilon^{JT} + \Pi_\epsilon^{TJ}) + B^2\Pi_\delta^{JJ} + B^2q^2\Pi_q^{JJ} - i\omega Bq^2(\Pi_{q\epsilon}^{JT} - \Pi_{q\epsilon}^{TJ})\right]$$
$$+\epsilon^{jl}\left[\omega^2\Pi_\epsilon^{TT} - i\omega B(\Pi_\delta^{JT} + \Pi_\delta^{TJ}) + B^2\Pi_\epsilon^{JJ} - \frac{i\omega B}{2}q^2(\Pi_q^{JT} + \Pi_q^{TJ})\right]$$
$$+q^jq^l\left[\omega^2\Pi_q^{TT} + 2i\omega B(\Pi_{q\epsilon}^{JT} - \Pi_{q\epsilon}^{TJ}) - B^2\Pi_q^{JJ}\right] \qquad (5.106)$$
$$+(\epsilon^{jm}q^l + \epsilon^{lm}q^j)q_m\left[\omega^2\Pi_{q\epsilon}^{TT} - \frac{i\omega B}{2}(\Pi_q^{JT} - \Pi_q^{TJ}) - B^2\Pi_{q\epsilon}^{JJ}\right]$$
$$= -i\omega\delta^{jl}q^2\eta + \epsilon^{jl}i\omega\left[B\rho - q^2\eta_H\right] - i\omega q^jq^l\zeta \,,$$

where we use (5.47) to derive the last line.

We use slightly different notations for the conductivities. $\Pi^{TT} = -i\omega\kappa, \Pi^{TJ} = -i\omega\alpha, \Pi^{JT} = -i\omega\bar{\alpha}, \Pi^{JJ} = -i\omega\sigma$, where $\kappa, \alpha, \bar{\alpha}, \sigma$ are thermal, thermoelectric, electric-thermo, and electric conductivity tensors. All these contributions are the direct generalizations of the familiar electric conductivity. We express these conductivity tensors explicitly

$$\tilde{G}^{0i0k} = -i\omega\left[\delta^{ik}\kappa_\delta + q^iq^k\kappa_q\right] - i\omega\left[\epsilon^{ik}\kappa_\epsilon + (\epsilon^{in}q_nq^k + \epsilon^{kn}q_nq^i)\kappa_{q\epsilon}\right],$$
$$\tilde{G}^{0i,k} = -i\omega\left[\delta^{ik}\alpha_\delta + q^iq^k\alpha_q\right] - i\omega\left[\epsilon^{ik}\alpha_\epsilon + (\epsilon^{in}q_nq^k + \epsilon^{kn}q_nq^i)\alpha_{q\epsilon}\right],$$
$$\tilde{G}^{i,0k} = -i\omega\left[\delta^{ik}\bar{\alpha}_{,\delta} + q^iq^k\bar{\alpha}_q\right] - i\omega\left[\epsilon^{ik}\bar{\alpha}_\epsilon + (\epsilon^{in}q_nq^k + \epsilon^{kn}q_nq^i)\bar{\alpha}_{q\epsilon}\right],$$
$$\tilde{G}^{ik} = -i\omega\left[\delta^{ik}\sigma_\delta + q^iq^k\sigma_q\right] - i\omega\left[\epsilon^{ik}\sigma_\epsilon + (\epsilon^{in}q_nq^k + \epsilon^{kn}q_nq^i)\sigma_{q\epsilon}\right].$$
$$(5.107)$$

Matching tensor structures, $\delta^{ik}, \epsilon^{jl}, q^iq^l, (\epsilon^{jm}q^l + \epsilon^{lm}q^j)q_m$, in (5.106), we find four independent Ward identities

$$\omega^2\kappa_\delta + i\omega B\left(\alpha_\epsilon + \bar{\alpha}_\epsilon + q^2[\alpha_{q\epsilon} - \bar{\alpha}_{q\epsilon}]\right) + B^2\left(\sigma_\delta + q^2\sigma_q\right) = q^2\eta \,,$$
$$\omega^2\kappa_\epsilon - i\omega B\left(\alpha_\delta + \bar{\alpha}_\delta + \frac{q^2}{2}[\alpha_q + \bar{\alpha}_q]\right) + B^2\sigma_\epsilon = -B\rho + q^2\eta_H \,,$$
$$\omega^2\kappa_q - 2i\omega B(\alpha_{q\epsilon} - \bar{\alpha}_{q\epsilon}) - B^2\sigma_q = \zeta \,,$$
$$\omega^2\kappa_{q\epsilon} + i\omega\frac{B}{2}(\alpha_q - \bar{\alpha}_q) - B^2\sigma_{q\epsilon} = 0 \,. \qquad (5.108)$$

These momentum space Ward identities are expressed in terms of the conductivity tensors in the presence of the translation invariance. We come back this equation when we seek the properties of Hall viscosity in the context of skyrmion motion.

If we do not require the translation invariance, various different contact terms come into play including the angular momentum. Of course, we also need to use the coordinates instead of the momentum variables. The Ward identity turns into

$$\int d^2x \Big\{ \delta^{jl} [\omega^2 \Pi^{TT}_\delta + i\omega B(\Pi^{jT}_\epsilon + \Pi^{Tj}_\epsilon) + B^2 \Pi^{jj}_\delta + B^2 \partial^2 \Pi^{jj}_{q^2} - i\omega B \partial^2 (\Pi^{jT}_{q^2\epsilon} - \Pi^{Tj}_{q^2\epsilon})]$$
$$+ \epsilon^{jl} [\omega^2 \Pi^{TT}_\epsilon - i\omega B(\Pi^{jT}_\delta + \Pi^{Tj}_\delta) + B^2 \Pi^{jj}_\epsilon - \frac{i\omega B}{2} \partial^2 (\Pi^{jT}_{q^2} + \Pi^{Tj}_{q^2})]$$
$$+ \partial^j \partial^l [\omega^2 \Pi^{TT}_{q^2} + 2i\omega B(\Pi^{jT}_{q^2\epsilon} - \Pi^{Tj}_{q^2\epsilon}) - B^2 \Pi^{jj}_{q^2}] \quad (5.109)$$
$$+ (\epsilon^{jm}\partial^l + \epsilon^{lm}\partial^j)\partial_m [\omega^2 \Pi^{TT}_{q^2\epsilon} + \frac{i\omega B}{2}(\Pi^{jT}_{q^2} - \Pi^{Tj}_{q^2}) - B^2 \Pi^{jj}_{q^2\epsilon}] \Big\}$$
$$= \int d^2x \Big\{ \delta^{jl}[-i\omega \partial^2 \bar{\eta}] + \epsilon^{jl} i\omega [B\rho - \partial^2 (\bar{\eta}_H + \frac{\ell_B}{2})] + \partial^j \partial^l [-p + BM - i\omega \bar{\zeta}] \Big\}.$$

Here ¯ indicates the coordinate expressions. We can also rewrite this coordinate Ward identity similar to (5.108) in terms of various conductivity tensor components. We note that some interesting combinations show up as groups. By comparing (5.109) and (5.106), we see that

$$\eta_H \to \bar{\eta}_H + \frac{\ell_B}{2},$$
$$i\omega \zeta \to i\omega \bar{\zeta} + p - BM, \quad (5.110)$$

when the translation symmetry is broken. This happens because these grouped quantities are linked by the same physical origins.

5.3.4 Galilean invariant Ward identities

In this section we consider Galilean invariant systems by taking a physically interesting limit of the Ward identity (5.109) and (5.106). In a Galilean invariant system with particles that have the same charge and mass m, the momentum density T^{0i} is proportional to the number current density j^i [107]. Specifically,

$$T^{0i} = mj^i, \quad (5.111)$$

where we consider the particles with unit charge. If one wants to have explicit expression with charge, one can have the substitution $m \to m/e$ with charge e. This relation has been used in the context of Ward identities in [43][98].

Then all the two spatial index Green's functions and equal time commutators given in (5.105) are proportional to each other: $G^{0m0n} = mG^{0m,n} = mG^{m,0n} = m^2 G^{mn}$ and $C^{0m0n} = mC^{0m,n} = mC^{m,0n} = m^2 C^{mn}$. The Ward identity given in (5.94) turns into

$$m^2 \left[\omega^2 \delta_n^j \delta_m^l - i\omega\omega_c e^j{}_n \delta_m^l + i\omega\omega_c e^l{}_m \delta_n^j + \omega_c^2 e^j{}_n e^l{}_m \right] G^{nm}$$
$$= \partial_n \partial'_m G^{njml} + m^2 \left(i\omega \delta_n^j \delta_m^l + \frac{\omega_c}{2} e^j{}_n \delta_m^l - \frac{\omega_c}{2} e^l{}_m \delta_n^j \right) C^{nm} \quad (5.112)$$
$$- \frac{1}{2} \left[\partial_n \partial_l \langle T^{nj} \rangle + \partial_j \partial_m \langle T^{ml} \rangle \right] \delta^{(2)}(\vec{x} - \vec{x}') ,$$

where the cyclotron frequency ω_c is defined as $\omega_c = B/m$. The terms in the last line comes from the three spatial index terms C^{nj0l} and C^{0jml}.

To have the terms in an explicitly symmetric form motivated by the combination (5.85), we introduce the tensors

$$\Omega^i{}_j = i\omega \delta^i_j + \omega_c e^i{}_j , \qquad (5.113)$$
$$\bar{\Omega}^i{}_j = -i\omega \delta^i_j + \omega_c e^i{}_j ,$$

which have the following properties

$$\Omega^i{}_k \bar{\Omega}^k{}_j = (\omega^2 - \omega_c^2) \delta^i_j , \qquad (5.114)$$
$$\bar{\Omega}^i{}_k \Omega^k{}_j = (\omega^2 - \omega_c^2) \delta^i_j .$$

We can rearrange the terms of (5.112) in more convenient form

$$m^2 \Omega^j{}_n \bar{\Omega}^l{}_m G^{nm} = \partial_n \partial'_m G^{njml}$$
$$+ \frac{m^2}{2} \left(\Omega^j{}_n \delta_m^l - \bar{\Omega}^l{}_m \delta_n^j \right) C^{nm} \quad (5.115)$$
$$- \frac{1}{2} \left[\partial_n \partial_l \langle T^{nj} \rangle + \partial_j \partial_m \langle T^{ml} \rangle \right] \delta^{(2)}(\vec{x} - \vec{x}') .$$

We arrive at the Ward identity with Galilean invariance in terms of Green's functions and equal time commutators.

5.3.4.1 Without spatial translation symmetry

When the spatial transition is not required, the contact terms in the last two lines of (5.115) can contribute. The equal time commutator gives

$$m^2 C^{nm} = C_B^{0n0m} + \frac{B}{2}\epsilon^n{}_s x^s C_B^{0,0m} + \frac{B}{2}\epsilon^m{}_s x'^s C_B^{0n,0}$$

$$= [\partial_n \langle T_B^{0m}\rangle - \partial_m \langle T_B^{0n}\rangle - \frac{B}{2}\epsilon^n{}_s x^s \partial_m \langle J^0\rangle + \frac{B}{2}\epsilon^m{}_s x'^s \partial_n \langle J^0\rangle]\delta^{(2)}(\vec{x}-\vec{x}')$$

$$= [\partial_n \langle T^{0m}\rangle - \partial_m \langle T^{0n}\rangle + B\epsilon^{nm}\langle J^0\rangle]\delta^{(2)}(\vec{x}-\vec{x}'), \quad (5.116)$$

where the subscript B indicates to use T_B given in (5.85), and we use $C^{00} = [J^0(\vec{x}), J^0(\vec{x})] = 0$ in the first line. With rotational invariance, the one point functions of the momentum density take the same form as in (5.57). Thus, the equal time commutator becomes

$$m^2 C^{nm} = [\frac{1}{2}(\epsilon^{ms}\partial_n - \epsilon^{ns}\partial_m)\partial_s \ell_T + \epsilon^{nm} B\langle J^0\rangle]\delta^{(2)}(\vec{x}-\vec{x}')$$

$$= \epsilon^{nm}[-\frac{1}{2}2\ell_T + B\langle J^0\rangle]\delta^{(2)}(\vec{x}-\vec{x}'). \quad (5.117)$$

This leads to the following form of the Ward identity.

$$m^2 \Omega^j{}_n \bar{\Omega}^l{}_m G^{nm} = \partial_n \partial'_m G^{njml}$$
$$+ \Omega^j{}_k \epsilon^{kl}[-\frac{1}{2}2\ell_T + B\langle J^0\rangle]\delta^{(2)}(\vec{x}-\vec{x}') \quad (5.118)$$
$$- \frac{1}{2}[\partial_n \partial_l \langle T^{nj}\rangle + \partial_j \partial_m \langle T^{ml}\rangle]\delta^{(2)}(\vec{x}-\vec{x}').$$

With appropriate contact terms, we can evaluate Ward identities with conductivity and viscosity tensors in either coordinate space or in momentum space depending on our physical interests. We consider a specific example below.

5.3.4.2 With spatial translation symmetry

With spatial translation symmetry, the Ward identity (5.115) is simplified to

$$m^2 \Omega^j{}_n \bar{\Omega}^l{}_m G^{nm}(\omega, \vec{x} - \vec{x}') = \partial_n \partial'_m G^{njml}(\omega, \vec{x} - \vec{x}') \\ + \Omega^j{}_n \epsilon^{nl} B\bar{n} \, \delta^{(2)}(\vec{x} - \vec{x}') \, , \quad (5.119)$$

which becomes, after a Fourier transformation in momentum space

$$m^2 \Omega^j{}_n \bar{\Omega}^l{}_m G^{nm}(\omega, \vec{q}) = q_n q_m G^{njml}(\omega, \vec{q}) \\ + \Omega^j{}_n \epsilon^{nl} B\bar{n} \, . \quad (5.120)$$

Here $\langle J^0 \rangle = \bar{n}$. The stress tensor correlator can be expanded as in (5.47). This leads to

$$m^2 \Omega^j{}_n \bar{\Omega}^l{}_m \tilde{G}^{nm}(\omega, \vec{q}) = -i\omega(q^2 \eta \delta^{jl} + \zeta q^j q^l + \eta_H q^2 \epsilon^{jl}) \\ + \Omega^j{}_n \epsilon^{nl} B\bar{n} \, . \quad (5.121)$$

To extract quantities with different tensor structures, we contract with $\bar{\Omega}^i{}_j$ and $\Omega^k{}_l$ and use (5.114) to get

$$m^2 (\omega^2 - \omega_c^2)^2 \tilde{G}^{ik}(\omega, \vec{q}) = (\omega^2 - \omega_c^2)(i\omega \epsilon^{ik} + \omega_c \delta^{ik}) B\bar{n} \\ - i\omega q^2 \Big[\eta \left((\omega^2 + \omega_c^2)\delta^{ik} + 2i\omega_c \omega \epsilon^{ik}\right) \\ + \eta_H \left((\omega^2 + \omega_c^2)\epsilon^{ik} - 2i\omega_c \omega \delta^{ik}\right) \\ + \zeta\left((\omega^2 - \omega_c^2)\frac{q^i q^k}{q^2} + \omega_c(\omega_c \delta^{ik} + i\omega \epsilon^{ik})\right)\Big]. \quad (5.122)$$

We identify the following form factors in the current correlator in terms of conductivities.

$$\tilde{G}^{ik} = -i\omega \big[\delta^{ik} \sigma_\delta + q^i q^k \sigma_q \\ + \epsilon^{ik} \sigma_\epsilon + (\epsilon^{in} q_n q^k + \epsilon^{kn} q_n q^i) \sigma_{q\epsilon}\big] \, . \quad (5.123)$$

By matching the tensor structures, we get

$$\sigma_\delta = \frac{1}{m^2(\omega^2 - \omega_c^2)} \left\{ \frac{i\omega_c}{\omega} B\bar{n} + \frac{q^2}{\omega^2 - \omega_c^2} \left[\eta(\omega^2 + \omega_c^2) - 2i\omega_c\omega\eta_H + \omega_c^2 \zeta \right] \right\},$$

$$\sigma_\epsilon = \frac{-1}{m^2(\omega^2 - \omega_c^2)} \left\{ B\bar{n} - \frac{q^2}{\omega^2 - \omega_c^2} \left[2i\omega_c\omega\eta + (\omega^2 + \omega_c^2)\eta_H + i\omega_c\omega\zeta \right] \right\},$$

$$\sigma_{q^2} = \frac{\zeta}{m^2(\omega^2 - \omega_c^2)},$$

$$\sigma_{q^2\epsilon} = 0. \tag{5.124}$$

We note that the conductivities σ_δ and σ_{q^2} are invariant under $B \to -B$ when we simultaneously change $\eta_H \to -\eta_H$, while σ_ϵ changes its sign. These are consistent with the Onsager relations $G^{ij}(B, \eta_H) = G^{ji}(-B, -\eta_H)$ discussed in §2.6.1.

We consider two different limits.

1. Zero magnetic field limit: When $B = \omega_c = 0$, the conductivity tensors reduce to

$$m^2\omega^2 \sigma_\delta = q^2 \eta,$$

$$m^2\omega^2 \sigma_\epsilon = q^2 \eta_H,$$

$$m^2\omega^2 \sigma_{q^2} = \zeta, \tag{5.125}$$

$$m^2\omega^2 \sigma_{q^2\epsilon} = 0.$$

When magnetic field B vanishes, we expect that all the contributions from the current J vanish as well. We can conform this by checking that the results in (5.125) are identical to (5.49) when $m^2\sigma = \kappa$. When there is no electric or magnetic fields, there are only heat or momentum transport.

2. Low frequency limit: We expand to linear order in ω (ζ and η may contain terms $\sim 1/\omega$)

$$\sigma_\delta \simeq -\frac{i\,\bar{n}}{\omega\,m} - \frac{q^2}{B^2}(\eta + \zeta),$$

$$\sigma_\epsilon \simeq \frac{\bar{n}}{B} + \frac{q^2}{B^2}\left\{\eta_H + 2i\frac{\omega}{\omega_c}\left(\eta + \frac{\zeta}{2}\right)\right\}, \qquad (5.126)$$

$$\sigma_{q^2} \simeq -\frac{\zeta}{B^2},$$

$$\sigma_{q^2\epsilon} = 0.$$

Note that the shear and bulk viscosities are related in the first equation, which is in contrast with the relations we have derived above, for example, (5.60) or (5.108). This happens as momentum density and current density are combined with antisymmetric tensor. In both limits (5.125) and (5.126), the results are in agreement with those presented in the reference [98]. These relations hold, for instance, for incompressible Hall fluids, and can be derived in several effective theory descriptions [108, 109].

Before moving on, let us briefly comment a possible generalization regarding the Galilean invariant formula (5.111). When using non-relativistic diffeomorphism invariance along with the broken parity symmetry, the relation (5.111) can be modified with an extra term [109].

$$T^{0i} = mJ^i - \frac{\gamma_0 - 2s}{4}\epsilon^{ij}\partial_j J^0, \qquad (5.127)$$

where γ_0 is the gyromagnetic ratio that determines the coupling to an external magnetic field and s determines the coupling to the spin connection. The term proportional to J^0 produce terms that depend on momentum due to the spatial derivative. Interested readers are encouraged to consult with the references in [99]. If you are also interested in non-relativistic diffeomorphism invariance, the original paper is an excellent source [46].

5.3.5 Ward identities with dissipative terms

Until now, we have assumed that momentum is conserved $\partial_\mu T^{\mu i} = 0$ or $\partial_\mu T^{\mu i}_B = 0$ in the presence of an external magnetic field. This is not necessarily the case in many systems of interest. For example, the microscopic theory is not translation invariant if there is a lattice or impurities with which the degrees of freedom that carry charge and momentum can scatter.

A simple way to model the momentum dissipation is by adding drag terms to the momentum conservation equation

$$\partial_\mu T^{\mu i} = B\epsilon^i{}_j J^j - \lambda_J J^i - \lambda_T T^{0i} . \tag{5.128}$$

This is a purely phenomenological characterization of a system where the effects of the scatterers are to change the momentum T^{0i} without changing the current J^i. The last two terms are motivated by the Drude model discussed in §2.6. This simple approximation is not expected to hold when the effects of the impurities become strong.

It is straightforward to repeat the same steps that we have used to derive the other Ward identities with the new terms. When $B = 0, J^i = 0$, (5.128) can be rewritten as

$$\partial_0 T^{0i} = -\partial_j T^{ji} - \lambda_T T^{0i} . \tag{5.129}$$

Then (5.39) turns into

$$\begin{aligned}\partial'_0 \partial_0 G^{0j0l}(x^\mu;x'^\mu) &= \partial_n \partial'_m G^{njml}(x^\mu;x'^\mu) - \lambda_T^2 G^{0j0l}(x^\mu;x'^\mu) \\ &\quad - \lambda_T \partial_0 G^{0j0l}(x^\mu;x'^\mu) - \lambda_T \partial'_0 G^{0j0l}(x^\mu;x'^\mu) \\ &\quad - \delta'(x^0 - x'^0) C^{0j0l}(x^\mu;x'^\mu) \\ &\quad - \delta(x^0 - x'^0)(\partial_0 - \partial'_0) C^{0j0l}(x^\mu;x'^\mu) ,\end{aligned} \tag{5.130}$$

where we use $\partial_j T^{ji} = -(\partial_0 + \lambda_T) T^{0i}$.

With the time translation symmetry, we can take the Fourier transform by performing the integral on both sides after multiplying $\int d(x^0 - x'^0) e^{i\omega(x^0 - x'^0)}$. Then the second line vanishes. For the homogeneous states, the contact terms vanishes. With the spatial translation symmetry, we also use Fourier transform (5.45). We arrive a simple result

$$(\omega^2 + \lambda_T^2)\tilde{G}^{0j0l}(\omega, \vec{q}) = q_n q_m \tilde{G}^{njml}(\omega, \vec{q}) \,. \tag{5.131}$$

In the presence of the drag term (5.129), the coefficient of the conductivity tensor is modified as $\omega^2 \to \omega^2 + \lambda_T^2$.

In the presence of magnetic field B and current J^i with (5.128), one can compute the Ward identities

$$\begin{aligned}
\omega^2 \tilde{G}^{0j0l}(\omega, \vec{q}) = {} & q_n q_m \tilde{G}_R^{njml}(\omega, \vec{q}) - \lambda_T^2 \tilde{G}_R^{0j0l}(\omega, \vec{q}) \\
& + (i\omega + \lambda_T)(B\epsilon^l{}_m - \lambda_J \delta^l_m)\tilde{G}_R^{0j,m}(\omega, \vec{q}) \\
& + (-i\omega + \lambda_T)(B\epsilon^j{}_n - \lambda_J \delta^j_n)\tilde{G}_R^{n,0l}(\omega, \vec{q}) \\
& - (B\epsilon^j{}_n - \lambda_J \delta^j_n)(B\epsilon^l{}_m - \lambda_J \delta^l_m)\tilde{G}_R^{nm}(\omega, \vec{q}) \,.
\end{aligned} \tag{5.132}$$

Here we consider physical systems with translation invariance and the corresponding contact terms vanish.

5.4 Hall viscosity from Ward identities

We have already mentioned that the Hall viscosity and angular momentum have the same tensor structures in (5.109). From the observation that angular momentum is generated by the operator S_2 (which is a commutator of S_1 and S_3)

$$\begin{aligned}
i[\frac{1}{2}\bar{\sigma}^1_{ij} \int d^2x x^i T^{0j}(\vec{x}), \frac{1}{2}\bar{\sigma}^3_{kl} \int d^2x'^k i T^{0l}(\vec{x}')] \\
= -\frac{1}{2}\bar{\sigma}^2_{ij} \int d^2x x^i T^{0j}(\vec{x}) \,,
\end{aligned} \tag{5.133}$$

given in (5.18)–(5.20), we can think about contracting indices to pick up only the relevant contributions from various Ward identities. We note $\bar{\sigma}^2_{ij} = \epsilon_{ij}$ as given in (5.10).

A non-zero angular momentum requires an expectation value of the momentum density of the form $\langle T^{0i}(\vec{x}) \rangle = (1/2)\epsilon^{ij}\partial_j \ell(\vec{x})$ given in (5.57). This is compatible with rotational symmetry when ℓ is a function of $r^2 = \vec{x} \cdot \vec{x}$. The angular momentum is

$$\langle L \rangle = \int d^2x \, \epsilon_{ij} x^i \langle T^{0j}(\vec{x}) \rangle = \int d^2x \, \ell(r^2) \,, \tag{5.134}$$

where we perform the integration by parts.

We are already familiar with this types of Ward identities as discussed in §5.3.2.3 and §5.3.2.2. To use the results already established in these sections, we introduce a tensor

$$S_{ijkl} = \frac{1}{4} \bar{\sigma}^1_{ij} \bar{\sigma}^3_{kl} \,, \tag{5.135}$$

where the Pauli matrices are given in (5.9) – (5.11). This tensor is motivated by (5.133) to explicitly pick the angular momentum and Hall viscosity contributions. There are several useful algebraic relations that we list here.

$$\begin{aligned}
&S_{ijkl} = S_{jikl} \,, \quad S_{ijkl} = S_{ijlk} \,, \\
&\delta^{ij} S_{ijkl} = 0 \,, \quad \delta^{kl} S_{ijkl} = 0 \,, \\
&S_{ijkl} \delta^{ik} = -\frac{1}{4} \epsilon_{jl} \,, \quad S_{ijkl} \delta^{jl} = -\frac{1}{4} \epsilon_{ik} \,, \\
&S_{ijkl} \epsilon^{ik} = -\frac{1}{4} \delta_{jl} \,, \quad S_{ijkl} \epsilon^{jl} = -\frac{1}{4} \delta_{ik} \,.
\end{aligned} \tag{5.136}$$

We note that S_{ijkl} is symmetric under the first two and last two indices as $\bar{\sigma}^1_{ij}$ is symmetric with the indices (ij), while $\bar{\sigma}^3_{kl}$ has only diagonal components. The identities in the second line come from the fact that the Pauli matrices are traceless.

We consider a neutral case with time translation invariance, whose Ward identity has been worked out in (5.50). Let us contract the tensor S_{ijkl} (5.135) with the equation (5.50). The contact terms in the last two lines yields

$$\int d^2x \, S_{ijkl} \, x^i x^k \left[i\omega e^{jl} \epsilon^n_m \partial_n \langle T^{0m}(\vec{x}) \rangle - \frac{1}{2} \left(\partial_l \partial_n \langle T^{nj}(\vec{x}) \rangle + \partial_j \partial_m \langle T^{ml}(\vec{x}) \rangle \right) \right]$$
$$= -\frac{1}{8} \int d^2x \, x^i x^k \left[i\omega \delta_{ik} \epsilon^n_m \epsilon^m_o \partial_n \partial_o \ell(\vec{x}) \right]$$
$$= \frac{1}{8} \int d^2x \, (\delta^i_n \delta^k_o + \delta^i_o \delta^k_n) \left[i\omega \delta_{ik} \delta^{no} \ell(\vec{x}) \right] \quad (5.137)$$
$$= \frac{i\omega}{2} \int d^2x \, \ell(\vec{x}) \,.$$

The contributions from $\langle T^{nj}(\vec{x}) \rangle$ and $\langle T^{ml}(\vec{x}) \rangle$ vanishes as their tensor structures are symmetric combinations of delta functions, for example $(\delta^{il}\delta^k_n + \delta^i_n\delta^{kl})\langle T^{nj}(\vec{x})\rangle = (\delta^{il}\delta^{kj} + \delta^{ij}\delta^{kl})p$, which vanishes when contracted with S_{ijkl}. To evaluate the first term, we use the last identity given in (5.136) and the relation (5.57), followed by the identity $\epsilon^n_m \epsilon^m_o = -\delta^n_o$ and integration by parts.

Then the Ward identity formula (5.50) becomes

$$0 = i\omega \frac{V_2 \bar{\ell}}{2} + S_{ijkl} \int_{V_2} d^2x \int_{V_2} d^2x' \, G^{ijkl}(\omega, \vec{x}, \vec{x}')$$
$$- \omega^2 S_{ijkl} \int_{V_2} d^2x \int_{V_2} d^2x' x^i x'^k \, G^{0j0l}(\omega, \vec{x}, \vec{x}') \,, \quad (5.138)$$

where we consider a system with a finite volume V_2 and define the average angular momentum as

$$\bar{\ell} = \frac{1}{V_2} \int_{V_2} d^2x \, \ell(\vec{x}) \,. \quad (5.139)$$

Now the tensor structure of the correlator $G^{ijkl}(\omega, \vec{x}, \vec{x}') = -i\omega \eta^{ijkl}(\omega, \vec{x}, \vec{x}')$ is the same as in (5.47) with coefficients that depend on the coordinates. We also define an average viscosity tensor as

$$\bar{\eta}^{ijkl}(\omega) = \frac{1}{V_2} \int_{V_2} d^2x \int_{V_2} d^2x' \, \eta^{ijkl}(\omega, \vec{x}, \vec{x}') \,. \quad (5.140)$$

If the translation invariance was unbroken $\bar{\eta}$ would be the same as the zero momentum viscosity tensor. We work out the contracted quantity $S_{ijkl}\bar{\eta}^{ijkl}$

$$S_{ijkl}\bar{\eta}^{ijkl} = S_{ijkl}\left[\frac{\bar{\eta}_H}{2}(\epsilon^{ik}\delta^{jl} + \epsilon^{il}\delta^{jk} + \epsilon^{jk}\delta^{il} + \epsilon^{jl}\delta^{ik})\right] \quad (5.141)$$
$$= -\bar{\eta}_H(\omega),$$

where only the Hall viscosity contributes as two delta functions contracted with S_{ijkl} vanishes.

Putting these together, we find the following relation between the average Hall viscosity and the average angular momentum density

$$\bar{\eta}_H(\omega) = -\frac{\bar{\ell}}{2} - i\omega S_{ijkl} \frac{1}{V_2} \int_{V_2} d^2x \int_{V_2} d^2x' x^i x'^k G^{0j0l}(\omega, \vec{x}, \vec{x}'). \quad (5.142)$$

This relation indicates that there is a definite relation among Hall viscosity, angular momentum, and a particular combination of conductivity tensor. When a system is gapless and there are low energy carriers, the conductivity tensors contributes. For the rest of the section, we focus on evaluating the last term especially when a system has a gap.

5.4.1 Spectral representation

To evaluate the last term in (5.142), we consider a generic quantum system. We label $|\alpha\rangle$ as one of the eigenstates of the Hamiltonian of the system. We also use $|0\rangle$ for the ground state. We can write the correlation function of two operators as

$$\langle 0|T^{0i}(t,\vec{x})T^{0j}(t',\vec{x}')|0\rangle = \sum_\alpha \langle 0|T^{0i}(t,\vec{x})|\alpha\rangle\langle\alpha|T^{0j}(t',\vec{x}')|0\rangle$$
$$= \sum_\alpha e^{i\epsilon_\alpha(t-t')}\langle 0|T^{0i}(\vec{x})|\alpha\rangle\langle\alpha|T^{0j}(\vec{x}')|0\rangle$$
$$= \int \frac{d\omega}{2\pi} e^{i\omega(t-t')} D^{ij}(\omega,\vec{x},\vec{x}'),$$
$$(5.143)$$

where α labels the complete set of quantum eigenstates of the system. ε_α is the energy of the state $|\alpha\rangle$ and also the difference between the state $|\alpha\rangle$ and the ground state. The Fourier transform of the two-point function is given by

$$D^{ij} = 2\pi \sum_\alpha \delta(\omega - \varepsilon_\alpha) \langle 0|T^{0i}(\vec{x})|\alpha\rangle \langle \alpha|T^{0j}(\vec{x}')|0\rangle \,. \tag{5.144}$$

To evaluate the retarded Green's function, we use the integral representation of step function given by

$$\theta(t - t') = i \int \frac{dk_0}{2\pi} \frac{e^{-ik_0(t-t')}}{k_0 + i\epsilon}, \tag{5.145}$$

where $\epsilon \to 0^+$. Then the time Fourier transform of the retarded correlator has the form

$$\begin{aligned} G^{0j0l}(\omega, \vec{x}, \vec{x}') &= \int \frac{dk_0}{2\pi} \frac{D^{jl}(k_0, \vec{x}, \vec{x}') - D^{lj}(k_0, \vec{x}', \vec{x})}{\omega - k_0 - i\epsilon} \\ &= 2i \int \frac{dk_0}{2\pi} \frac{1}{\omega - k_0 - i\epsilon} \rho^{jl}(k_0, \vec{x}, \vec{x}'). \end{aligned} \tag{5.146}$$

Note that the last term only picks up the imaginary part as

$$\begin{aligned} \rho^{ij}(k_0, \vec{x}, \vec{x}') = 2\pi \sum_\alpha \delta(\omega - \varepsilon_\alpha) \\ \times \operatorname{Im}\left(\langle 0|T^{0i}(\vec{x})|\alpha\rangle \langle \alpha|T^{0j}(\vec{x}')|0\rangle\right). \end{aligned} \tag{5.147}$$

Note that, when $|\alpha\rangle = |0\rangle$, the expectation value of T^{0i} is real, so the ground state contribution drops out of the sum.

We want to evaluate the integral (5.146). For the limit $\omega \to 0$, there are no divergences coming from the integral over k_0 as long as

$$\lim_{\omega \to 0} \rho^{ij}(\omega, \vec{x}, \vec{x}') < \infty. \tag{5.148}$$

This can be checked using the decomposition of the pole in the principal value part and a delta function

$$\frac{1}{k_0 - \omega + i\epsilon} = \mathcal{P} \frac{1}{k_0 - \omega} - i\pi \delta(k_0 - \omega). \tag{5.149}$$

Then, we find that

$$\int \frac{dk_0}{2\pi} \frac{\rho^{jl}(k_0,\vec{x},\vec{x}')}{\omega - k_0 - i\epsilon}$$
$$= \frac{i}{2}\rho^{jl}(\omega,\vec{x},\vec{x}') - \mathcal{P}\int \frac{dk_0}{2\pi} \frac{\rho^{jl}(k_0,\vec{x},\vec{x}')}{k_0 - \omega}. \quad (5.150)$$

Both the principal value and the imaginary term, depending on the spectral density $\rho^{ij}(k_0,\vec{x},\vec{x}')$, contribute to the Hall viscosity. We define

$$F(k_0) = \frac{1}{V_2} S_{ijkl} \int_{V_2} d^2\vec{x} \int_{V_2} d^2\vec{x}' \, x^i x'^k \rho^{jl}(k_0,\vec{x},\vec{x}'). \quad (5.151)$$

Then, the Hall viscosity takes a simple form

$$\bar{\eta}_H(\omega) = -\frac{\bar{\ell}}{2} - 2\omega \mathcal{P}\int \frac{dk_0}{2\pi} \frac{F(k_0)}{k_0 - \omega} + i\omega F(\omega), \quad (5.152)$$

where we consider the neutral case without a magnetic field.

When $\omega F(\omega) \to 0$ as $\omega \to 0$, only the angular momentum density will contribute to the Hall viscosity. This happens if there is an energy gap in the spectrum. We see this as the spectral function has been formally expanded as a sum over energy eigenstates as given in (5.147). Clearly, the energy difference between the states $|\alpha\rangle$ and $|0\rangle$ is non-zero, $\varepsilon_\alpha \neq 0$. Thus, the function $\rho^{ij}(\omega,\vec{x},\vec{x}')$ in (5.147) vanishes at $\omega = 0$ as the delta function is not supported there. Note that there are no special requirements on the form of the spectrum above the gap.

The situation is different at finite temperature, where the spectral function has more complicated form depending on Maxwell's distribution function.

$$\rho^{ij}(\omega,\vec{x},\vec{x}')_T = \pi \sum_{\alpha,\beta;\varepsilon_\alpha \neq \varepsilon_\beta}$$
$$\left[e^{-\frac{\varepsilon_\beta}{T}} \delta(\omega - (\varepsilon_\alpha - \varepsilon_\beta)) \, \text{Im}\left(\langle\beta|V_a^i(\vec{x})|\alpha\rangle\langle\alpha|V_b^j(\vec{x}')|\beta\rangle\right) \right. \quad (5.153)$$
$$\left. + e^{-\frac{\varepsilon_\alpha}{T}} \delta(\omega + (\varepsilon_\alpha - \varepsilon_\beta)) \, \text{Im}\left(\langle\beta|V_b^j(\vec{x}')|\alpha\rangle\langle\alpha|V_a^i(\vec{x})|\beta\rangle\right) \right].$$

We see that if the spectrum is discrete, the spectral function will vanish at zero frequency even at non-zero temperature. Thus the Hall viscosity is proportional to the angular momentum. However, for a continuous spectrum this does not need to be true in general.

If a theory does not have a gap, the relation between the Hall viscosity and the angular momentum depends on the matrix elements of T^{0i}. In a theory with spontaneous symmetry breaking, we can have massless Goldstone bosons separated by an energy gap from other kind of excitations. In such a case the energy-momentum tensor at low energies will be proportional to derivatives of the Goldstone field ϕ

$$T^{0i} \simeq \partial^0 \phi \partial^i \phi, \tag{5.154}$$

in which case one expects the matrix element of the momentum density to be proportional to the energy of the eigenstates

$$\langle 0|T^{0i}|\alpha\rangle \simeq i\varepsilon_\alpha \langle 0|\phi \partial^i \phi|\alpha\rangle. \tag{5.155}$$

Even though the continuous excitations of the Goldstone bosons reach zero energy, this factor would prevent them from contributing to the Hall viscosity.

We confirm that, in the absence of a magnetic field, the static Hall viscosity at zero temperature will be given by Read's formula for any field theory with an energy gap.

$$\bar{\eta}_H = -\frac{\bar{\ell}}{2}. \tag{5.156}$$

This is a universal result and have been attracted much attention in the context of the quantum Hall systems. We see below that this Hall viscosity also appears in the physical systems with skyrmions.

5.4.2 Fermions in magnetic field

In this section we consider a concrete model by taking the Dirac fermions ψ in the presence of a background gauge field A_μ. The system is described by the action $\mathcal{S} = \int d^3x \mathcal{L}$ with the Lagrangian

$$\mathcal{L} = \bar{\psi}(i\hbar c\gamma^\mu D_\mu - mc^2)\psi, \tag{5.157}$$

where the covariant derivative is $D_\mu \psi = (\partial_\mu - ieA_\mu)\psi$ and $D_\mu \bar{\psi} = (\partial_\mu + ieA_\mu)\bar{\psi}$. Here we take $\hbar = c = e = 1$.

The stress energy tensor and current operators are straightforward to evaluate by Noether's procedure explained in §1.4.1.

$$T^\mu_{\ \nu} = -\frac{i}{2}\bar{\psi}\gamma^\mu \overset{\leftrightarrow}{D}_\nu \psi + \frac{1}{2}\delta^\mu_\nu \left(i\bar{\psi}\gamma^\sigma \overset{\leftrightarrow}{D}_\sigma \psi - 2m\bar{\psi}\psi\right), \tag{5.158}$$

$$J^\mu = \bar{\psi}\gamma^\mu \psi.$$

Using the equations of motion

$$(i\gamma^\mu D_\mu - m)\psi = 0,$$
$$iD_\mu \bar{\psi}\gamma^\mu + m\bar{\psi} = 0, \tag{5.159}$$

and the algebra of the gamma matrices

$$\{\gamma^\mu, \gamma^\nu\} = -2\eta^{\mu\nu}\mathbf{1}, \tag{5.160}$$

one can check the conservation equations

$$\partial_\mu T^\mu_{\ \nu} = F_{\nu\sigma}J^\sigma. \tag{5.161}$$

In order to compute the equal time commutators we will use the following identity for the commutator of composite operators.

$$[AB, CD] = A\{B, C\}D - AC\{B, D\} + \{A, C\}DB - C\{A, D\}B, \tag{5.162}$$

and the equal time anti-commutator of two fermions

$$\{\psi_\alpha(\vec{x}), \bar{\psi}_\beta(\vec{y})\} = \delta^{(2)}(\vec{x} - \vec{y})\gamma^0_{\alpha\beta},$$
$$\{\psi_\alpha(\vec{x}), \psi_\beta(\vec{y})\} = \{\bar{\psi}_\alpha(\vec{x}), \bar{\psi}_\beta(\vec{y})\} = 0. \tag{5.163}$$

Let us first compute the equal time commutator between two currents

$$[J^\mu(\vec{x}), J^\nu(\vec{y})] = \bar{\psi}(\vec{x}) \left(\gamma^\mu \gamma^0 \gamma^\nu - \gamma^\nu \gamma^0 \gamma^\mu \right) \psi(\vec{x}) \, \delta^{(2)}(\vec{x} - \vec{y}) \,. \quad (5.164)$$

If any of the currents is the time component J^0 the commutator vanishes, as expected.

The equal time commutator with the momentum density is

$$x^i [T^0_j(\vec{x}), J^\mu(\vec{y})] = i \left(x^i \partial_j J^\mu(\vec{x}) + \delta^i_j J^\mu(\vec{x}) \right) \delta^{(2)}(\vec{x} - \vec{y}) \,. \quad (5.165)$$

Then, for

$$S^0_{Ba}(\vec{x}) = \frac{(\bar{\sigma}_a)^j_i}{2} x^i \left[T^0_j - \frac{B}{2} \epsilon_{jn} x^n J^0 \right] \,, \quad (5.166)$$

the equal time commutator with the current is

$$i \left[S^0_{Ba}(\vec{x}), J^\mu(\vec{y}) \right] = -\frac{(\bar{\sigma}_a)_{ij}}{2} x^i \partial_j J^\mu(\vec{x}) \delta^{(2)}(\vec{x} - \vec{y}) \,. \quad (5.167)$$

Note that for the current S^0_a has the same equal time commutator as S^0_{Ba}. However, the action over the fermionic fields is different. Using

$$[AB, C] = A\{B, C\} - \{A, C\}B, \quad (5.168)$$

we find

$$i \left[S^0_{Ba}(\vec{x}), \psi(\vec{y}) \right] = -\frac{(\bar{\sigma}_a)_{ij}}{2} \left[x^i D_j - \frac{iB}{2} x^i \epsilon_{jn} x^n \right] \psi(\vec{x}) \delta^{(2)}(\vec{x} - \vec{y}) \,. \quad (5.169)$$

The term proportional to B would be absent in the commutator with S^0_a. This term is necessary in order to satisfy the right $SL(2, \mathbb{R})$ algebra. It is easily seen in the symmetric gauge

$$A_i = -\frac{B}{2} \epsilon_{in} x^n \,, \quad (5.170)$$

where the commutator reduces to the usual shear transformation

$$i \left[S^0_{Ba}(\vec{x}), \psi(\vec{y}) \right] = -\frac{(\bar{\sigma}_a)_{ij}}{2} x^i \partial_j \psi(\vec{x}) \, \delta^{(2)}(\vec{x} - \vec{y}) \,. \quad (5.171)$$

Here we used the canonical energy-momentum tensor for simplicity, in principle the shear transformations can be generalized for the symmetric stress energy tensor.

To understand the energy spectrum for the fermions in a magnetic field, the Hamiltonian has the form

$$\mathcal{H} = \frac{\Pi_i^2}{2m}, \qquad \Pi_i = p_i - A_i. \tag{5.172}$$

The 'kinetic' momentum operators satisfy the commutation relations

$$[\mathcal{H}, \Pi_i] = -i\omega_c \epsilon_i^j \Pi_j, \tag{5.173}$$

where we use $[\Pi_i, \Pi_j] = i\epsilon_{ij}B$, which can be easily verified.

In the rotationally invariant gauge the single-particle wavefunctions in the lowest Landau level can be expanded in a basis

$$\psi_n(z) = N_n z^n e^{-B|z|^2/4}, \tag{5.174}$$

where we use the complex coordinate $z = x + iy$.

For the usual definition of the angular momentum operator $L_{xy}^p = xp_y - yp_x$ with p_i the canonical momentum operators, these wavefunctions carry n units of angular momentum. The total momentum of N fermions in the lowest Landau level will be then of order N^2. However, L_{xy}^p is not gauge-invariant and has no direct physical interpretation. A gauge-invariant definition involves the kinetic momentum operator $L_{xy}^\Pi = x\Pi_y - y\Pi_x$. For this operator, the angular momentum is independent of n and is actually -1. For the $\mathcal{N}$th Landau level, the single particle states have angular momentum $-(2\mathcal{N}+1)$. In the case of ν filled Landau levels we can use the fact that each Landau level is equally degenerate, so the average value is

$$\frac{L_{xy}^\Pi}{N} = -\frac{1}{\nu}\sum_{\mathcal{N}=0}^{\nu-1}(2\mathcal{N}+1) = -\nu. \tag{5.175}$$

These are the values that determine the shift.

In our analysis T^{0i} is a gauge-invariant operator, we can see that it is indeed related to the kinetic momentum operators in quantum mechanics. In the presence of the magnetic field the conservation equation is $\partial_\mu T^{\mu i} = B\epsilon^i{}_j J^j$. In addition, in a theory with Galilean invariance $T^{0i} = mJ^i$, in which case we can write the conservation equation as

$$\partial_\mu T^{\mu i} = \omega_c \epsilon^i{}_j T^{0j}. \tag{5.176}$$

The momentum operators $P^i = \int d^2x T^{0i}$ then satisfy

$$\partial_t P^i = \omega_c \epsilon^i{}_j P^j \;\Rightarrow\; i[\mathcal{H}, P^i] = \omega_c \epsilon^i{}_j P^j, \tag{5.177}$$

where $\mathcal{H}$ is the Hamiltonian. This agrees with the commutation relation for the kinetic momentum operators Π_i. Therefore, the angular momentum $L_{xy} = \int d^2x \epsilon_{ij} x^i T^{0j}$ corresponds to L_{xy}^Π and should capture the right value of the shift.

We see that in the presence of a magnetic field not only the angular momentum contributes but there is a term which would be divergent in the infinite volume limit if the density remains constant. This divergence is related to the static Hall conductivity. In the presence of the magnetic field we may extract a contribution from the current correlator of the form

$$G_R^{ij}(\omega;\vec{x},\vec{x}') = i\omega \epsilon^{ij} \frac{\bar{n}}{B} \delta^{(2)}(\vec{x} - \vec{x}') + \hat{G}_R^{ij}(\omega;\vec{x},\vec{x}'), \tag{5.178}$$

where $\bar{n}$ is the average charge density. This leads to

$$\begin{aligned}
\tilde{I}_{13}(\epsilon) = &\int d^2\vec{x}\, e^{-i2\epsilon\cdot\vec{x}} i\omega \left[\langle S_2^0(\vec{x})\rangle - \frac{B}{4} x^2 \left(\langle J^0(\vec{x})\rangle - \bar{n} \right) \right] \\
&+ \int d^2\vec{x}\, d^2\vec{x}'\, e^{-i\epsilon\cdot\vec{x} - i\epsilon\cdot\vec{x}'} S_{ijkl} x^i x'^k \\
&\times \Big[\partial_n \partial'_m G_R^{njml}(\omega;\vec{x},\vec{x}') - i\omega B \epsilon^j{}_n G_R^{n,0l}(\omega;\vec{x},\vec{x}') \\
&\quad + i\omega B \epsilon^l{}_m G_R^{0j,m}(\omega;\vec{x},\vec{x}') - B^2 \epsilon^j{}_n \epsilon^l{}_m \hat{G}_R^{nm}(\omega;\vec{x},\vec{x}') \Big].
\end{aligned} \tag{5.179}$$

So the contact term vanishes when the density is constant as $\langle J^0(\vec{x})\rangle = \bar{n}$. There can also be a diamagnetic term in the current correlator $G_R^{ij} \sim \bar{n}\delta^{ij}$, but it will drop after contracting with S_{ijkl}. Thus the contributions from the density vanish.

Then, the Ward identity for the average Hall viscosity can be written as

$$\bar{\eta}_H(\omega) = -\frac{\bar{k}}{2} + S_{ijkl}\frac{1}{V_2}\int_{V_2} d^2\vec{x}\int_{V_2} d^2\vec{x}'\, x^i x'^k M^{ab}\left\langle V_a^j V_b^l\right\rangle_R, \quad (5.180)$$

where $\bar{k}$ is the full contact term ($\bar{k} = \bar{\ell}$ if $B = 0$). We also define the vector operators

$$V_1^i = T^{0i}, \qquad V_2^i = \epsilon^i{}_n J^n, \quad (5.181)$$

and the matrix

$$M^{ab} = \begin{pmatrix} -i\omega & -B \\ B & -i\frac{B^2}{\omega} \end{pmatrix}. \quad (5.182)$$

Similar to the computation in the previous section, we get

$$\bar{\eta}_H(\omega) = -\frac{\bar{k}}{2} + i\omega F_{11}(\omega) + B\left(F_{12}(\omega) - F_{21}(\omega)\right) + \frac{iB}{\omega}F_{22}(\omega),$$

$$- 2\mathcal{P}\int \frac{dk_0}{2\pi}\left[\omega\frac{F_{11}(k_0)}{k_0 - \omega} - iB\frac{F_{12}(k_0) - F_{21}(k_0)}{k_0 - \omega} - \frac{2B}{\omega}\frac{F_{22}(k_0)}{k_0 - \omega}\right],$$

$$(5.183)$$

where

$$F_{ab}(k_0) = \frac{1}{V_2}S_{ijkl}\int_{V_2} d^2\vec{x}\int_{V_2} d^2\vec{x}'\, x^i x'^k \rho_{ab}^{jl}(k_0, \vec{x}, \vec{x}'), \quad (5.184)$$

with the $\rho_{ab}^{jl}(k_0, \vec{x}, \vec{x}')$ is defined as (5.147) by generalizing T^{0i} as V_a^i. Surely, in the absence of a magnetic field, the formula for the Hall viscosity reduces to (5.152). Thus the discussion for the relation between the Hall viscosity and angular momentum is still valid in the presence of a magnetic field.

In quantum Hall systems and other topological states such as chiral superfluids, the Hall viscosity is proportional to the shift $\mathcal{S}$ [49, 42, 50, 48]. More precisely,

$$\eta_H = \frac{\mathcal{S}}{4}\bar{n}, \quad (5.185)$$

where $\bar{n}$ is the average particle number density. When put on a curved space, the shift determines the change in the number of particles relative to flat space

$$N = \nu^{-1} N_\phi - (1-g)\mathcal{S}, \qquad (5.186)$$

where g is the genus of the two-dimensional surface, ν is the filling fraction for a Hall system (for chiral superfluids $\nu^{-1} = 0$) and N_ϕ is the number of magnetic flux quanta. In the superfluid $\mathcal{S}$ is the orbital angular momentum of the Cooper pair. For free non-relativistic fermions in a magnetic field, it is a mean orbital angular momentum per particle, defined as $\mathcal{S} = 2E_0/\omega_c$, where E_0 is the energy of the ground state and ω_c the cyclotron frequency [98]. We mention that the static Hall viscosity was argued to be exactly (5.185) in a system with a mass gap and no magnetic field in [98].

5.5 Ward identity with topological charge

Until now we have developed the Ward identity for the physical systems without any topological non-trivial configurations. Here we generalize the Ward identity in the presence of such topological objects, especially the skyrmions with broken parity symmetry. The discussion is general and can be applied to the systems with(out) parity symmetry and other symmetries with appropriate modifications.

We first introduce a topological object as a central extension of momentum-momentum commutator. This formulation provides a pathway to include the topological charge in the Ward identities. After introducing the general Ward identities, we specialize two different cases involved with different mutually compatible observables in the next chapter. Along the way, we explain necessary physical quantities, such as topological and skyrmion Hall effects and skyrmion Seebeck effect in Chapter 6.

5.5.1 Topological charge as a central extension

When quantum mechanics came along, the deterministic idea of physics, that all the physical quantities can be completely determined simultaneously to the perfect precision, has shattered. Precise and simultaneous measurements of a particle's position and momentum are not possible at the very fundamental level. This is referred to the uncertainty relation

$$\Delta x^i \Delta p^j \geq \frac{\hbar}{2} \delta^{ij} , \qquad (5.187)$$

that originates from the commutation relations

$$[x^i, p^j] = i\hbar \delta^{ij} , \qquad [x^i, x^j] = [p^i, p^j] = 0 , \qquad (5.188)$$

where $i, j = 1, ..., d$ for d-dimensional space. To present the material in a simpler manner, we set $\hbar = 1$.

In some situations, there even exist further limitations for determining the momenta in different coordinate directions. This is realized by the momentum-momentum commutation relation that can be modified by a central extension [110]. In quantum field theories, we can formulate this as

$$[P^i, P^j] = iC^{ij} , \qquad (5.189)$$

where $P^i = \int d^d x \, T^{0i}(\vec{x})$ is the momentum operator formed from the local energy momentum tensor $T^{0i}(\vec{x})$.

There are several obstructions to have a non-zero C^{ij} due to the Jacobi identities, which are necessarily satisfied. Let us start with the definition of the momentum operator that generates translations

$$[P^i, T^{0j}(t, \vec{x})] = -i\partial_i T^{0j}(t, \vec{x}) . \qquad (5.190)$$

Upon integrating this expression on both sides, we find that the right-hand side is a total derivative. To circumvent this, we need either finite boundary contributions from the boundary or some singularities in T^{0j}. We provide some concrete realizations of these in the later part of this section.

As we are interested in the underlying symmetries, we consider the consistency conditions through the required Jacobi identity. For simplicity, let us consider the set of symmetry generators satisfying the Poincaré symmetry without any non-trivial central extension such as the skyrmion charge. The Poincaré generators satisfy the commutation relations,

$$i[J^{\mu\nu}, J^{\rho\sigma}] = \eta^{\nu\rho} J^{\mu\sigma} - \eta^{\mu\rho} J^{\nu\sigma} - \eta^{\sigma\mu} J^{\rho\nu} + \eta^{\sigma\nu} J^{\rho\mu},$$
$$i[P^{\mu}, J^{\rho\sigma}] = \eta^{\mu\rho} P^{\sigma} - \eta^{\mu\sigma} P^{\rho}, \qquad (5.191)$$
$$[P^{\mu}, P^{\nu}] = 0,$$

where $\mu, \nu, \rho, \sigma = 0, 1, 2, \cdots, d$ and $i, j, k = 1, 2, \cdots, d$ for $d+1$ dimensional space-time, $\eta^{\mu\nu} = \mathrm{diag}\{-1, 1, 1, 1\}$. The Poincaré generators include the homogeneous Lorentz generators $J^{\mu\nu}$, that includes the rotation generators $J_i = (1/2)\epsilon_{ijk} J^{jk}$ and the boost generators $K^i = J^{i0} = -J^{0i}$, and the inhomogeneous translation generators P^{μ}.

When the Jacobi identity involves a boost operator denoted as J^{0i}, energy and a momentum operator,

$$J(J^{0i}, P^0, P^j)$$
$$\equiv [J^{0i}, [P^0, P^j]] + [P^0, [P^j, J^{0i}]] + [P^j, [J^{0i}, P^0]] \qquad (5.192)$$
$$= i[P^j, P^i] = -C^{ji} = C^{ij},$$

where we use $[J^{0j}, P^{\mu}] = i(\delta^{\mu 0} P^j - \delta^{\mu j} P^0)$ and also (5.189). Thus the Jacobi identity is violated by the central extension C^{ij}. In the presence of a boost symmetry, either Lorentz or Galilean, the left-hand side vanishes and thus C^{ij} is not compatible with the boost symmetry. Thus to have the central extension C_{ij} in the system of our interest, the boost symmetry should be broken.

We may also consider the Jacobi identity involving a rotation operator J^{ij} and two momentum operators, P^k and P^l. We also contract the indices l and j to have a simple result

$$\begin{aligned}
\delta_{lj} J(J^{ij}, P^k, P^l) &= \delta_{lj}([J^{ij}, [P^k, P^l]] + [P^k, [P^l, J^{ij}]] + [P^l, [J^{ij}, P^k]]) \\
&= i\delta_{lj}([J^{ij}, C^{kl}] + C^{lj}\delta^{ki} + C^{ki}\delta^{lj} - C^{li}\delta^{kj} - C^{kj}\delta^{li}) \\
&= i(d-2)C^{ki} ,
\end{aligned} \qquad (5.193)$$

where we use $[M^{ij}, P^k] = i(\delta^{ki} P^j - \delta^{kj} P^i)$ and d is the number of spatial dimensions. The left-hand side vanishes in the presence of rotation symmetry. We have a couple of options to satisfy (5.193). We either discard the rotation symmetry or focus on two spatial dimensions. By combining these two results using the Jacobi identity, we choose to consider the physical system with a rotational symmetry, set $d = 2$ henceforth, and discard the boost symmetry.

There is a well known example for the physical system with a central extension we already encountered in this book. We consider a constant background magnetic field B with a modified energy momentum tensor as

$$T_B^{0j} = T^{0j} - \frac{B}{2} \epsilon_n^j x^n J^0 , \qquad (5.194)$$

where J^0 is a charge density operator. This modification turns out to be equivalent to a minimally coupled momentum operator.

$$[P_B^i, T_B^{0j}(t, \vec{x})] = -i\partial_i T_B^{0j}(t, \vec{x}) - iq\epsilon^{ijk} B^k J^0(t, \vec{x}) . \qquad (5.195)$$

There exists a central extension due to the last term, for $J^0 \neq 0$. Boost symmetry is broken by the background magnetic field.

Actually there is more interesting physical system that allows a central extension: a spin system with magnetic skyrmions that is described by a continuous spin configuration $\vec{n}(t, \vec{x})$.

$$\vec{n} = (\sin\Theta(\rho)\cos\Phi(\phi), \sin\Theta(\rho)\sin\Phi(\phi), \cos\Theta(\rho)) , \qquad (5.196)$$

which automatically satisfy $\vec{n}^2 = 1$. We adapt the coordinate system $\vec{x} = (\rho\cos\phi, \rho\sin\phi, z)$.

The Lagrangian density for $\vec{n}$ that admits skyrmions, even without external magnetic field, is given by

$$\mathcal{L} = \dot{\Phi}(\cos\Theta - 1) - \frac{J}{2}\partial_i \vec{n} \cdot \partial_i \vec{n}, \tag{5.197}$$

where $\dot{\Phi}$ is the time derivative of Φ, whose Dirac commutator with conjugate momentum $p_\Phi(t,\vec{x}) = \cos\Theta(t,\vec{x}) - 1$ is given by

$$[\Phi(t,\vec{x}), p_\Phi(t,\vec{x}')] = i\delta^2(\vec{x} - \vec{x}'). \tag{5.198}$$

The energy momentum tensor can be computed as $T^{0i}(t,\vec{x}) = p_\Phi(t,\vec{x})\partial_i\Phi(t,\vec{x})$. Then [111][20]

$$[P^i, T^{0j}(t,\vec{x})] = -i\partial_i T^{0j}(t,\vec{x}) + i\epsilon^{ij}(\epsilon^{kl} p_\Phi(t,\vec{x})\partial_k \partial_l \Phi(t,\vec{x})). \tag{5.199}$$

This system has a central extension in the commutation relations because the field Φ has a vortex singularity at $\Theta = \pi$. The singularity also breaks boost symmetry. Note that the conservation equation $\partial_\mu T^{\mu\nu} = 0$, with $\mu,\nu = 0,1,2$, can be shown to be satisfied explicitly.

For our purpose, it is convenient to use a local version of (5.199)

$$[T^{0i}(t,\vec{x}), T^{0j}(t,\vec{x}')] = i\left(-\partial_i T^{0j} + \partial_j T^{0i} + i\epsilon^{ij} c\right)\delta^2(\vec{x} - \vec{x}'), \tag{5.200}$$

where c is topological charge density. While we have presented this relation in the context of a particular spin model, it holds for skyrmion systems in general, independent of the details such as the form of Lagrangian given in (5.197).

The total skyrmion charge is the integral of the charge density c. To see this, let us consider a particular spin configuration that is composed of up-spins ↑ at infinity, $\cos\Theta(\rho = \infty) = 1$, and down-spins ↓ at the center, $\cos\Theta(\rho = 0) = -1$. We further specify $\Phi(\phi) = m\phi + \gamma$. m represents the winding number along the ϕ direction and γ parameterizes the skyrmions with different phases. See, for example [112]. The topological skyrmions charge is integral of the density

$C^{ij} = \epsilon^{ij} C.$

$$C = \int d^2x\, \vec{n} \cdot \left[\frac{\partial \vec{n}}{\partial x} \times \frac{\partial \vec{n}}{\partial y}\right] = \int_0^\infty d\rho \int_0^{2\pi} d\phi \frac{d\Theta}{d\rho}\frac{d\Phi}{d\phi} \sin\Theta$$

$$= \cos\Theta(\rho)\Big|_0^\infty \Phi(\phi)\Big|_0^{2\pi} = 2 \cdot 2\pi m$$

$$= 4\pi Q\,. \tag{5.201}$$

Here m is the number of 2π rotations of the spin between the infinity and the center, and Q is the skyrmion number. Only $Q = m = \pm 1$ are stable skyrmions.

Magnetic skyrmions of this sort with $\gamma = 1/2$ have been first observed in MnSi using neutron scattering experiment [113] and also in thin films with a real space Lorentz transmission electron microscope in FeCoSi [114]. They are named as Bloch skyrmions. The skyrmions with $\gamma = 0, \pi$ have been also discovered on the Fe monolayer on the Ir(111) by direct image of spin-polarized scanning tunneling microscopy [115]. They are know as Néel skyrmions. Since then skyrmions have been observed in many different materials such as conductor MnSi, semi-conductor $Fe_xCo_{1-x}Si$ and insulator Cu_2OSeO_3 with different geometry including 3 dimensional bulk, 2 dimensional thin film (monolayer) and 1 dimensional nanowires [112].

5.5.2 Topological Ward identity

We are going to present a simple quantum field theory Ward identities including the topological charge given in (5.200) [116]. We use the conservation equation $\partial_\mu T^{\mu\nu} = 0$ as Ward identity is the results of symmetry and conserved equations.

Following the constructions above §5.3.1, we consider the same retarded Green's function (5.38), $G^{0j0l}(x^\mu; x'^\mu) = i\theta(x^0 - x'^0)\langle [T^{0j}(x^\mu), T^{0l}(x'^\mu)]\rangle$. We take the same steps done by taking two time derivatives on this retarded function and evaluate

(5.39). We pause for a moment with the following step

$$\partial_0'\partial_0 G^{0j0l}(x^\mu;x'^\mu) = \partial_n\partial_m' G^{njml}(x^\mu;x'^\mu)$$
$$- [\delta'(x^0 - x'^0) + \delta(x^0 - x'^0)(\partial_0 - \partial_0')]C^{0j0l}(x^\mu;x'^\mu),$$
(5.202)

to note that he contact term

$$C^{0j0l}(x^0,\vec{x},\vec{x}') = i\langle [T^{0j}(x^0,\vec{x}), T^{0l}(x^0,\vec{x}')]\rangle \tag{5.203}$$

has an additional contribution proportional to the topological charge density c in (5.200) compared to the previous cases.

Assuming time translation invariance, we perform a Fourier transform in the time direction by multiplying $\int d(x^0 - x'^0)e^{i\omega(x^0-x'^0)}$ on both sides of (5.202), we arrive at our general result in this simplest case with the skyrmion charge.

$$\omega^2 G^{0j0l}(\omega,\vec{x},\vec{x}') = \partial_n\partial_m' G^{njml}(\omega,\vec{x},\vec{x}')$$
$$+ i\omega e^{jl}\epsilon^n_m \partial_n \langle T^{0m}(\vec{x})\rangle \delta(\vec{x}-\vec{x}')$$
$$- \frac{1}{2}[\partial_l\partial_n\langle T^{nj}(\vec{x})\rangle + \partial_j\partial_m\langle T^{ml}(\vec{x})\rangle]\delta(\vec{x}-\vec{x}')$$
$$- i\omega c^{jl}\delta(\vec{x}-\vec{x}').$$
(5.204)

This is abstract, yet general Ward identities with a topological charge, topological Ward identities. Here $G^{0j0l}(\omega,\vec{x},\vec{x}')$ and $G^{njml}(\omega,\vec{x},\vec{x}')$ are two spatial and four spatial index Green's functions, while $\langle T^{0m}(\vec{x})\rangle$, $\langle T^{nj}(\vec{x})\rangle$, and $\langle T^{ml}(\vec{x})\rangle$ are one point functions. These quantities are specified further in the following sections, depending on the situations in the physical systems of interest. We note the presence of the central charge c^{jl}, that arises from the commutation relations among energy momentum operators [116] based on [99][22]. As advertised before, there are two independent and exclusive cases we can consider depending on symmetries. For simplicity, we focus on the case with translation and rotation invariance in the rest of the section.

5.5.2.1 Rotation and translation symmetries

Starting from the general Ward identity (5.204), we focus on the physical system with the rotation and translation symmetries and examine the consequences of the topological Ward identities. The translation symmetry does not allow any non-trivial spatial dependence for the one point functions. Thus all the contact terms in the second and the third lines in (5.204) vanish. This makes the Ward identity particularly simple. Then, (5.204) becomes

$$\omega^2 G^{0j0l}(\omega, \vec{x}, \vec{x}') = \partial_n \partial'_m G^{njml}(\omega, \vec{x}, \vec{x}') - i\omega e^{jl} c \delta(\vec{x} - \vec{x}'), \tag{5.205}$$

where we use $c^{jl} = e^{jl}c$ in the absence of the parity (mirror) symmetry.

In the presence of the translation symmetry, we perform a Fourier transform using $G^{\mu\nu\alpha\beta}(\omega, \vec{x} - \vec{x}') = 1/(4\pi^2) \int d^2\vec{q}\, e^{i\vec{q}\cdot(\vec{x}-\vec{x}')} \tilde{G}^{\mu\nu\alpha\beta}(\omega, q)$. Then, we further use the rotation symmetry to organize the most general tensor structures for the retarded Green's functions in momentum space as

$$\begin{aligned}\tilde{G}^{0i0k} &= -i\omega \big[\delta^{ik}\kappa_\delta + \epsilon^{ik}\kappa_\epsilon + q^i q^k \kappa_q + (\epsilon^{in} q_n q^k + \epsilon^{kn} q_n q^i)\kappa_{q\epsilon}\big], \\ \tilde{G}^{njml} &= -i\omega[\eta(\delta^{nm}\delta^{jl} + \delta^{nl}\delta^{mj} - \delta^{nj}\delta^{ml}) + \zeta \delta^{nj}\delta^{ml} \\ &\quad + \frac{\eta_H}{2}(\epsilon^{nm}\delta^{jl} + \epsilon^{nl}\delta^{jm} + \epsilon^{jm}\delta^{nl} + \epsilon^{jl}\delta^{nm})], \end{aligned} \tag{5.206}$$

where κ_δ, κ_q, κ_ϵ, and $\kappa_{q\epsilon}$ are the symmetric and antisymmetric thermal (momentum) conductivities, while η, ζ, and η_H are the viscosities. In the presence of the parity symmetry, κ_q, κ_ϵ, and η_H vanish. More explanations have been presented in §5.3.

After a little algebra for (5.205) with the tensor structures (5.206), we obtain

$$\omega^2 [\delta^{jl}\kappa_\delta + \epsilon^{jl}\kappa_\epsilon + q^j q^l \kappa_q + (\epsilon^{jn} q_n q^l + \epsilon^{ln} q_n q^j)\kappa_{q\epsilon}] = \delta^{jl} q^2 \eta + \epsilon^{jl}(c + q^2 \eta_H) + q^j q^l \zeta. \tag{5.207}$$

While this equation seems to give a single Ward identity, there are actually 4 independent tensor structures, and thus 4 independent

Ward identities.

$$\begin{aligned} \omega^2 \kappa_\delta &= q^2 \eta\,, \\ \omega^2 \kappa_\epsilon &= c + q^2 \eta_H\,, \\ \omega^2 \kappa_q &= \zeta\,, \\ \omega^2 \kappa_{q\epsilon} &= 0\,. \end{aligned} \tag{5.208}$$

Thus thermal conductivities are directly related to viscosities except the topological charge. This is simple topological Ward identities for the physical systems with translation and rotation symmetry in momentum space.

The second equation in (5.208) is particularly interesting as it involves with the skyrmion charge density c. We come back to this identity in the following chapter after extending the Ward identity with a conserved current.

5.5.2.2 Rotation and angular momentum

Finally, we discuss the topological Ward identities in the presence of the rotation symmetry and the angular momentum without the translation symmetry. When the system of interest is not translationally invariant, there are additional contributions to the general topological Ward identity (5.204) as the spatial derivatives of the one point functions no longer vanish.

To facilitate our discussion, we consider the contact terms, $\langle T^{0i} \rangle$ and $\langle T^{ij} \rangle$, that appear in (5.204). A particularly interesting one arises with the spontaneously generated angular momentum ℓ [20, 21] in the systems with broken parity. Without the spatial translation symmetry, the momentum generator can develop an expectation value

$$\langle T^{0i} \rangle = \frac{1}{2} \epsilon^{ik} \partial_k \ell\,, \tag{5.209}$$

which is described in (1.148). It turns out that the structure of the term associated with the angular momentum yields a tensor similar to η_H. As a result, η_H in (5.208) is modified as [22]

$$\eta_H \to \eta_H + \frac{\ell}{2}\,. \tag{5.210}$$

The other contact term actually provides $\langle T^{ij}(\vec{x})\rangle = \delta^{ij} p(\vec{x}^2)$, where p is pressure, another universal contribution to the thermodynamic description. Similar to the angular momentum, due to the particular tensor structure, the contribution would replace ζ in the general Ward identities by the combination

$$\zeta \to \zeta - \frac{i}{\omega} p. \tag{5.211}$$

Thus the Ward identity for the case without translation invariance would get these two extra contributions due to the contact terms in (5.204).

Here we consider, for example, the counter part of the second Ward identity given in (5.208) without translation invariance. As the translation symmetry in broken, we keep the spatial dependence and also keep the spatial derivatives in this subsection. Then the Ward identity takes the following form

$$\omega^2 \bar{\kappa}_e = c + \partial^2 \left(\bar{\eta}_H + \frac{\bar{\ell}}{2} \right), \tag{5.212}$$

where $\partial^2 = \partial_i \partial_i$ with $i = 1, 2$. Of course one can take different form as done in §5.3.2.2. Here $\bar{}$ represents the quantity in the coordinate space representation.

Before closing this chapter, we comment on the generalization of the Ward identity in the presence of the external magnetic field B and conserved current J^μ. Then there are more contributions

$$\begin{aligned}\eta_H &\to \eta_H + \frac{\ell}{2}, \\ \zeta &\to \zeta - \frac{i}{\omega}(p - BM),\end{aligned} \tag{5.213}$$

where M is magnetization that can be defined as $\langle J^i \rangle = \epsilon^{ik} \partial_k M$. More details can be found in [116]. Of course, these changes are additional to the charge density ρ.

Chapter 6

Skyrmion Transport in Magnets

As mentioned in the earlier parts of this book, we have a high degree of manipulating power for individual skyrmions in research labs. In particular, we can create, move around, and annihilate individual skyrmions. In this chapter we want to consider conducting and insulating magnetic materials that host skyrmions and demonstrate their transport properties, such as conductivity and viscosity. We first present interesting physical phenomena that have been reported in literature through experimental facts and numerical results based on phenomenological models. We revisit them from the Ward identity point of view to check whether we can offer something new and useful to the community. There are gaps between the advanced experimental advancements and the theoretical Ward identity approaches, which are still under active developments. We comment on the latter throughout this chapter.

6.1 Conducting magnets

Understanding skyrmion motion is essential for revealing their fundamental physical properties, not to mention for using them for any practical applications such as skyrmion racetrack devices.

Skyrmions and Hall Transport
Bom Soo Kim
Copyright © 2023 Jenny Stanford Publishing Pte. Ltd.
ISBN 978-981-4968-34-8 (Hardcover), 978-1-003-37253-0 (eBook)
www.jennystanford.com

Figure 6.1 A schematic figure that explains basic elements of the skyrmion motion under the influence of electric currents through the spin transfer torque. The thick green arrow represents the motion of conduction electrons, while the thick gray arrow does that of skyrmion. The red and blue arrows are emergent magentic and electric fields, explained in the text.

In particular, realizing next-generation storage devices such as skyrmion racetrack requires high precision as skyrmions move at a high speed. One of the early developments for the skyrmion motion is involved with electric currents through the spin transfer torque (STT) introduced in §4.2.1.

Individual skyrmion is made of a bunch of spins and thus is an electrically neutral object. Nevertheless, electric currents turn out to be a useful way to manipulate the skyrmion motion due to the interaction, so-called Hund's rule coupling, between the spins of the skyrmion and moving electrons passing through it. Majority of the experiments on skyrmion motion have been focused on this aspect. It turns out that skyrmions can be moved by an ultra low current density which is five or six orders of magnitude smaller than that is required for domain wall (DW) motion in ferromagnets [117].

With the parity symmetry breaking, skyrmions reveal not only the longitudinal motion, but also the transverse motion called Hall effect. Skymion Hall effect is similar to that of the electron Hall effect described by the Lorentz force law under the magnetic field. With this motivation, we start with the Hall effect measurement of skyrmions. There are two parts to it, topological Hall effect and skyrmion Hall effect. This part of skyrmion Hall effect is formulated in terms of the Thiele equation in §4.4.2.1.

6.1.1 Topological Hall effect

Mobile spins interact with localized spins and vice versa. The extended spin configuration of the skyrmions produces non-trivial physical effects through a coupling with the spins of the conduction electrons in conducting materials. For simplicity, we assume strong Hund's rule coupling that the conduction electron spin is coupled strongly to, and forced to be parallel to, the localized spin at each atomic site [112]. The resulting motion due to the interaction is called spin transfer torque discussed in §4.2.1. This produces several interesting physical consequences. For example, this coupling of the skyrmion spin and that of the conduction electrons leads to the emergent electromagnetic fields $\vec{e}$ and $\vec{b}$ depicted in Fig. 6.1. See also [79][81][118][80].

$$e_i = E_i^e = \vec{n} \cdot (\partial_i \vec{n} \times \partial_t \vec{n}),$$
$$b_i = B_i^e = \frac{1}{2}\epsilon_{ijk}\vec{n} \cdot (\partial_j \vec{n} \times \partial_k \vec{n}),$$
(6.1)

where i,j are spatial indices and $\vec{n} = \vec{M}/M_s$. These emergent electromagnetic fields measure the solid angle for an infinitesimal loop in space time that is captured by the quantum mechanical Berry phase covered by $\vec{n}$. We also have mentioned these explicit higher derivative contributions in the spin torque considered above in (4.37), which produce prominent effects on the magnetic metals.

Here we illustrate the emergent electromagnetic fields by working on a small part of a Bloch skyrmion with up-spins at the edge and down-spins in the middle, illustrated in Fig. 6.2. To figure out the emergent magnetic field $\vec{b}$, we consider a triangle labeled by three points (a, b, c) in the left panel of the figure. As we work on two spatial dimensions, $\partial_z \cdots$, any derivative involved with z coordinate, vanishes as we cannot take a variation along z coordinate. Thus $\vec{b} = b\hat{z}$ according to (6.1). To understand the sign of b, we consider the spins at three corners of the triangle. The corresponding spins are

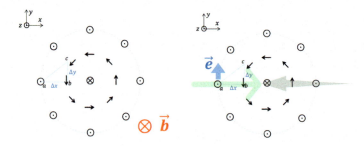

Figure 6.2 Illustrations of the emergent electromagnetic fields. Left: the magnetic field, depicted in red ⊗ symbol, is proportional to topological skyrmion charge. Right: the electric field, in blue arrow, arises due to the relative motion between electron and skyrmion.

$\vec{n}_a = \hat{z}, \vec{n}_b = -\hat{y}$, and $\vec{n}_c = \frac{1}{\sqrt{2}}(-\hat{x} - \hat{y})$. Then,

$$b_z = \epsilon_{zyx}\vec{n} \cdot (\partial_y \vec{n} \times \partial_x \vec{n}) \approx \epsilon_{zyx}\vec{n}_a \cdot \left(\frac{\vec{n}_c - \vec{n}_b}{\Delta y} \times \frac{\vec{n}_b - \vec{n}_a}{\Delta x}\right)$$

$$\propto -\hat{z} \cdot \left(\left[-\frac{1}{\sqrt{2}}\hat{x} + (1 - \frac{1}{\sqrt{2}})\hat{y}\right] \times \left[-\hat{y} - \hat{z}\right]\right) \quad (6.2)$$

$$= -\frac{1}{\sqrt{2}},$$

where we use $\Delta x, \Delta y > 0$. Thus the emergent magnetic field has $-\hat{z}$ direction, indicated as ⊗ in the left panel of Fig. 6.2. One can also check to have the same conclusion for the other parts of the skyrmion as well. This direction is tied with the fact that the depicted Bloch skyrmion is actually an antiskyrmion (with a negative topological charge) according to our definition. The magnetic field can be written directly in terms of the topological charge as

$$\vec{b} = \hat{z}\frac{hc}{e}\frac{1}{4\pi}\epsilon_{zij}\vec{n} \cdot (\partial_i \vec{n} \times \partial_j \vec{n}) = \hat{z}\frac{hc}{e}Q, \quad (6.3)$$

where we restored the fundamental constants with Planck constant h, and Q is the topological charge density. We do not pay much attention whether a skyrmion has a positive or negative topological charges unless that plays an important role. One can also explicitly work out the direction of the emergent magnetic field for the Néel skyrmions.

Conduction electrons passing through the emergent magnetic field is subject to the Lorentz force law, and thus the trajectory bends. This is depicted in the right panel of Fig. 6.2. This is called "topological Hall effect" as the emergent magnetic field $\vec{b}$ is nothing but the topological skyrmion charge.

Through the Hund's rule spin-spin coupling, skyrmion also moves by picking up momentum that comes from moving electrons. In ideal situation, all the momentum lost by electrons will be transferred to skyrmions. Skyrmions have, not only longitudinal motion along the direction of electron motion, but also transverse motion that is opposite to the electron. This transverse motion of skyrmions, as a consequence of momentum conservation, is part of the "skyrmion Hall effect" illustrated in Fig. 6.1.

To illustrate the emergent electric field, we consider an electron moving horizontally with a velocity, $\vec{v}_s = v_s \hat{x}$, with respect to a stationary skyrmion. The time derivative of spin vector can be rewritten in terms of the spatial derivative along the electron moving direction, $\partial_t \vec{n} = (\vec{v}_s \cdot \vec{\nabla})\vec{n} = v_s \partial_x \vec{n}$. From this, we can work out the direction of $\vec{e}$ according to (6.1). It is straightforward to see $e_z = 0$ as the z-derivative vanishes. The electric field along the moving direction turns out to vanish as

$$e_x = \vec{n} \cdot (\partial_x \vec{n} \times \partial_t \vec{n}) \propto \vec{n} \cdot (\partial_x \vec{n} \times \partial_x \vec{n}) = 0 , \qquad (6.4)$$

while the emergent electric field perpendicular to the motion gives

$$\begin{aligned} e_y &= \vec{n} \cdot (\partial_y \vec{n} \times \partial_t \vec{n}) \approx \vec{n}_a \cdot \left(\frac{\vec{n}_c - \vec{n}_b}{\Delta y} \times \frac{\vec{n}_b - \vec{n}_a}{\Delta t} \right) \\ &\propto \hat{z} \cdot \left([-\frac{1}{\sqrt{2}}\hat{x} + (1 - \frac{1}{\sqrt{2}})\hat{y}] \times [-\hat{y} - \hat{z}] \right) \qquad (6.5) \\ &= \frac{1}{\sqrt{2}} . \end{aligned}$$

Thus $\vec{e} = e\hat{y}$ and the electron is pushed toward $-\hat{y}$ direction, which is the same direction of Lorentz force due to $\vec{b}$ and provide an additional Hall contribution [80]. This is illustrated in the right panel of Fig. 6.2.

We can understand this either from the reference frame of electron with the skyrmion moving as the transparent thick gray arrow or from the skyrmion reference frame. $\partial_t \vec{n}$ gives the same results as it only depends on the difference between the final and initial spin vectors. The magnitude of electric field is proportional to the relative velocity $|\vec{v}_s - \vec{v}_d|$, where $\vec{v}_s, \vec{v}_d$ are electron and skyrmion velocities in observer's reference frame. This is still the case even when the electron has a complicated velocity profile. This is illustrated in Fig. 6.1. Emergent electric field is perpendicular to the relative motion between the electron and skyrmion.

Without relative motion between electron and skyrmion, we are able to observe neither topological Hall effect nor skyrmion Hall effect according to (4.51). There is a close relation between these fields, that can be written as

$$\vec{\nabla} \times \vec{e} = -\frac{\partial \vec{b}}{\partial t}. \tag{6.6}$$

This is nothing but one of the Maxwell's equation in electrodynamics. Here the associated velocity is the relative velocity $\vec{v}_d - \vec{v}_s$ between the skyrmion and the electron.

6.1.1.1 Experimental verification of topological Hall effect

The conduction electrons moving to the right (namely along the $\hat{x}$) pass through the emergent magnetic field $\vec{b} = -b\hat{z}$ in Fig. 6.1. This Magnus force makes the electrons bend toward $-\hat{y}$ direction due to the topological Hall effect. Experimentally, this effect has been confirmed to add extra contribution to the Hall resistivity or conductivity due to an external magnetic field $\vec{B}$ [119, 120, 118, 80, 121, 122].

Experimental data for the topological Hall effect are presented in Fig. 6.3. It is highly unusual to see a rectangular-shape contribution as in the the left panel of Fig. 6.3. It turns out that these contributions only exist in the skyrmion phase, which we have considered with Fig. 1.10 in §1.3.3. To see this more clearly, we follow a line labeled by "7" in the left panel of Fig. 6.3 that has an additional constant contribution for a range of an external magnetic field $0.1 \text{ T} < \mu_0 H < 0.42 \text{ T}$ with a unit Tesla (T). The range of the magnetic field for a fixed temperature is identical to skyrmion

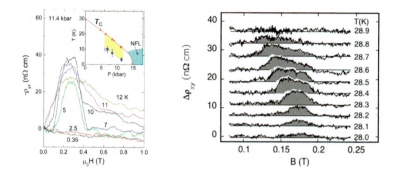

Figure 6.3 Experimental data for the topological Hall effects. Left: a rectangular function-like contribution exists only in the skyrmion phase confirming the existence of skyrmions. Right: only the topological Hall resistivity data are displayed after subtracting Lorentz and anomalous Hall resistivity. The ranges of the magnetic field and temperature correspond to those of skyrmion phase (A-phase) of MnSi. Reproduced with permission from [119][120].

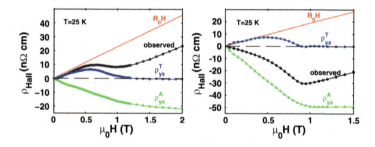

Figure 6.4 Experimental data for three different Hall resistivity. Left: 10nm MnSi think film. Right: 50nm MnSi think film. Reproduced with permission from [122].

phase in the phase diagram. The location of skyrmion phase is slightly different from those reported in [114] due to the applied pressure. The left panel of Fig. 6.3 shows the data points that include other temperature. This extra contribution to the electric resistivity confirms the existence of the emergent magnetic field $\vec{b}$ and thus the existence of the skyrmions.

The right panel of Fig. 6.3 is a slightly different presentation of the topological Hall effect. The Hall resistivity data in the left panel of the figure contains all the contributions as

$$\rho_H = \rho_{nH} + \rho_H^A + \rho_H^T, \tag{6.7}$$

where $\rho_{nH} = R_0 B = R_0 H$ is the normal resistivity term with the Hall coefficient R_0 from the Lorentz force term, $\rho_H^A \propto M$ is the anomalous Hall resistivity induced by the spontaneous magnetization M. The last term ρ_H^T represents the topological Hall resistivity due to the skyrmions, which is presented in the right panel of Fig. 6.3 by subtracting the known resistivity contributions from the total resistivity, $\rho_H^T = \rho_H - \rho_{nH} - \rho_H^A$. In Fig. 6.4, these three different resistivity components are individually plotted as a function of the external magnetic field at a fixed temperature, $T = 25K$ for the same material MnSi. If interested, you are encouraged to read the details in [122].

You might be also curious about mismatches of this temperature $T = 25K$ in Fig. 6.4 compared to the temperature range around $T \sim 28K$ given in the right panel of Fig. 6.4 and also to those in the left panel of Fig. 6.3. The sample for the latter measurements were under extra pressure. The skyrmion phase turns out to extend to almost $T = 0K$ for a thin film reported in [122] and also in other materials [112]. We also nota that there are two different contributions to the topological Hall effect in the metallic thin films due to the emergent magnetic and electric fields as discussed in this section and also in [80].

6.1.2 Skyrmion Hall effect

In §6.1, we have introduced skyrmion Hall effect on the skyrmions in addition to the topological Hall effect on conduction electrons, two related yet different transverse motions. Here we look into the experimental verification of skyrmion Hall effect done in [123][72]. These two papers use different spin torques to reveal the skyrmion Hall effects: spin Hall torque (SHT) of the current in the heavy metal layer [123] and spin-orbit torque (SOT) in the multilayer ferromagnetic system with in-layer current [72]. These spin torque techniques have been explained in §4.2.

Here we start our discussion with the celebrated Thiele equation (4.48) that describes the translation motion of the skyrmion center without macroscopic deformation of its structure. The equation has been introduced in §4.4. In the context of spin Hall torque adapted in [123], it has the following 2 dimensional in-plane form (4.62)

$$\vec{G} \times \vec{v}_d + \alpha \mathcal{D} \cdot \vec{v}_d + 4\pi \mathcal{B} \cdot \vec{J}_{HM} = 0, \tag{6.8}$$

where $\vec{v}_d = (v_{dx}, v_{dy})^T$ is an in-plane drift velocity of the skyrmion, $\vec{J}_{HM}$ is the current in the heavy metal layer (without a current in the ferromagnet layer). Thus, the in-plane spin current due to conduction electrons vanishes, $\vec{v}_s = 0$. As explained in §4.4, the integral of gyromagnetic coupling vector is nothing but the skyrmion charge, $\vec{G} = -4\pi Q \hat{z}$ pointing perpendicular to the plane. The first term in (6.8) captures the Magnus force that results in transverse motion of the skyrmion. This term acts similar to the Lorentz force for electric charge and produces transverse skyrmion motion.

The first term in (6.8) is a vector. To make sense of the other terms, the quantities $\mathcal{D}$ and $\mathcal{B}$ should be second rank tensors. They can have the following general forms in 2 dimensions

$$\mathcal{D} = \begin{pmatrix} D_{xx} & D_{xy} \\ D_{yx} & D_{yy} \end{pmatrix}, \quad \mathcal{B} = \begin{pmatrix} B_{xx} & B_{xy} \\ B_{yx} & B_{yy} \end{pmatrix}. \tag{6.9}$$

The dissipative force tensor $\mathcal{D}$ is related to the damping torque. The tensor $\mathcal{B}$ is proportional to spin Hall angle and quantifies the efficiency of the spin Hall spin torque over the 2-dimensional spin texture of the skyrmion and is in general non-zero as discussed in (4.61). We compute the tensors $\mathcal{D}$ and $\mathcal{B}$ explicitly using a continuous skyrmion model in the next two subsections.

6.1.2.1 A continuous skyrmion model

Let us consider an isolated Néel-type skyrmion

$$\vec{n} = \sin\Theta(\rho)\hat{\rho} + \cos\Theta(\rho)\hat{z}, \tag{6.10}$$

where we set $\Phi = 0$ for Néel-type skyrmion in the general parametrization of the magnetization vector $\vec{n} = \sin\Theta\cos\Phi\hat{\rho} + \sin\Theta\sin\Phi\hat{\rho} + \cos\Theta\hat{z}$. The skyrmion configuration has rotational symmetry and is independent of ϕ in 2-dimensional space (ρ, ϕ) with

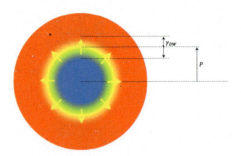

Figure 6.5 Illustration of the continuous skyrmion model with spin-up (red) in outer region, spin-down (blue) in inner region, and domain wall (yellow) interpolating between them. The domain wall has in-plane spins that are pointing outwards.

$\rho = \sqrt{x^2 + y^2}$ or $x = \rho \cos\phi, y = \rho \sin\phi$. To be specific, we consider [123]

$$\Theta(\rho) = \begin{cases} \pi, & \rho - P < -\frac{\gamma_{DW}}{2}, \\ \frac{\pi}{2} - \frac{\rho - P}{\gamma_{DW}}\pi, & -\frac{\gamma_{DW}}{2} \leq \rho - P \leq \frac{\gamma_{DW}}{2}, \\ 0, & \rho - P > \frac{\gamma_{DW}}{2}, \end{cases} \qquad (6.11)$$

where P and γ_{DW} represent the position and width of the domain wall (DW) illustrated in Fig. 6.5. This DW describes the interpolating region of the skyrmion with spin-up vector $\vec{n} = \hat{z}$ in the outer region $\rho > P + \gamma_{DW}/2$ and with spin-down vector $\vec{n} = -\hat{z}$ in the inner region $\rho < P - \gamma_{DW}/2$.

With the continuous skyrmion model given in (6.11), we can compute the dissipative force tensor defined in (4.47)

$$D_{ij} = -\int d^2x \left(\frac{\partial \vec{n}}{\partial x^i} \cdot \frac{\partial \vec{n}}{\partial x^j}\right), \qquad (6.12)$$

where the dot product runs for the components of the vector $\vec{n}$. First, we compute

$$\frac{\partial \vec{n}}{\partial x} = \left(\frac{\partial \rho}{\partial x}\right)\left(\frac{\partial \Theta}{\partial \rho}\right)(\cos\Theta(\rho)\hat{\rho} - \sin\Theta(\rho)\hat{z}),$$

$$\frac{\partial \vec{n}}{\partial x} \cdot \frac{\partial \vec{n}}{\partial x} = \left(\frac{\partial \rho}{\partial x}\frac{\partial \Theta}{\partial \rho}\right)^2 = \frac{\pi^2}{\gamma_{DW}^2}\cos^2\phi, \quad (6.13)$$

where the contribution only survives in the DW region, $-\gamma_{DW}/2 \leq \rho - P \leq \gamma_{DW}/2$. Thus,

$$D_{xx} = -\int_{P-\frac{\gamma_{DW}}{2}}^{P+\frac{\gamma_{DW}}{2}} d\rho\rho \int_0^{2\pi} d\phi \cos^2\phi \left(\frac{\pi^2}{\gamma_{DW}^2}\right)$$

$$= -\pi^3 \frac{P}{\gamma_{DW}}. \quad (6.14)$$

For D_{yy}, the result is the same as $\cos^2\phi$ is replaced by $\sin^2\phi$ whose integral yields the same result. The off diagonal components vanish as $D_{xy} = D_{yx} = 0$ because it has $\sin\phi\cos\phi$ which vanishes upon ϕ integral. Thus,

$$D = D_{xx} = D_{yy} = -\pi^3 \frac{P}{\gamma_{DW}}. \quad (6.15)$$

We note only the diagonal components give non-zero contributions.

6.1.2.2 Computing spin Hall torque

The skyrmion is under the influence of a current $\vec{J}_{HM}$ that is flowing in an adjacent heavy metal layer along the x-direction. Its contribution to the Thiele equation is captured by the last term in (6.8). We already described this term in (4.61) in §4.4. In component expressions,

$$B_{ij} = B_1 \int d^2x \epsilon_{zlj}\epsilon_{lmn}\frac{\partial n_m}{\partial x^i}n_n$$

$$= B_1 \int d^2x \left(\frac{\partial n_j}{\partial x^i}n_z - \frac{\partial n_z}{\partial x^i}n_j\right), \quad (6.16)$$

where the sign of B_1 depends on the spin Hall angle θ_{SH} of a specific heavy metal.

We again use the continuous skyrmion model given in (6.11) to compute B_{ij}. We starts with B_{xx}.

$$\begin{aligned} B_{xx} &= B_1 \int d^2x \left[\frac{\partial n_x}{\partial x} n_z - \frac{\partial n_z}{\partial x} n_x \right] \\ &= B_1 \int d^2x \left[\left(\frac{\partial \rho}{\partial x} \frac{\partial \Theta}{\partial \rho} \right) \cos\phi - \sin\Theta \cos\Theta \left(\frac{\partial \phi}{\partial x} \right) \sin\phi \right] \\ &= B_1 \int_{P-\frac{\gamma_{DW}}{2}}^{P+\frac{\gamma_{DW}}{2}} d\rho\, \rho \int_0^{2\pi} d\phi \cos^2\phi \left(-\frac{\pi}{\gamma_{DW}} \right) \\ &= -\pi^2 P B_1 \equiv B_0\,, \end{aligned} \qquad (6.17)$$

where we use $n_x = \sin\Theta(\rho)\cos\phi$ from $\vec{n} = \sin\Theta(\rho)[\cos\phi\,\hat{x} + \sin\phi\,\hat{y}] + \cos\Theta(\rho)\hat{z}$ in the first line. The second term in the second line can be shown to vanish as follows:

$$\begin{aligned} -\int d^2x \sin\Theta \cos\Theta \left(\frac{\partial \phi}{\partial x} \right) \sin\phi &= \pi \int d\rho \sin\Theta\cos\Theta \\ &= -\gamma_{DW} \int_{DW} \sin\Theta\, d(\sin\Theta) = -\frac{\gamma_{DW}}{2}\left(1-\cos^2\Theta\right)\Big|_{\Theta=\pi}^{\Theta=0} \\ &= 0\,, \end{aligned} \qquad (6.18)$$

where $\partial\phi/\partial x = -\sin\phi/\rho$ and the integral over ϕ gives π. We also used $d(\sin\Theta) = \cos\Theta \times (d\Theta/d\rho) = (-\pi/\gamma_{DW})\cos\Theta$, that only contributes to the integral in the domain wall region.

From the symmetry of the problem and explicit computation, we can easily see $B_{xx} = B_{yy} = B_0$. Similar computations show that the off-diagonal components, $B_{xy} = B_{yx}$, vanish. For example,

$$\begin{aligned} B_{xx} &= B_1 \int d^2x \left[\frac{\partial n_y}{\partial x} n_z - \frac{\partial n_z}{\partial x} n_y \right] \\ &= B_1 \int d^2x \left[\left(\frac{\partial \rho}{\partial x} \frac{\partial \Theta}{\partial \rho} \right) \sin\phi - \sin\Theta\cos\Theta \left(\frac{\partial \phi}{\partial x} \right) \cos\phi \right] \quad (6.19) \\ &\propto \int d\phi\, \sin\phi \cos\phi = 0\,, \end{aligned}$$

where we use $n_y = \sin\Theta(\rho)\sin\phi$. Thus we see that only diagonal components survive for the tensors $\mathcal{D}$ and $\mathcal{B}$.

6.1.2.3 Skyrmion Hall angle

Now we are ready to solve the Thiele equation (6.8) to see the skyrmion Hall motion quantitatively. In the matrix form, the equation reads

$$\begin{pmatrix} \alpha D & -4\pi Q \\ 4\pi Q & \alpha D \end{pmatrix} \begin{pmatrix} v_{dx} \\ v_{dy} \end{pmatrix} = 4\pi \begin{pmatrix} B_0 & 0 \\ 0 & B_0 \end{pmatrix} \begin{pmatrix} J_0 \\ 0 \end{pmatrix} \tag{6.20}$$

where we use the current in the heavy metal layer along x direction, $\vec{J}_{HM} = J_0 \hat{x}$. We also use $\vec{G} \times \vec{v}_d = 4\pi Q(v_{dy}\hat{x} - v_{dx}\hat{y})$, combine the first two terms, and arrange them so that we can solve it with a matrix algebra.

By inverting the matrix, we get

$$\begin{pmatrix} v_{dx} \\ v_{dy} \end{pmatrix} = 4\pi \begin{pmatrix} \alpha D & -4\pi Q \\ 4\pi Q & \alpha D \end{pmatrix}^{-1} \begin{pmatrix} B_0 & 0 \\ 0 & B_0 \end{pmatrix} \begin{pmatrix} J_0 \\ 0 \end{pmatrix}$$

$$= \frac{4\pi B_0}{\alpha^2 D^2 + (4\pi)^2 Q^2} \begin{pmatrix} \alpha D J_0 \\ -4\pi Q J_0 \end{pmatrix}. \tag{6.21}$$

Thus we solve the Thiele equation to get the drift velocities for skyrmion motion. Both components are proportional to the spin Hall torque and to the current of the adjacent heavy fermion layer.

One illuminating and convenient quantity is the so-called skyrmion Hall angle, the relative angle of the transverse velocity compared to the longitudinal velocity. It is given by the ratio of these two in-plane velocities. Thus,

$$\Theta_{SkH} = \tan^{-1}\left(\frac{v_{dy}}{v_{dx}}\right), \quad \frac{v_{dy}}{v_{dx}} = \frac{-4\pi Q}{\alpha D}. \tag{6.22}$$

Note that this is independent of B_0 and thus the spin Hall angle θ_{SH}. Here α is the Gilbert damping parameter.

6.1.2.4 Experimental verification of skyrmion Hall effect

The top right panel of Fig. 6.6 demonstrates the skyrmion Hall effect [123] with the transverse component of velocity under the electric current in the heavy metal layer. The skyrmion Hall effect is further confirmed when the magnetic field is reversed. The reversal of the magnetic field changes the magnetization direction of the underlying

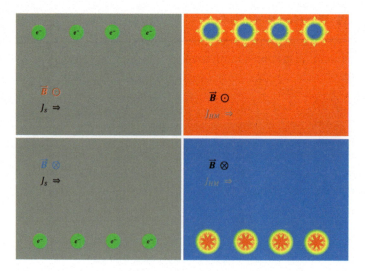

Figure 6.6 An experimental verification of the skyrmion Hall effect through the accumulation of the skyrmions in one side of the film. Left panel illustrates the Hall effect for electrons with electronic charge $-e$ that accumulate at the opposite edges of the film upon reversal of the magnetic field direction. Reversing the magnetic field does not change the sign of the carrier. Right panels illustrate the skyrmion Hall effect, for which the reversal of the magnetic field direction reverses the sign of the topological charge from $Q = -1$ to $Q = +1$ leading to the accumulation of skyrmions of opposite topological charges at the opposite edge.

ferromagnetic material. Note also the change of the spin directions in the interpolating region between up- and down-spins. Thus the sign of the topological charge is changed. Then the skyrmion Hall angle given in (6.22) changes its sign, and the direction of the transverse skyrmion motion is reversed. This happens because the topological charge is odd function of the magnetization $\vec{n}$ which is clear in (5.201).

The Thiele equation (6.8) and the corresponding analysis done in this section include neither possible pinning effects due to the material imperfections, nor the internal degrees of freedom that can be excited and modify the dynamics of skyrmions. We note here that these results in [123] are associated with the Néel-type skyrmion in an asymmetric trilayer material, Ta(5 nm)/Co$_{20}$Fe$_{60}$B$_{20}$(CoFeB)

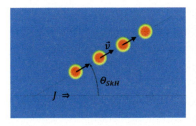 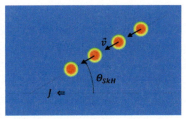

Figure 6.7 Real time dynamic imaging of the skyrmion motion. It is clear to observe the longitudinal and transverse motion of Skyrmions. The Hall angle Θ_{SkH} given in (6.22) can be measured directly.

(1.1 nm)/TaO$_x$ (3 nm), which shows a considerable creep motion influenced by pinning potential of randomly distributed defects. The skyrmions are electrically generated and do not show skyrmion Hall effect with low current densities ($J_0 < 1 \times 10^9 A/m^2$ that is normalized by the total thickness of Ta (5 nm) and CoFeB (1.1 nm)). Interested readers are encouraged to read the reference [123], especially Fig. 3, which contains the systematic skyrmion Hall angle data for two different applied magnetic fields as a function of an applied current J_0.

It is worthwhile to mention that there is another independent skyrmion Hall measurement [72] using a different technique, spin orbit torque (SOT) introduced in §4.2.3. Here, the corresponding skyrmion is also a Néel-type one. Thus we can work out the detailed computations similar to (6.20) of the skyrmion motion using (4.59) as well as the field-like SOT contribution described by $\mathcal{T}_{ij}^2$ in (4.58). We leave its explicit verifications to the interested readers.

Here we mention several interesting aspects of the findings in [72]. The experiment is done for a thin multi-layers of [Pt(3.2 nm)/CoFe(0.7 nm)/MgO (1.4 nm)]$_{15}$ with a current density $4.2 \times 10^{11} A/m^2$ for the Pt and CoFe layers. This skyrmion exists at room temperature. The Hall effect is precisely measured by direct dynamics imaging technique, that reveals the dynamic velocities and displacement direction in real time. See the illustration, Fig. 6.7, of actual measurement of the skyrmion Hall angle Θ_{SkH} defined in (6.22).

It is particularly interesting to note that the skyrmion motion in this material is very efficient so that the measured velocities are those predicted for the perfect pinning-free systems. As mentioned in (4.27) in §4.2.3, field-like SOT plays an important role for skyrmion motion. The skyrmion mass scale is inversely related with the rigidity of the spin structure, which turns out to be very high due to the strong Dzyaloshinskii–Moriya interaction (DMI). The effective skyrmion mass of the material is small, and the velocity of the skyrmions follows the AC current without any noticeable delay. Thus, this material is shown to exhibit excellent homogeneity and a very low pinning effect. Skyrmions without pinning or impurity effects have been also discussed in [84]. To have efficient performances in controlling the skyrmion motion, such as next-generation skyrmion racetrack devices, it is desirable to have this type of clean materials.

6.2 Ward identity for conducting magnets

After discussing some basic experimental facts on conducting magnets with skyrmions, we consider the topological Ward identities with rotation and translation symmetries considered in §5.5.2.1. As there are conduction electrons that interact with skyrmions, we need to extend the Ward identities by including the current contributions similar to §5.3.3. In the presence of the topological charge, the generalization of the Ward identity can be done in a straightforward manner [116]. This generalization is explained in §6.2.1. Then, we discuss the consequences of the resulting Ward identity in the following subsections §6.2.2 at zero momentum and §6.2.3 at finite momentum, where we propose a way to measure the Hall viscosity.

6.2.1 With magnetic field and electric current

The topological Ward identities (5.204) in an abstract form and (5.207) with translation and rotation symmetries in momentum space only capture the conservation equations related to the energy momentum tensor. Here we generalize this Ward identity by including both a background magnetic field and an electric current.

Electric currents turn out to be useful for controlling the skyrmion motion due to a strong interaction between the skyrmion spins and electron spin.

In the presence of the magnetic field B and the electric current J^μ, the conservation equations and momentum generators are modified as follows in 2+1 dimensions as described in (5.84):

$$\partial_\mu T^{\mu i} = B\epsilon^i{}_j J^j, \qquad (6.23)$$

where the equation is from the general expression $\partial_\mu T^{\mu\nu} = F^{\nu\rho} J_\rho$ for both energy momentum tensor and current. The time component of the conserved generator can be refined as in (5.85).

$$T_B^{0j} = T^{0j} - \frac{B}{2}\epsilon^j{}_n x^n J^0 . \qquad (6.24)$$

This can be thought as a minimal coupling substitution in the presence of an electromagnetic vector potential.

These modifications produce several changes for the Ward identities. When one uses the conservation equation $\partial_0 T^{0i} = -\partial_m T^{mi} + B\epsilon^i{}_m J^m$, the momentum-momentum correlator $G^{0j0l} \sim \langle [T^{0j}, T^{0l}]\rangle$ on the left-hand side of (5.205) also includes $G^{0j,m} \sim \langle [T^{0j}, J^m]\rangle$, $G^{n,0l} \sim \langle [J^n, T^{0l}]\rangle$, and $G^{nm} \sim \langle [J^n, J^m]\rangle$. Similar to the thermal conductivities, all these two spatial index Green's functions in the momentum space are the conductivities: thermal conductivities $\kappa \sim \langle [T, T]\rangle$, thermoelectric conductivities $\alpha \sim \langle [T, J]\rangle$ and $\alpha^* \sim \langle [J, T]\rangle$, and also the familiar electric conductivities $\sigma \sim \langle [J, J]\rangle$. On the right-hand side of (5.205), we have only one change, the addition of $B\langle J^0\rangle$, which is nothing but the magnetic field times the electric charge density $\langle J^0\rangle = \rho$.

While the computation is slightly more complicated, it is straightforward. After some algebra, we obtain the topological Ward identities in the presence of the translation and rotation symmetry.

$$\begin{aligned}\omega^2 G^{0j0l} - i\omega B\epsilon^j{}_n G^{n,0l} + i\omega B\epsilon^l{}_m G^{0j,m} + B^2 \epsilon^j{}_n \epsilon^l{}_m G^{nm} \\ = \partial_n \partial'_m G^{njml} - i\omega \epsilon^{jl}\left[c - B\langle J^0\rangle\right]\delta^{(2)}(\vec{x} - \vec{x}') .\end{aligned} \qquad (6.25)$$

After performing the Fourier transform and using the index structures similar to (5.46) and (5.47), we obtain the Ward identities using the thermoelectric conductivities.

$$\begin{aligned}
&\delta^{jl}\left[\omega^2\kappa_\delta + i\omega B\left(\alpha_\epsilon + \alpha_\epsilon^* + q^2[\alpha_{q\epsilon} - \alpha_{q\epsilon}^*]\right) + B^2\left(\sigma_\delta + q^2\sigma_q\right)\right] \\
&+ \epsilon^{jl}\left[\omega^2\kappa_\epsilon - i\omega B\left(\alpha_\delta + \alpha_\delta^* + \frac{1}{2}q^2[\alpha_q + \alpha_q^*]\right) + B^2\sigma_\epsilon\right] \\
&+ q^j q^l\left[\omega^2\kappa_q - 2i\omega B(\alpha_{q\epsilon} - \alpha_{q\epsilon}^*) - B^2\sigma_q\right] \\
&+ (\epsilon^{jo}q^l + \epsilon^{lo}q^j)q_o\left[\omega^2\kappa_{q\epsilon} + \frac{i}{2}\omega B(\alpha_q - \alpha_q^*) - B^2\sigma_{q\epsilon}\right] \\
&= \epsilon^{jl}\left[c - B\rho + q^2\eta_H\right] + \delta^{jl}q^2\eta + q^j q^l\zeta\,.
\end{aligned} \qquad (6.26)$$

This is the general topological Ward identity in the presence of charge density $\langle J^0\rangle = \rho$ and the magnetic field B. While the expressions are a little complicated, the result is quite illuminating. The left-hand side has four tensor structures that are nicely packaged along with all possible combinations of the allowed conductivities depending on the tensor structures. Thus there are four independent Ward identities that can be readily extracted from the expression. For the Ward identity with the parity breaking hydrodynamics and without the topological charge, $c = 0$, we have a consistency check that these results reduce to (5.108).

To go a little further, we isolated the momentum independent parts of the Ward identities (6.26). Then only terms with the tensor structures δ^{jl} and ϵ^{jl} contribute. Thus,

$$\begin{aligned}
\omega^2\kappa_\delta^{(0)} + i\omega B(\alpha_\epsilon^{(0)} + \alpha_\epsilon^{*(0)}) + B^2\sigma_\delta^{(0)} &= 0\,, \\
\omega^2\kappa_\epsilon^{(0)} - i\omega B(\alpha_\delta^{(0)} + \alpha_\delta^{*(0)}) + B^2\sigma_\epsilon^{(0)} &= c - B\rho\,,
\end{aligned} \qquad (6.27)$$

where the superscript $^{(0)}$ denotes the momentum independent part. When we take $B \to 0$, these expressions reduce to

$$\omega^2\kappa_\delta^{(0)} = 0\,, \qquad \omega^2\kappa_\epsilon^{(0)} = c\,, \qquad (6.28)$$

which we have derived in (5.208) and (5.207). In the opposite limit $B \to \infty$ without the topological charge, we have

$$\sigma_\delta^{(0)} = 0\,, \qquad \sigma_\epsilon^{(0)} = -\frac{\rho}{B}\,. \qquad (6.29)$$

This result is consistent with the known results of the Hall conductivity from the dyonic black hole with an appropriate identification of the sign convention using the AdS/CFT correspondence [22][104].

For the non-zero momentum, there are four independent relations connecting viscosities and conductivities as in the neutral case. Here we only consider the parity odd contributions with the tensor structure ϵ^{jl} that are related to the Hall transports and the Hall viscosity.

$$q^2 \eta_H = \omega^2 \bar{\kappa}_\epsilon + B^2 \bar{\sigma}_\epsilon$$
$$- i\omega B \left[\bar{\alpha}_\delta + \bar{\alpha}_\delta^* + q^2 \frac{\alpha_q + \alpha_q^*}{2} \right], \qquad (6.30)$$

where the bar $^-$ indicates the momentum dependent part that is defined as, for example, $\bar{\kappa}_\epsilon = \kappa_\epsilon - \kappa_\epsilon^{(0)} = q^2 \kappa_\epsilon^{(2)} + q^4 \kappa_\epsilon^{(4)} + \cdots$. Thus, when we further take the limit $q^2 \to 0$ for (6.30), we get

$$\eta_H = \omega^2 \kappa_\epsilon^{(2)} + B^2 \sigma_\epsilon^{(2)}$$
$$- i\omega B \left[\alpha_\delta^{(2)} + \alpha_\delta^{*(2)} + \frac{\alpha_q^{(0)} + \alpha_q^{*(0)}}{2} \right]. \qquad (6.31)$$

This Ward identity indicates that the Hall viscosity is related to several different thermoelectric conductivities, not to mention its dependence on the frequency ω. In the following subsections, we revisit this to find simpler ways to seek the signatures of the presence of the Hall viscosity.

6.2.2 Modeling the Hund's rule coupling

The Ward identities (6.26) have both the momentum and charge responses through the stress energy tensor and conserved current operator. These two are connected by the relations (6.24). Nevertheless, the interactions between the skyrmion spins and conduction electron spins are not incorporated into the Ward identity. In this section we model this by using the Hund's rule coupling.

We have discussed the surprising experimental measurements in the presence of the topological skyrmions in §6.1.1.1, where the electric Hall conductivity or Hall resistivity shows a unique rectangular-shape signature in the so-called A-Phase of MnSi [119][120]. Similar experimental results have been confirmed later in [118].

How does this happen? Skyrmions are electrically neutral objects because they are made from bunch of spins tightly arranged. Then why do they display electric responses? This happens through the interactions between the spins of skyrmions and those of conduction electrons that are modeled by the ferromagnetic spin coupling [80][112]. In the strong coupling limit, the spin wave functions of the conduction electrons are identified with that of the localized spins $\vec{n}(x^\mu)$ of the skyrmions. This limit is described by a tight binding model with Hund's rule coupling.

There is an equivalent way to say this. The skyrmionic spin configurations create an emergent magnetic field $\vec{b}$ with magnitude $b = c/2$ for the conduction electrons due to the tight binding interactions between their spins, where c is the topological charge density of skyrmions.

This Hund's rule coupling is not built in the Ward identities. One can model this coupling by combining the energy momentum tensors and electric currents through the modification of the parameters existing in the topological Ward identities. This produces the effects of the interaction between the thermal and charge responses. In particular, we note that the skyrmion charge density c produces an emergent magnetic field $b = c/2$ [112], which can change the dynamics of conduction electrons. This is also the case for the magnetic field B, which is modified in the presence of $\boldsymbol{b}$. For simplicity, we assume that the emergent magnetic field is homogeneous and constant, which is true for most of practical measurements. For the flip side, the motion of the conduction electrons will also influence the thermal response of the skyrmions due to the tight binding.

After taking these effects into account, the second equation of the Ward identity (6.27) at the vanishing momentum becomes

$$\omega^2 \kappa_e^{(0)} - i\omega B_b(\boldsymbol{\alpha}_\delta^{(0)} + \boldsymbol{\alpha}_\delta^{*(0)}) + B_b^2 \sigma_e^{(0)} = c_b - B_b \rho \,. \tag{6.32}$$

This Ward identity is of the same form as (6.27), with the modifications $B \to B_b \equiv B + b$, which contributes to the charge response, and $c \to c_b \equiv c + c_{el}$, which incorporates an additional contribution to the thermal response from the conduction electrons c_{el}. Although there are these modifications, the topological charge density c and the background magnetic field B do not alter. The quantities c and b are constants and independent of B, while c_{el} is expected to be proportional to B and depends on the tight binding strength. We expect that b, c, c_{el} can be readily measured experimentally. In particular, b can be identified from a rectangular-shape signature in the Hall conductivity σ_e [119][120][118][121][122], as one passes into and out of the A-Phase, in which the skyrmions develop a finite density c. Such behavior will also confirm the presence of a nonzero density c, which will likewise produce a similar rectangular-shape contribution in the thermal Hall conductivity κ_e (with an additional B-dependent c_{el}) when experimental setup sweeps the magnetic field B or the temperature T independently.

If the ferromagnetic binding between the skyrmion and conduction electron spins would not exist, the electric Hall conductivity would only pick up contributions from the conduction electrons, and B_b would reduce to B. On the other hand, the thermal Hall conductivity would include both contributions, c and c_{el}, with the latter being independent of B. Thus the tight binding and the corresponding strength are readily verifiable.

6.2.3 Hall viscosity in conducting materials

Once the experimental setup described in §6.2.2 is well established, one can move on to the momentum dependent Ward identities that is described in (6.30). With this equation and proper modifications considered in §6.2.2, we propose a simple way to measure the Hall viscosity in the conducting materials.

We would like to remove the ω dependence from the Ward identity relation. To do so, let us divide equation (6.30) by the second equation of (6.27), followed by substituting $B \to B_b, c \to c_b$ to include the effects of the Hund's rule coupling as discussed in the previous section. The expression is still complicated. To make things more clear, we take the limit $\omega/B_b \to 0$ and the zero momentum limit $q^2 \to 0$. We expect these approximations are straightforward and reliable as discussed the results related to (6.29). Then, we obtain a simple expression.

$$\eta_H = (c_b - B_b\rho) \frac{\sigma_e^{(2)}}{\sigma_e^{(0)}} . \tag{6.33}$$

Once the electric Hall conductivity σ_e is measured as a function of q^2 in the presence of the rotation symmetry, the Hall viscosity is nothing but the modified skyrmion density $c_b - B_b\rho$ multiplied by the ratio between the slope $\sigma_e^{(2)}$ and σ_e-intercept $\sigma_e(q^2 = 0)$. See Fig. 6.8. We note that one can apply this to the physical systems without topological objects, such as quantum Hall systems, where $c_b = c = 0$ and $B_b = B$. This is quite different from the neutral case, which is not applicable in the absence of the topological charge as the right side of the second equation of (6.27) vanishes. Indeed, similar relation between the Hall viscosity and Hall conductivity for the Galilean invariant qantum Hall systems has been proposed in [43].

We come back to the neutral case in §5.4. It turns out that in the presence of the topological charge, the neutral case at zero temperature has much cleaner picture that is described in (6.84) in §6.4.1.

Let us consider the opposite limit $B_b/\omega \to 0$ after dividing equation (6.30) by the second equation of (6.27), followed by substituting $B \to B_b, c \to c_b$. Then η_H reduces to

$$\eta_H = (c_b - B_b\rho) \frac{\kappa_e^{(2)}}{\kappa_e^{(0)}} . \tag{6.34}$$

Here κ_e is the thermal Hall conductivity. Note that this identification of η_H can also be applied to the systems without skyrmions due to the presence of the magnetic field and density $B_b\rho$. This is different from (6.84) with the modification $c \to c_b - B_b\rho$ due to the presence of the conduction electrons and the magnetic fields B. There can be another figure for this by replacing σ to κ in Fig. 6.8. It will be interesting and exciting to verify the Hall viscosity signature from the electric or thermal Hall conductivity data. Again we mention that the neutral case considered in (6.84) has much simpler result as there is no conduction electrons.

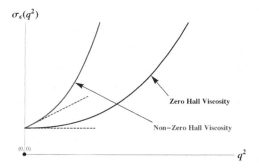

Figure 6.8 Confirmation of the existence of the Hall viscosity in the conducting magnetic materials. Intercept of electric Hall conductivity $\sigma_\epsilon^{(0)}$ is non-zero and is proportional to the skyrmion charge density. Non-vanishing slope $\sigma_\epsilon^{(2)}$ as a function of momentum squared q^2 confirms the existence of the Hall viscosity.

Before moving on, we briefly mention a new possibility to have a definite relation between the thermal and charge responses by imposing a relation between the stress energy tensor and charge current as

$$T^{0i} = \mu J^i, \tag{6.35}$$

which is similar to the Galilean invariant case. Then it is straightforward to get the corresponding Ward identity by using the results in §5.3.4 with rotation and translation invariance. In particular, there are definite relations between various conductivity tensors as

$$\kappa = \mu\alpha = \mu\alpha^* = \mu^2\sigma. \tag{6.36}$$

This leads the distinct results on the conductivity. In the language of the electric conductivity σ, we have

$$\sigma_\delta^{(0)} = -i\frac{\omega_c}{\omega}\frac{c - B\rho}{\mu^2(\omega^2 - \omega_c^2)},$$

$$\sigma_\epsilon^{(0)} = \frac{c - B\rho}{\mu^2(\omega^2 - \omega_c^2)}, \tag{6.37}$$

where $\omega_c = B/\mu$. For small magnetic field $\omega_c \ll \omega$, the Hall conductivity is directly related to the topological charge density

$\sigma_\delta^{(0)} \approx 0, \sigma_\varepsilon^{(0)} \approx c/\mu^2\omega^2$. In the opposite limit with large magnetic field $\omega_c \gg \omega$, $\sigma_\delta^{(0)} \approx -i\rho/\mu\omega, \sigma_\varepsilon^{(0)} \approx 0$. Such distinctive behaviors can be easily measured.

6.3 Insulating magnets

Here we consider the skyrmion motion in the electrically insulating magnets. There exist low energy excitations, called magnons, in magnetic materials that we need to take into account to properly describe the skyrmion motion. As magnons excitations always present above the temperature where they can be excited, the skyrmion dynamics in insulating magnets, in some ways, are more fundamental than those in conducting magnets, where skyrmion motion is dominated by its interaction with electrons. Nevertheless, skyrmion motion in insulating magnets is less developed as we don't understand magnons as much as we do electrons. Fortunately, understanding magnon dynamics has been under active developments along with the advancement of the field of spintronics. Below we describe some surprising skyrmion motions in insulating magnets.

6.3.1 Magnon Hall effect

A ferromagnetic material has a ground state with all the spins are aligned in one direction as in the top panel of Fig. 6.9. At finite temperature, there exists a low energy excited states called spin waves. Magnon, quantized spin wave, is a quasi-particle or a collective excitation of the spin structure in a crystal lattice. It is illustrated as a wave of neighboring electron spins precessing at the same frequency but with a different phase as shown in bottom panel of Fig. 6.9. They arise because neighboring electron spins interact strongly in a magnet, making it energetically favorable to excite the collective magnon mode rather than flipping a single spin as in the middle panel.

6.3.1.1 Magnon and electric currents

Before focusing on insulating magnet, we describe two different spin currents and also spin pumping mechanism using inverse spin Hall

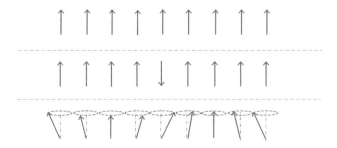

Figure 6.9 Excitations of a ferromagnetic material. Top: the configuration represents the ground state of a ferromagnet. Middle: A spin flip in the middle represents a high energy excitation. Bottom: illustration of a spin wave or quantized magnon that has a lower energy than a spin flip. The spins are precessing around their equilibrium positions.

effect introduced in §4.2.5.1. In solid there are two different types of non-equilibrium spin currents as illustrated in Fig. 6.10 [124][125]. The first is the conduction electron spin current, the flow of net spin angular momentum that is carried by the conduction electrons. For example, in ferromagnetic metal layer, the electrons effectively have the same polarization due to the Hund's rule coupling, and their motion subject to a bias voltage results in charge and spin currents. In a thin slab geometry, a bias voltage along the long side of slab derives a longitudinal charge current as well as a spin current along the transverse direction due to the spin Hall effect, as illustrated in the right panel of Fig. 4.11 in §4.2.5.1. It is also possible to observe only the spin current without charge current through the ballistic motion of electrons as depicted in the right panel of Fig. 6.10, consisting of spin-up electrons traveling in one direction and spin-down electrons traveling in the opposite direction [126][127].

The second is the spin wave spin current. The spin angular momentum is carried by the spin wave, magnon, as depicted in the right panel of Fig. 6.10. This spin wave spin current (magnon current that exists even in magnetic insulators) has attracted much attention recently as it has been shown to persist much greater distances than conduction electron spin currents, which disappear within a very short distance, typically hundreds of nanometers [128]. For instance,

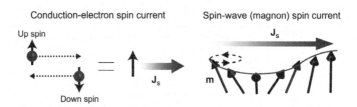

Figure 6.10 Two different spin currents. Left: Illustration of conduction electron spin current that is carried by the electron diffusion, spin-up electrons traveling to the right and spin-down electrons traveling to the left. Right: Spin wave spin current that is carried by collective magnetic-moment precession. Reproduced with permission from [125].

in the ferromagnetic insulator $Y_3Fe_5O_{12}$ (YIG), the spin-wave decay length can be several centimeters and thus the waves are propagated over a relatively long distance. Thus, this material can be served as an ideal conductor for spin-wave spin currents, while it is an insulator for electric currents.

By combining these two spin currents together, one can convert the magnon spin angular momentum in the insulating ferromagnet layer to the conduction electron spins in the conducting layer, called spin pumping, and vice versa [124]. In particular, the spin wave spin current can be converted into electric current (voltage). This is illustrated in the left panel of Fig. 6.11. The spin wave in the $Y_3Fe_5O_{12}$ (YIG) layer transfers spins to the Pt layer that generates the spin current $\vec{J}_s$. This spin current is converted into electric current through inverse spin Hall effect as illustrated in Fig. 4.12. The direction of the resulting electric field is perpendicular to the field $\vec{H}$. The magnetization vector of the spin wave process around the field $\vec{H}$. The spin pumping is inverse process of the spin transfer torque we discussed in §4.2.1, where spin is transferred from a conducting layer to a ferromagnetic layer. We can transfer information from one Pt layer to another by combining the spin transfer torque (from a Pt layer to a ferromagnetic layer) and the spin pumping (from the ferromagnetic layer to an another Pt layer where a signal can be detected electrically). As the magnon can propagate a long distance, the two Pt layers can be well separated. The direction of magnon propagation is transverse to the charge current in the Pt layer. This

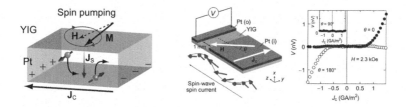

Figure 6.11 Left: Illustration of spin pumping, through which the spin wave spin current can be converted into electric current and electric voltage (transverse to the field direction $\vec{H}$) through the inverse spin Hall effect. Middel & Right: Experimental setup for transferring the applied charge current from PT (i) layer to another charge current in Pt (o) layer through the spin wave spin current. Reproduced with permission from [125].

is illustrated in the middle panel of Fig. 6.11. Note that there is a threshold electric current $J_c^{critical}$ = 0.6 GA/m^2 with GA =10^{9}A. This confirms that electric signal transmission in the YIG is activated via the spin wave spin currents.

Before moving on to consider the driving mechanism of magnon for experimental setups [124][125]. For the case illustrated in the middle panel of Fig. 6.11, the YIG sample is placed at the center of TE$_{011}$ microwave cavity, where the magnetic-field component $\vec{h}$ of the microwave is maximized while the electric field component $\vec{e}$ is minimized. An external static in-plane magnetic field $\vec{H}$ is applied, so that precession of magnetization is induced when $\vec{H}$ fulfills the ferromagnetic spin-wave resonance conditions. For the illustration in the right panel of Fig. 6.11, the applied electrically charged current $\vec{J}_c$ in the Pt (i) layer induce the magnetization oscillation through the spin transfer torque across the Pt/Y$_3$Fe$_5$O$_{12}$ interface to produce a measurable signal in the Pt (o) layer depending on the angle θ between the inplane field $\vec{H}_{\text{eff}}$ and the applied current $\vec{J}_e$. This magnetization oscillation is the source of spin wave spin current (magnon) and it is expected for the magnon spreads from the source isotropically when its layer geometry allows. In the following subsection, we consider the response of magnon under the temperature gradient.

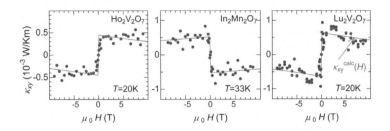

Figure 6.12 Thermal Hall conductivity κ_{xy} measurements for the $Ho_2V_2O_7$, $In_2Mn_2O_7$, and $Lu_2V_2O_7$ at the low temperature insulating phases. Reproduced with permission [131]. (The data set given for $Lu_2V_2O_7$ are essentially the same as given in [129].)

6.3.1.2 Magnon motion under temperature gradient

Here we consider insulating magnets. Under a temperature gradient, magnons diffuse from the hotter side to the colder side. It turns out that there exists a transverse motion, magnon thermal Hall effect, that has been reported for $Lu_2V_2O_7$ in [129] based on a previous theoretical study [130], in addition to this longitudinal motion along the temperature gradient.

Novel feature of the observation is its profile as a function of applied magnetic field $H(T)$ at the temperature well below the curie temperature. The thermal Hall conductivity steeply increases and saturates in the low magnetic field region. See Fig. 6.12. There are a few interesting observations to make. First, the profile is not related to the normal Hall effect that is proportional to magnetic field. The anomalous Hall effect, steeply rising and saturating Hall conductivity at low magnetic field, is affected by the spontaneous magnetization. Second, the conductivity curve gradually decreases with magnetic field after saturation (happening in the low-temperature region for all three data sets in Fig. 6.12) can be explained by the magnon gap induced by the magnetic field [129].

For insulating magnets at low temperature, there is still another low energy excitation, phonons, in addition to magnons. Thermal phonon Hall effect has been reported in $Tb_3Ga_5O_{12}$ through the spin-phonon interaction [132]. Here, the phonon mean free path is expected to increase with magnetic field as a result of reduced scattering by magnetic fluctuations. Thus, the decreasing behavior of the thermal Hall conductivity in the high-field region cannot be explained by the phonon Hall effects.

This is similar to the Leduk-Righi effect, which is related to the generation of the transverse temperature gradient in the presence of a temperature gradient and magnetic field (without charge current in metal) discussed in (2.131) in §2.6.3.

6.3.2 Skyrmion Seebeck effect

Seebeck effect is the generation of electric field in the presence of thermal gradient due to the accumulation of charge carriers in one side of a conducting sample and has been introduced in §2.6.2. For the insulating materials, manipulations using electric field is not effective. Instead, temperature gradient is a useful tool. Spin Seebeck effect has been also introduced in §4.5.1. In the following subsections, we introduce the magnon-driven spin Seebeck and skyrmion Seebeck effects.

6.3.2.1 Magnon-driven spin Seebeck effect

As explained in §4.5.1, spin Seebeck effects have been observed in conducting material [89], where its physical explanation has been provided in terms of the electrochemical difference for up-spin and down-spin conduction electrons. Subsequently, the role of conduction electrons was questioned due to their short spin-flip diffusion length [93]. Through further investigations of spin Seebeck effects in a ferromagnetic insulator $Y_3Fe_5O_{12}$ (YIG) [124] (explained in the previous section §6.3.1) and a direct observation of the spin Seebeck effect in insulator $LaY_2Fe_5O_{12}$ [90], it has been accepted that magnon can be a carrier for the spin Seebeck effect [92]. This magnon induced spin Seebeck effect is a major spin transport mechanics for a large (order of millimeter) length scale even in the conducting magnets.

Here we briefly introduce the theory of the magnon induced spin Seebeck effect following [93]. Interested readers for technical details are encouraged to read this illuminating paper. Its setup is illustrated in Fig. 6.13. The left panel describes the interface between a ferromagnetic block (F) with magnetization pointing horizontal direction, $\vec{M} \propto \hat{z}$, at temperature T_F and a normal metal block (N) at

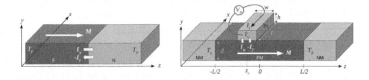

Figure 6.13 Illustration of magnon induced spin Seebeck effect. Left: Interface between ferromagnetic block (F) and normal metal (N). The interface exchange the spin pumping current I_{sp} and fluctuating spin current I_{fl}. Right: A ferromagnetic block in contact with two thermal reservoirs of normal metal reveals spin currents through inverse spin Hall effects in the overlaid Pt layer. Reproduced with permission from [93].

$T_N \neq T_F$. We assume that three different subsystems, magnon (m), conduction electron (e), and phonon (p), can be described by their local temperatures: $T_F^{m,e,p}$ and $T_N^{e,p}$ along with additional assumption that electron and phonon have interactions so that $T_F^e = T_F^p = T_F$ and $T_N^e = T_N^p = T_N$. (When one considers ferromagnetic insulators, the conduction electron subsystem in the ferromagnet can be removed.) The key ingredient is the fact that the magnon temperature can deviate from the electron/phonon temperature, $T_F^m \neq T_F$ without thermal equilibrium between the two blocks.

At the interface between F and N, a spin current noise I_{sp} is transferred to N due to the spin pumping discussed in §6.3.1.1. On the other hand, the thermal noise in N also induces a fluctuating spin current I_{fl}. Thus the spin current flowing through the interface is given by the combination of these two currents. The average dc current component along the temperature gradient is computed to give

$$\langle I_z \rangle \propto (T_F^m - T_N), \tag{6.38}$$

where $\langle \cdots \rangle$ is ensemble average, and the proportionality constant is an interfacial spin Seebeck coefficient that is omitted. This equation tells that the spin pumping current it proportional to the temperature difference between the magnon and electron/phonon temperatures. The other spin pumping current components along $\hat{x}$ and $\hat{y}$ vanish.

In the right panel of Fig. 6.13, the ferromagnetic block F is sandwiched between two normal metals with temperature T_L and T_R. By including the spin pumping currents and the fluctuating spin currents on both interfaces, the magnon-phonon temperature difference in the ferromagnetic block at a fixed location z (The middle point of F is set as the origin $z = 0$) is computed and given by [93]

$$T_F^m(z) - T_F^p(z) \propto \frac{\sinh(z/\lambda_m)}{\sinh(L/2\lambda_m)}(T_L - T_R), \qquad (6.39)$$

where $\lambda_m^2 \propto \tau_m \tau_{mp}$ with the magnon scattering time τ_m and the magnon-phonon thermalization (or spin-lattice relaxation) time τ_{mp}.

The equation (6.39) tells that the deviation of the magnon temperature from the lattice (phonon) temperature is proportional to applied temperature bias $T_L - T_R$ and decays from the interfaces with the length scale λ_m. In particular, the temperature difference vanishes at $z = 0$ and has different sign for the two regions $-L/2 < z < 0$ and $0 < z < L/2$. From this point of view, at $z = 0$, there is a crossover of temperature profile between the constant magnon temperature $T_F^m(z)$ and the phonon temperature $T_F^p(z)$, which is a linear interpolation between two interfaces. (This explanation is in contrast to the crossover of the electrochemical between spin-up and spin-down conduction electrons as discussed in Fig. 4.17 in §4.5.1.) The Pt block is placed on top of F at different location in z coordinate to measure the spin current through F using the inverse spin Hall effect, which can be measured by the charge current I_c or emf V_H. It is assumed that the measurement does not affect substantially the spin current in F when the spin currents exchanged in the interface between F and the Pt block. The linear dependence of spin current in F as a function of z is depicted in the middle panel of Fig. 4.18 in §4.5.1.

6.3.2.2 LLG equation with magnon interaction

The discussion in the previous subsection tells that magnon can play important roles in insulating magnets, especially when there is temperature gradient. Thus, here we provide a quantitative formulation of domain wall or skyrmion motion in the presence of magnons [133]. The derivation is technical and a little long.

Nevertheless, the final result is simple, illuminating and analogous to that with the conduction electron.

We split the magnetization vector into a slow one $\vec{m}_s$ and a fast one $\vec{m}_f$ in a similar spirit of the reference [63], which is used to derive the spin transfer torque. The slow magnetization vector describes the motion of domain wall or skyrmions. We use

$$\vec{m} = (1 - m_f^2)^{1/2}\,\vec{m}_s + \vec{m}_f. \tag{6.40}$$

Here the two vectors $\vec{m}_s$ and $\vec{m}_f$ are orthogonal to each other, $\vec{m}_s \cdot \vec{m}_f = 0$. Thus we can write $\vec{m}_f = \vec{m}_s \times \hat{n}$, which describes the motion processing around the slow magnetization $\vec{m}_s$. We have used similar descriptions earlier in this book. This decomposition is designed to ensure $\vec{m}^2 = \vec{m}_s^2 = 1$ as they are supposed to be. Now we expand the vector $\vec{m}$ for a small fast one $\vec{m}_f$ as

$$\vec{m} \approx \vec{m}_s + \vec{m}_f - \frac{1}{2}(m_f^2)\vec{m}_s, \tag{6.41}$$

where we discard $\mathcal{O}(m_f^3)$.

At finite temperature we use the stochastic LLG equation

$$\dot{\vec{m}} = -\gamma \vec{m} \times (\vec{H}_{\text{eff}} + \vec{h}) + \alpha \vec{m} \times \dot{\vec{m}}, \tag{6.42}$$

where γ and α are the gyromagnetic ratio and Gilbert damping parameter. The influence of temperature is encoded in the fluctuating magnetic field $\vec{h}$ that satisfies the local fluctuation-dissipation theorem

$$\langle h_i(\vec{x},t)h_j(\vec{x}',t')\rangle = \frac{\alpha a^2 k_B T}{\gamma}\delta_{ij}\delta(\vec{x}-\vec{x}')\delta(t-t'), \tag{6.43}$$

where a is a length scale representing the lattice spacing and also the fluctuation of the fast magnetization vector $\vec{m}_f$.

In the rest of this subsection, we derive the equation that governs the motion of the slow magnetization $\vec{m}_s$. We start with $\vec{H}_{\text{eff}}$. The following Hamiltonian $H = \int d^2x \mathcal{H}$ hosts skyrmions

$$\mathcal{H} = \frac{J}{2}(\partial_l \vec{m})^2 + \frac{D}{a}\vec{m} \cdot (\vec{\nabla} \times \vec{m}) - \frac{1}{a^2}\vec{H} \cdot \vec{m}, \tag{6.44}$$

with the external magnetic field $\vec{H} \propto \hat{z}$. We can compute the effective field $\vec{H}_{\text{eff}}$.

$$\begin{aligned}\vec{H}_{\text{eff}} = -\frac{\delta H}{\delta \vec{m}} &= Ja^2 \partial_l^2 \vec{m} - 2Da \vec{\nabla} \times \vec{m} + \vec{H} \\ &\approx Ja^2 \partial_l \big(\partial_l \vec{m}_s + \partial_l \vec{m}_f - m_{fo}(\partial_l m_{fo})\vec{m}_s - \frac{1}{2} m_f^2 \partial_l \vec{m}_s\big) \\ &- 2Da\big(\vec{\nabla} \times \vec{m}_s + \vec{\nabla} \times \vec{m}_f - m_{fo}(\vec{\nabla} m_{fo}) \times \vec{m}_s - \frac{1}{2} m_f^2 \vec{\nabla} \times \vec{m}_s\big) \\ &+ \vec{H},\end{aligned} \quad (6.45)$$

where we evaluate the expression by plugging in (6.42). At this point we compare these expressions to pick up dominant contributions. $\vec{m}_s$ is slowly changing over the length scale Ja/D, while $\vec{m}_f$ changes rapidly over the scale a. Thus spatial derivative of $\vec{m}_f$ dominates over that of $\vec{m}_s$, $\partial_l \vec{m}_f \gg \partial_l \vec{m}_s$. As the DM terms have only one derivative and are proportional to D, they are less important compared to the exchange interaction term proportional to J. Similarly, the effect of $\vec{H}$ can be ignored. Then we are left with the terms $Ja^2 \big(\partial_l^2 \vec{m}_f - (\partial_l m_{fm})^2 \vec{m}_s\big)$. Here we note that the last term vanishes when we apply $\vec{m} \times$ on this expression after ignoring m_f^3.

Then, the corresponding term in the LLG equation turns into

$$-\gamma \vec{m} \times \vec{H}_{\text{eff}} \approx -J\gamma a^2 \big[\vec{m}_s \times (\partial_l^2 \vec{m}_f) + \vec{m}_f \times (\partial_l^2 \vec{m}_f)\big]. \quad (6.46)$$

To go further, we consider the time dependence. The fast magnetization vector precesses around $\vec{m}_s$, for example $\vec{m}_f \sim \sin(w_0 t)$ with a characteristic frequency w_0. Thus the term that is linear in $\vec{m}_f$ vanishes when averaged over time. Thus we have just one term in the end. We slightly change the form as $\vec{m}_f \times (\partial_l^2 \vec{m}_f) = \partial_l [\vec{m}_f \times (\partial_l \vec{m}_f)]$. This term turns out to be the magnon current tensor we are looking for.

$$\begin{aligned} \mathbf{j}_l^m &= \vec{m}_f \times \partial_l \vec{m}_f, \\ \mathbf{j}_{al}^m &= \epsilon_{abc}(m_f)_b \partial_l (m_f)_c, \end{aligned} \quad (6.47)$$

where we explicitly express the tensor structure of the magnon spin current with indices a, b, c in spin space and l in coordinate space. In the following, we are going to put this into a slightly different form.

We employ the expression $\vec{m}_f = \vec{m}_s \times \hat{n}$ introduced above to satisfy the orthogonality condition $\vec{m}_s \cdot \vec{m}_f = 0$. Then, we compute the term $\vec{m}_f \times (\partial_l \vec{m}_f)$.

$$\begin{aligned}\vec{m}_f \times (\partial_l \vec{m}_f)|_a &= (\vec{m}_s \times \hat{n}) \times [(\partial_l \vec{m}_s) \times \hat{n} + \vec{m}_s \times (\partial_l \hat{n})]|_a \\ &= \epsilon_{abc}(\vec{m}_s \times \hat{n})_b[(\partial_l \vec{m}_s) \times \hat{n} + \vec{m}_s \times (\partial_l \hat{n})]_c \\ &= \epsilon_{abc}\epsilon_{b\alpha\beta}\epsilon_{c\gamma\delta}(m_{s\alpha}n_\beta)[(\partial_l m_{s\gamma})n_\delta + m_{s\gamma}(\partial_l n_\delta)] \\ &= \epsilon_{\alpha\gamma\delta}m_{s\alpha}n_a\partial_l(m_{s\gamma})n_\delta - \epsilon_{\beta\gamma\delta}m_{s\alpha}n_\beta m_{s\gamma}(\partial_l n_\delta) \\ &= -\hat{n}_a(\partial_l \vec{m}_s) \cdot (\vec{m}_s \times \hat{n}) + \vec{m}_{sa}(\partial_l \hat{n}) \cdot (\vec{m}_s \times \hat{n}),\end{aligned}$$ (6.48)

where we use $\epsilon_{abc}\epsilon_{b\alpha\beta} = \delta_{c\alpha}\delta_{a\beta} - \delta_{c\beta}\delta_{a\alpha}$. The first term can be discarded as $\partial_l \hat{n} \gg \partial_l \vec{m}_s$ that comes from $\partial_l \vec{m}_f \gg \partial_l \vec{m}_s$. Now we define the magnon current vector as

$$\begin{aligned}\vec{j}^m &= (\vec{\nabla}\hat{n}) \cdot (\vec{m}_s \times \hat{n}) \\ &= \epsilon_{abc}(\vec{\nabla}\hat{n}_a)(\vec{m}_{sb}\hat{n}_c),\end{aligned}$$ (6.49)

where we explicitly indicate that the magnon current vector index is associated with the spatial derivative. Apparently, $\vec{j}^m$ lost some information compared to the tensor form (6.47). In this form, the magnon spin current is polarized along the direction of $\vec{m}_s$ as the magnon current tensor is written as a direct product, $\mathbf{j}^m_{al} = \vec{m}_{sa}j^m_l$. Then

$$\begin{aligned}-\gamma \vec{m} \times \vec{H}_{\text{eff}} &\approx -J\gamma a^2 \partial_l[\vec{m}_f \times (\partial_l \vec{m}_f)] \\ &= -J\gamma a^2[(\partial_l \vec{m}_s)j^m_l + \vec{m}_s \partial_l j^m_l] \\ &\to -J\gamma a^2 j^m_l (\partial_l \vec{m}_s).\end{aligned}$$ (6.50)

In the last line, we consider the case with steady state current $\partial_l j^m_l = 0$.

We can also evaluate other terms in the LLG equation (6.42).

$$\gamma \vec{m} \times \vec{h} \approx \gamma \vec{m}_s \times \vec{h},$$ (6.51)

as the time average of $\vec{m}_f$ vanishes. Finally

$$\alpha \vec{m} \times \dot{\vec{m}}$$
$$\approx \alpha[(1 - \frac{1}{2}m_f^2)\vec{m}_s + \vec{m}_f] \times [(1 - \frac{1}{2}m_f^2)\dot{\vec{m}}_s + (1 - \vec{m}_s(\vec{m}_f))\dot{\vec{m}}_f]$$
$$= \alpha[(1 - \frac{1}{2}m_f^2)^2 \vec{m}_s \times \dot{\vec{m}}_s + \vec{m}_f \times \dot{\vec{m}}_f + \vec{m}_s \times \dot{\vec{m}}_f + \vec{m}_f \times \dot{\vec{m}}_s]$$
$$= \alpha \vec{m}_s \times \dot{\vec{m}}_s, \tag{6.52}$$

where we use the fact that the time averages of $\vec{m}_f$, $\dot{\vec{m}}_f$ and $\vec{m}_f \times \dot{\vec{m}}_f \sim \sin(w_0 t)\cos(w_0 t)$ vanish. Here we also use $m_f^2 \ll 1$ and $\vec{m}_s \times \vec{m}_s = 0$.

We finally arrive at the LLG equation for slow magnetization vector $\vec{m}_s$ as

$$\dot{\vec{m}}_s = -J\gamma a^2 (j_l^m \partial_l)\vec{m}_s - \gamma \vec{m}_s \times \vec{h} + \alpha \vec{m}_s \times \dot{\vec{m}}_s. \tag{6.53}$$

This is a particularly simple and interesting LLG equation that contains the magnon current j_l^m.

At this point, we remind of the magnetization equation with the spin transfer torque (STT) developed in (4.16) in §4.2.1. By identifying $\vec{M} \propto \vec{m}_s$, we rewrite (4.16) as

$$\dot{\vec{m}}_s = [\beta_e + \tilde{\beta}_e \vec{m}_s \times](j_i^s \partial_i)\vec{m}_s + \alpha \vec{m}_s \times \dot{\vec{m}}_s, \tag{6.54}$$

where $\vec{j}^s$ is the spin current of the conduction electrons in the conducting magnets. The terms proportional to β_e and $\tilde{\beta}_e$ are the adiabatic and the non-adiabatic spin torque terms, respectively. These coefficients β_e and $\tilde{\beta}_e$ can be estimated theoretically and also fit with experimental data for a given material.

Now back to the equation (6.53), we can also include the analogue of non-adiabatic term for magnon current as well. Thus,

$$\dot{\vec{m}}_s = -[\beta_m + \tilde{\beta}_m \vec{m}_s \times](j_l^m \partial_l)\vec{m}_s - \gamma \vec{m}_s \times \vec{h} + \alpha \vec{m}_s \times \dot{\vec{m}}_s, \tag{6.55}$$

where the cross product × acts on the magnetization vector $\vec{m}_s$. The front factors $J\gamma a^2$ in (6.53) is absorbed into β_m or the magnon current j_l^m. The terms with β_m and $\tilde{\beta}_m$ are the adiabatic and the non-adiabatic magnon spin torques. The non-adiabatic term is expected to come

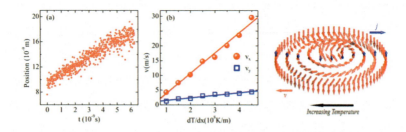

Figure 6.14 Numerical studies for the skyrmion motion in the insulating magnets. Left: The positions of skyrmions, which move from the cold to the hot region with a constant velocity. Right: The skyrmions exhibit the transverse motion (square data points) in addition to the longitudinal motion (ball data points). They increase as a function of temperature gradient. Reproduced with permission from [133].

from the DMI term proportional to D in (6.45) that we omitted in the above derivation. We cannot neglect the non-adiabatic term when magnon wavelength becomes comparable to skyrmion size as skyrmion's internal structure is deformed through the interaction with magnon [134][135]. Here we are only concerned with the adiabatic term with β_m.

6.3.2.3 Skyrmion Seebeck effect

In addition to the discovery of the skyrmion crystals in metallic materials, they have been also discovered in insulating materials, such as Cu_2OSeO_3 [136][137] and $BaFe_{1-x-0.05}Sc_xMg_{0.05}O_{19}$ [138].

One clear advantage of the skyrmions in insulating materials is the absence of dissipation due to the conduction current. (Depending on your view, this can be viewed as a disadvantage as you miss the relatively well known ability to control skyrmions using electrons.) While the skyrmion Hall effect has been experimentally measured as discussed in §6.1.2, understanding the motion of the skyrmions in insulating materials shed insights on their interactions with magnons, the low energy excitations of magnetization. Here we consider the numerical and analytical studies of the skyrmion Hall effect in the insulating magnets based on recent development of the spin Seebeck effect [133][134][135].

Surprisingly, when the insulating magnets hosting skyrmions are under the temperature gradient, the skyrmions move towards the high temperature region according to these numerical studies. This is illustrated in the left panel of Fig. 6.14. This is against our usual intuition based on the Brownian diffusion, particles like electrons move toward the colder regions as illustrated in Fig. 2.3 while we discuss the Seebeck effect in §2.6.2.

To understand this counterintuitive skyrmion motion, a magnon-assisted theory is employed in [133]. The result shows that skyrmions move steadily with a constant velocity as in the left panel of Fig. 6.14. The middle panel illustrates the velocity as a function of temperature gradient. When the temperature gradient is increased along $\hat{x}$, the velocity along that direction, v_x increases linearly as indicated with ball-shaped data points. There is also a transverse velocity v_y with squared data points, the thermal Hall conductivity, that is also linearly increasing.

The right panel of Fig. 6.14 illustrates the interaction between the skyrmion and magnon. Magnon is the low energy quanta with the up-spin fluctuations in an up-spin ferromagnetic layer, while the skyrmion has a down-spin center with up-spins at the edge in the up-spin ferromagnet as in the schematic left panel of Fig. 6.15. We look into this more closely following [133] based on a previous work [139]. When the ferromagnetic layer is polarized along $\hat{z}$, the spin's deviation that describes a magnon can be described by (n_x, n_y). The magnon's creation operator and its number operator are given by

$$a^\dagger = \frac{1}{\sqrt{2}}(n_x - in_y),$$
$$\rho = a^\dagger a = \frac{1}{2}(n_x^2 + n_y^2). \tag{6.56}$$

The spin component along the ferromagnetic equilibrium configuration is

$$n_z = \sqrt{1 - (n_x^2 + n_y^2)} \approx 1 - \rho. \tag{6.57}$$

This indicates that each magnon carries spin one polarized antiparallel to the equilibrium point. Due to this anti-parallel alignment of magnon spins to skyrmions that is opposite to the spin

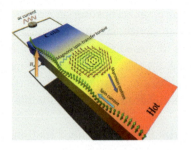
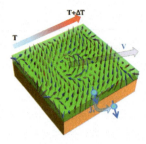

Figure 6.15 Left: Skyrmions move from the cold to the hot region, while magnons diffuse from the hot to the cold region. The skyrmions exhibit the transverse motion in addition to the longitudinal motion. Reproduced with permission from [134]. Right: An additional non-magnetic layer, such as Pt, can be used to detect the spin pumping resulting from skyrmion motion. Due to mostly out-of-plane magnetization configuration of the ferromagnet, the ordinary spin Seebeck effect would be suppressed. Reproduced with permission from [135].

interaction between skyrmion and conduction electron, magnon current provides a negative transfer torque on the skyrmion. Through the conservation angular momentum, the magnon and skyrmion move in opposite directions. Thus the right panel of Fig. 6.14 does not contradict to left panel of Fig. 6.15.

Before going to explicitly consider the Thiele equation with magnon contributions, we mention the results of skyrmion center motion without deformation [133] using

$$\vec{m}_s = \vec{m}_s^0(\vec{x} - \vec{X}(t)), \qquad \vec{X}(t) = \vec{v}_d t. \tag{6.58}$$

By inserting this equation to (6.53) and integrating over the ground state configuration, the average velocity of skyrmions has been obtained in [133].

$$v_{dx} = \gamma J a^2 j_x^m - \frac{\gamma}{\pi Q^2} \alpha \tilde{\eta} a^2 k_B \frac{dT}{dx} \equiv v_x^m - v^B,$$

$$v_{dy} = 2\alpha\tilde{\eta} v_x^m, \tag{6.59}$$

where $\tilde{\eta}$ is a skyrmion form factor that is order one. The magnon current j_x^m along the $\hat{x}$ direction provides the longitudinal velocity component $v_x^m = \gamma J a^2 j_x^m$ to skyrmion. This pushes the skrmion toward hot region, while the Brownian contribution proportional to dT/dx pushes the skyrmion to cold region. Typically the magnon contribution dominates, and thus skyrmions move toward the hot region. On the other hand, the Brownian motion along the $\hat{y}$ direction vanishes on the average. We revisit this result with the Thiele equation.

In the previous subsection §6.3.2.2, we derive the LLG equation for skyrmion in the pressence of magnon current. We can read off the Thiele equation for this case by looking at the Thiele equation with the electric current, which has been developed earlier in (4.55). Instead of carrying all the details, we present the result in a simple form using the effective magnon $\vec{v}_m$ and electron $\vec{v}_e$ velocities normalized compared to the skyrmion drift velocity $\vec{v}_d$. The resulting Thiele equation gives [134][135]

$$\vec{G} \times (\vec{v}_d + \vec{v}_m - \vec{v}_e) + \mathcal{D} \cdot (\alpha \vec{v}_d + \tilde{\beta}_m \vec{v}_m - \tilde{\beta}_e \vec{v}_e) = 0 \,. \tag{6.60}$$

In this form, it is straightforward to check that the skyrmion interaction with the magnons is different from that with electrons. Explicitly, the skyrmion interactions with conduction electrons and magnons have opposite signs in the Thiele equation. We also mention that the skyrmion and antiskyrmion have opposite Hall angles under these interactions as $\vec{G}$ has opposite signs for them.

For the insulating magnets, $\vec{v}_e = 0$.

$$-4\pi Q \hat{z} \times (\vec{v}_d + \vec{v}_m) + D(\alpha \vec{v}_d + \tilde{\beta}_m \vec{v}_m) = 0 \,, \tag{6.61}$$

where we use $\vec{G} = -4\pi Q \hat{z}$. The damping torque tensor $\mathcal{D}$ is assumed to have only the diagonal components, and thus $D = -\int d^2x (\partial_i \vec{m}_s)^2$ that is called the skyrmion form factor explicitly computed in §6.1.2.1. Having only the magnon current along the direction of the temperature gradient ($\vec{v}_m = v_{mx} \hat{x}$), the equation (6.61) gives

$$\begin{pmatrix} \alpha D & 4\pi Q \\ -4\pi Q & \alpha D \end{pmatrix} \begin{pmatrix} v_{dx} \\ v_{dy} \end{pmatrix} = v_{mx} \begin{pmatrix} D\tilde{\beta}_m \\ 4\pi Q \end{pmatrix} \,. \tag{6.62}$$

Solving the equation gives

$$v_{dx} = \frac{(\alpha\tilde{\beta}_m D^2 - (4\pi Q)^2) v_{mx}}{\alpha^2 D^2 + (4\pi Q)^2},$$

$$v_{dy} = \frac{(\alpha + \tilde{\beta}_m)(4\pi Q) D v_{mx}}{\alpha^2 D^2 + (4\pi Q)^2}. \quad (6.63)$$

The longitudinal motion described by v_{dx} indicates that there is a competition between the drag term and the skyrmion charge. If $\alpha\tilde{\beta}_m D^2 - (4\pi Q)^2 < 0$, skyrmions move toward the high temperature region against the motion of magnon. For $\alpha\tilde{\beta}_m D^2 - (4\pi Q)^2 > 0$, skyrmions move toward the cold region as the drift Brownian motion dominates the force exerted by magnons. This result is logical and consistent with the picture presented above. This result depends on Q^2, and thus is independent of skyrmion or antiskyrmion.

The transverse motion, Hall motion, of skyrmion depends on the topological charge Q as can be seen in v_{dy}. We can compute the Hall angle Θ_{SkX}.

$$\Theta_{SkX} = \frac{v_{dy}}{v_{dx}} = \frac{(\alpha + \tilde{\beta}_m)(4\pi Q) D}{(\alpha\tilde{\beta}_m D^2 - (4\pi Q)^2)}. \quad (6.64)$$

This equation tells that the Hall angle has four different regimes, not two, based on charge Q. For example, assuming $D < 0$ based on its definition above, the Hall angle is positive either 1) for skyrmion $Q > 0$ when skyrmion moves toward the high temperature region or 2) for antiskyrmion $Q < 0$ when it moves toward the cold temperature region with Brownian motion.

It has been proposed to detect temperature induced skyrmion dynamics by employing spin pumping into the neighboring non-magnetic metallic layer, such as Pt [135]. The ordinary spin Seebeck effect should be suppressed since the polarization of the moving electron in the Pt layer is parallel to the magnetization of the ferromagnet layer which is pointing out of plane. See the right panel of Fig. 6.15. Tunable magnon thermal Hall effect in skyrmion crystal of ferrimagnets in the vicinity of the angular momentum compensation point has been proposed recently [140]. We come back to the Hall angle for ferri-magnets in the next chapter when we model skyrmion Hall viscosity by generalizing the Thiele equation.

6.3.3 Rotational motion of skyrmions

As the last part of the skyrmion motion in insulting magnets, we would like to include the topic of the experimental realization of the skyrmion rotational motion under temperature gradient or magnetic field gradient.

Rotation motion of skyrmions happens universally under the temperature gradient in the conducing materials such as MnSi [117][141][83] and insulating materials Cu_2OSeO_3 [141][137]. In experimental and numerical studies [141], the rotation motions of the skyrmions are systematically studied, for both conducting and insulating materials, to find that the rotational motion can be driven by the thermal gradient. In the left panel of Fig. 6.16 [83], various different force components acting on skyrmion lattice are explained in the presence of current along the vertical direction and the linear temperature gradient either horizontal or vertical. For a static and non-moving skyrmion without internal deformations, the red horizontal arrows correspond to the Magnus force from the interaction with electric currents, while the green vertical arrows to dissipative forces. In the presence of a temperature or field gradient, these forces change smoothly across a domain, thereby inducing rotational torques which depend sensitively on the relative orientation of current and gradient (and on the direction in which the skyrmion lattice moves). Small black arrows show the local orientation of the magnetization projected into the plane perpendicular to the magnetic field $\vec{B}$. In each unit cell the magnetization winds once around the unit sphere.

Under the temperature gradient along the radial direction in a plane perpendicular to the magnetic field, skyrmions shows clockwise rotation motion that has been observed by Lorentz transmission electron microscopy (LTEM) image in the chiral-lattice magnets MnSi [141]. Same rotation is also verified by numerical simulation for the thermally driven motion for skyrmion microcrystal [141]. Moreover, rotational motion is also observed under a radial magnetic field gradient in the plane along the z-direction with the maximum in the center [142]. Charge-coupled device (CCD) camera snapshots have been taken while the field gradient-induced skyrmion lattice rotations happen as can be seen in

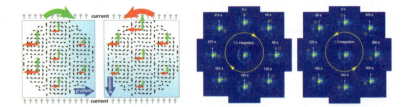

Figure 6.16 Left: Illustration of various force components acting on skyrmion lattice in the presence of current and temperature gradient. There is visible rotation motion not to mention translation motion of skyrmion in conducting magnet. Reproduced with permission from [83]. Right: Illustration of rotational motion under a radial magnetic field gradient in the plane along the z-direction with the maximum in the center. There is skyrmion rotational motion both with temperature gradient or field gradient for the insulating magnets. Credit: Nature Communications [142].

the right panel of Fig. 6.16. To see the rotation motion clear, one of the magnetic satellites is marked by an orange circle. When the direction of the applied magnetic field changes, the direction of rotation is also reversed.

6.3.3.1 Rotation motion from LLG equation

In this subsection, we formulate the rotation motion of rigid skyrmions without the deformation of their internal structures. First we summarize the Thiele equation as the derivation is similar. To start, let us consider the LLG equation with the spin transfer torque given in (4.50) in §4.4.2.1

$$M_s \vec{H}_{\text{eff}} = M_s \vec{m} \times (\partial_t + \vec{v}_s \cdot \vec{\nabla}) \vec{m} \\ + M_s (\alpha \partial_t + \beta \vec{v}_s \cdot \vec{\nabla}) \vec{m} , \tag{6.65}$$

where $\vec{m} = \vec{M}/M_s$. In case we want to include emergent fields, we can modify the parameters α and β as in §4.4.2.1. To get the Thiele equation, we use the parametrization of the center of skyrmion motion as

$$\vec{m} = \vec{m}(\vec{x} - \vec{v}_d t) , \tag{6.66}$$

which turns time derivative into spatial one with velocity $\vec{v}_d$. Then we project the LLG equation (6.65) to the translation mode by $\partial_i \vec{m}$ and integrate over a unit cell (UC).

$$\vec{\mathcal{G}} \times (\vec{v}_s - \vec{v}_d) + \mathcal{D} \times (\beta \vec{v}_s - \alpha \vec{v}_d) = 0 , \tag{6.67}$$

where

$$\begin{aligned}\mathcal{G}_i &= \frac{1}{2} \int_{UC} d^2x M \, \epsilon_{ijk} \vec{m} \cdot (\partial_j \vec{m} \times \partial_k \vec{m}) , \\ \mathcal{D}_{ij} &= \int_{UC} d^2x M \, \partial_i \vec{m} \partial_j \vec{m} .\end{aligned} \tag{6.68}$$

This result has been derived in (4.48) for $\vec{v}_s = 0$ and also in (4.55) in §4.4.

Now we are ready to understand the skyrmion rotation motion. Here we neglect the macroscopic deformation of the skyrmion internal structures. Due to the weak spin-orbit coupling, rotational invariance on its own is broken. Thus, rotational torque due to the current can be balanced by a counter torque of the underlying crystal lattice. Similar discussion has been given in §1.4.3. The rotation motion of skyrmion in the plane perpendicular to $\hat{z}$ (along the external magnetic field if there is any) can be described by the parametrization [143][83] introduced in §4.4.3 as

$$\vec{m}'(\vec{x}, t) = R_{\hat{z}}(\phi) \cdot \vec{m}\left[R_{\hat{z}}^{-1}(\phi) \cdot (\vec{x} - \vec{v}_d t)\right] , \tag{6.69}$$

where $R_{\hat{z}}(\phi(t))$ is the rotation matrix and $\vec{v}_d$ is the linear drift velocity of skyrmion. $(R \cdot \vec{m})_i = R_{ij} m_j$ is a vector, so is its argument. For an infinitesimal rotation $\phi \ll 1$, the matrix has the form

$$R_{\hat{z}}(\phi(t))_{ij} = \delta_{ij} - \phi \epsilon_{ijk} \hat{z}_k + \mathcal{O}(\phi^2) . \tag{6.70}$$

Infinitesimal change of the vector $\vec{m}$ can be rewritten as

$$\begin{aligned}\delta_R \vec{m} &= \vec{m}' - \vec{m} \\ &= \phi \left(\hat{z} \times \vec{m} - \left[\hat{z} \cdot ((\vec{x} - \vec{v}_d t) \times \vec{\nabla})\right] \vec{m}\right) + \mathcal{O}(\phi^2) \\ &\equiv \phi \left(\hat{G}_{rot} \vec{m}\right) + \mathcal{O}(\phi^2) .\end{aligned} \tag{6.71}$$

We can verify this in component form as

$$\begin{aligned}
m'_i &= R_{ij} m_j [R_{kl}^{-1} x_k] \\
&= (\delta_{ij} - \phi \epsilon_{ijn} \hat{z}_n) m_j [x_k + \phi \epsilon_{klm} \hat{z}_m x_l] + \mathcal{O}(\phi^2) \\
&= m_i - \phi \epsilon_{ijn} \hat{z}_n m_j + \phi \delta_{ij} \epsilon_{klm} \hat{z}_m x_l \partial_k m_j + \mathcal{O}(\phi^2) \\
&= m_i + \phi(\hat{z} \times \vec{m})_i - \phi[\hat{z} \cdot (\vec{x} \times \nabla)](\vec{m})_i + \mathcal{O}(\phi^2) ,
\end{aligned} \quad (6.72)$$

where we use $\epsilon_{klm} \hat{z}_m x_l \partial_k = -[\hat{z} \cdot (\vec{x} \times \nabla)]$, which is a scalar. The argument of m_j is also a vector, whose independent index l in the first line contributes independently in the argument.

Similar to the translation motion, to describe the rotational motion along $\hat{z}$ direction, we project LLG equation (6.65) by multiplying $(\hat{G}_{rot} \vec{m}) \equiv \hat{z} \times \vec{m} - [\hat{z} \cdot ((\vec{x} - \vec{v}_d t) \times \vec{\nabla})] \vec{m}$ as a scalar product and integrating over a unit cell. Then we arrive

$$\vec{\mathcal{P}}_R \cdot (\vec{v}_s - \vec{v}_d) + \vec{\mathcal{P}}_D \cdot (\tilde{\beta} \vec{v}_s - \tilde{\alpha} \vec{v}_d) = \tau , \quad (6.73)$$

where

$$\begin{aligned}
\mathcal{P}_R^i &= \int_{UC} d^2 x \, (\vec{m} \times \partial_i \vec{m}) \cdot (\hat{G}_{rot} \vec{m}) , \\
\mathcal{P}_D^i &= \int_{UC} d^2 x \, (\partial_i \vec{m}) \cdot (\hat{G}_{rot} \vec{m}) .
\end{aligned} \quad (6.74)$$

The reactive rotational coupling vector $\mathcal{P}_R^i$ comes from the Berry phase picked up by spin currents in the presence of a non-trivial topological object like skyrmions. The dissipative rotational coupling vector $\mathcal{P}_D^i$ comes from the dissipative term. These vectors have only components in plane as there is no non-zero z-directional derivative. These two terms describe the rotational torques created by the applied current.

These rotational torques due to the current is balanced by the angular momentum transferred between the skyrmion lattice and the underlying atomic lattice. The expression is given by

$$\tau = \int_{UC} d^2 x \, \vec{H}_{\text{eff}} \cdot (\hat{G}_{rot} \vec{m}) . \quad (6.75)$$

The theoretical study of this term has been explored in [143][83]. Interested readers are encouraged to read the references. Instead we present an analytical model that include the rotation motion along with the influence of magnons.

6.3.3.2 An analytical model for skyrmion rotation motion

Systematic study through experiments and numerics for the conducting and insulating magnets [141] has concluded that the rotation motion is solely a thermal effect. It was also confirmed that even in conducting magnets the rotation is not related to the conduction electrons as their interaction strength with skymions are not enough to create the skyrmion motion.

The universal rotation was modeled and explained by the interaction between the skyrmions and magnons. Magnons can be treated as particles, and thus the rotational motion of the skyrmion interacting with the magnons can be understood similarly. To include the rotational motion and the interaction with the magnons, the following magnetization vector was introduced for the Hamiltonian (6.44) as

$$\vec{m}(t,\vec{x}) = \vec{m}_s(\vec{x} - \vec{R}(t)) + \vec{m}_d(\partial_t \vec{R}(t)) + \vec{m}_f, \qquad (6.76)$$

where the magnetization $\vec{m}$ is separated into the slow part $\vec{m}_s$ for the skyrmion motion from the fast part $\vec{m}_f$ for the magnon motion. $\vec{R}(t) = (X(t), Y(t))$ is the coordinate for the center of the skyrmion.

The middle term $\vec{m}_d$ captures the deformation of a moving skyrmion involved with $\partial_t \vec{R}(t)$. Once we include this term, the Thiele equation has another contribution proportional to the mass

$$\mathcal{M} = \frac{1}{2} \int d^2x \vec{m}_s \cdot \left[(\partial \vec{m}_d / \partial \dot{R}_i) \times (\partial \vec{m}_s / \partial r_i) \right], \qquad (6.77)$$

which is multiplied by $\partial_t^2 \vec{R}(t) = (\ddot{X}(t), \ddot{Y}(t))$. Then, by projecting the LLG equation to the rotational motion, the analogue of the Thiele equation becomes

$$\begin{aligned}
\mathcal{M}\ddot{X} + G(\dot{Y} + J_y^M) + \alpha D\dot{X} &= -\frac{\partial U}{\partial X}, \\
\mathcal{M}\ddot{Y} - G(\dot{X} + J_x^M) + \alpha D\dot{Y} &= -\frac{\partial U}{\partial Y},
\end{aligned} \qquad (6.78)$$

where the force has the form $\vec{F} = (F_x, F_y) = (-GJ_x^M - \partial U/\partial X, GJ_y^M - \partial U/\partial X)$ with the reaction force $\vec{F}_M$ due to the magnon currents $\vec{J}^M$ and the other force $\vec{\nabla} U$ due to the repulsion of skyrmions from disk edge, magnetic fields, and impurities [141][80].

Under the temperature gradient along the radial coordinate r, $T(r)$, the skew scattering of the magnons off skyrmions gives rise to the topological magnon Hall effect.

$$J_r^M = \kappa_{xx}^{Magnon}\left(-\frac{dT}{dr}\right), \quad J_\theta^M = \kappa_{xy}^{Magnon}\left(-\frac{dT}{dr}\right). \quad (6.79)$$

The latter is transverse to the temperature gradient, corresponding to counter-clockwise rotation of magnon gas. Conservation of momentum makes the skyrmions rotate along the clockwise direction. Interested readers are encouraged to look into more on this motion described in [141].

Theoretical computation of the thermal Hall conductivity of the magnon under the influence of the emergent magnetic fields due to the skyrmion crystals in insulating magnets [144]. Skyrmion rotation under the influence of conduction electrons through the spin transfer torque was studied previously [143].

6.4 Ward identity for insulating magnets

After discussing the experimental facts on the skyrmion transport for insulating magnets, especially the interactions with the low energy excitations, magnons, we briefly discuss the corresponding situation for the Ward identity approach. We consider the skyrmion transport at zero temperature, where we can neglect the contributions from the magnons.

6.4.1 A way to measure Hall viscosity

In the presence of translation and rotation symmetries, the topological Ward identity with the stress energy tensors has been constructed as (5.207) and (5.208) in §5.5.2.1.

$$\omega^2[\delta^{jl}\boldsymbol{\kappa}_\delta + \epsilon^{jl}\boldsymbol{\kappa}_\epsilon + q^j q^l \boldsymbol{\kappa}_q + (\epsilon^{jn}q_n q^l + \epsilon^{ln}q_n q^j)\boldsymbol{\kappa}_{q\epsilon}] \\ = \delta^{jl}q^2\eta + \epsilon^{jl}(c + q^2\eta_H) + q^j q^l \zeta \,, \quad (6.80)$$

where the thermal conductivities are given by $\boldsymbol{\kappa}$, while the viscosities are η, ζ, and η_H. We also include the topological skyrmion charge c. Due to the rotational invariance, linear momentum dependent terms are not allowed.

We can further refine the Ward identities by considering the momentum dependence. When we take the zero momentum limit, $q^2 \to 0$, we get two different identities.

$$\omega^2 \boldsymbol{\kappa}_\delta^{(0)} = 0 \,, \\ \omega^2 \boldsymbol{\kappa}_\epsilon^{(0)} = c \,, \quad (6.81)$$

where the superscript $^{(0)}$ denotes only the momentum independent contribution. Intuitively, the reason $\boldsymbol{\kappa}_\delta^{(0)}$ vanishes, while $\boldsymbol{\kappa}_\epsilon^{(0)}$ does not, is the fact that skyrmions are associated with spontaneously broken translation symmetry along with the broken parity. The imprints of the broken parity symmetry can only enter through the parity odd parts of the conductivity at zero momentum. More precisely, the second identity predicts that the formation of a single skyrmion results in the creation of a unit of thermal Hall conductivity $\boldsymbol{\kappa}_\epsilon^{(0)}$ in units of the quantized topological charge density. The frequency dependence is a consequence of the pole structure of the Goldstone boson that manifests itself in the retarded momentum correlator.

In the presence of disorder, the behavior $\boldsymbol{\kappa}_\epsilon^{(0)} = c/\omega^2$ could, in principle, be lifted. However, numerical simulations have confirmed that skyrmion motions are unaffected by impurities, in contrast to the case of domain walls [84]. Furthermore, recent experiments confirmed that the skyrmions in some materials move very efficiently with velocities reaching the predicted for perfect pinning free systems [72]. The thermal Hall conductivity $\boldsymbol{\kappa}_\epsilon$ is dissipationless and exists even at zero temperature. While our Ward identity relations can be valid at finite temperatures as far as we ignore the magnon contributions, measurements will be cleaner at very low temperatures, where additional dissipative contributions are suppressed. Another interpretation of (6.81) is that the skyrmions carrying the thermal current propagate in an effective magnetic field given by the skyrmion charge density c_{ij}, leading to a thermal Hall effect.

The momentum dependent Ward identities of (6.80) can be worked out to be

$$\omega^2 \bar{\kappa}_\delta = q^2 \eta ,$$
$$\omega^2 \bar{\kappa}_\epsilon = q^2 \eta_H , \qquad (6.82)$$
$$\omega^2 \kappa_q = \zeta ,$$

where the bar ~ signifies the momentum dependent part. For example, $\bar{\kappa}_\epsilon = \kappa_\epsilon - \kappa_\epsilon^{(0)} = q^2 \kappa_\epsilon^{(2)} + q^4 \kappa_\epsilon^{(4)} + \cdots$. Thus, thermal conductivities are directly connected to the viscosities of the system, which are previously confirmed in [22]. Furthermore, it follows from (5.208) that $\kappa_{q\epsilon} = 0$. As $q^2 \to 0$, these Ward identities can be recast in a slightly different form as

$$\eta(q^2 \to 0) = \omega^2 \kappa_\delta^{(2)} ,$$
$$\eta_H(q^2 \to 0) = \omega^2 \kappa_\epsilon^{(2)} , \qquad (6.83)$$
$$\zeta(q^2 \to 0) = \omega^2 \kappa_q(q^2 \to 0) .$$

While these directly connect the viscosities in terms of the conductivities, actual verifications can be difficult as the time dependent measurements depending on ω are more difficult than time independent cases.

To remove the ω dependence, we can divide two Ward identities when they do not vanish. It turns out that this is useful in the presence of the topological skyrmion charge. By dividing the second equations of (6.82) and (6.81),

$$\eta_H(q^2 \to 0) = \lim_{q^2 \to 0} \frac{c \bar{\kappa}_\epsilon}{q^2 \kappa_\epsilon^{(0)}} = c \frac{\kappa_\epsilon^{(2)}}{\kappa_\epsilon^{(0)}} , \qquad (6.84)$$

where we take the limit $q^2 \to 0$. Once the thermal Hall conductivity κ_ϵ is measured as a function of q^2, the Hall viscosity is nothing but the skyrmion density multiplied by the ratio between the slope and κ_ϵ-

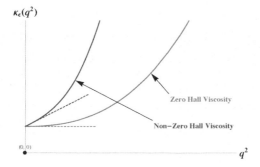

Figure 6.17 Confirmation of the existence of the Hall viscosity. Intercept of thermal Hall conductivity $\kappa_\epsilon^{(0)}$ is non-zero and is proportional to the skyrmion charge density. Non-vanishing slope $\kappa_\epsilon^{(2)}$ as a function of momentum squared q^2 confirms the existence of the Hall viscosity.

intercept $\kappa_\epsilon(q^2 = 0)$. Note that this is only applicable in the presence of nonzero skyrmion density. See Fig. 6.17. This provides a simple way to measure the Hall viscosity or to confirm its existence.

As discussed in §6.3, the interaction between skyrmions and magnons are universal at finite temperature. Thus it is important to include the magnon contributions in Ward identity approach. It will be also interesting to model their interactions in Ward identities.

6.5 Outlook: Measuring Hall viscosity?

Before ending this chapter, we would like to speculate a possible way to measure Hall viscosity in skyrmion systems neither with electric response in the presence of the electric currents nor with thermal response in the presence of the temperature gradients.

As introduced in §2.3, the Hall viscosity is transverse and dissipationless component of the shear tensor given in (2.38). Pictorially, the direction of the Hall viscosity is given in Fig. 2.2. Thus we can think about a similar experimental setup with an insulating magnet even at finite temperature, while it will be better at low temperature where the thermal excitations are suppressed. Possible magnon contributions can be isolated as we see below. Of course there is no conduction electrons in the system. The initial setup

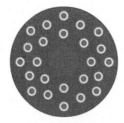 

Figure 6.18 Illustration of possible measurement of the Hall viscosity of skyrmion system in contact with a rotating cylinder without electric current or thermal gradient. Left: Initial setup with the random population of skyrmions. Right: The rotation of the cylinder in the ferromagnet produces the radially outward motion of skyrmions. This produces the spin imbalance that can be measured by an overlaid non-magnet metal through inverse spin Hall effect. Reproduced with permission from [85].

is depicted in the left panel of Fig. 6.18 [85]. The skyrmions are populated with an appropriate magnetic field and temperature. Blue color represents spin-up background of the ferromagnetic material. Red blobs are the skyrmions. In the middle of the ferromagnetic material, there is a solid cylinder that can be rotated.

When the cylinder at the center rotates, the skyrmions move radially inward or outward depending on the rotation direction due to the Hall viscosity as depicted in the right panel of Fig. 6.18. This will produce the spin imbalance between the inner boundary and the outer boundary. These spin imbalances can be measured by the inverse spin Hall effect by attaching a non-magnetic layer, for example a Pt layer, that is represented as a green rectangle. Because majority of the spin background in the ferromagnetic layer is up-spin, the magnon contribution for the inverse spin Hall effect of the transport will be suppressed. Thus what we observe for the inverse spin Hall effect on the Pt layer is from the skyrmions motion due to Hall viscosity!

Chapter 7

Modeling Hall Viscosity

Transverse skyrmion motion has been one of the central components of recent investigations [123][72] as it pushes skyrmion transverse to the skyrmion's moving direction, possibly resulting in its annihilation and information loss at devices' edges in the context of skyrmion racetrack devices. The conventional skyrmion Hall effect is sensitive to its charge, especially its sign depending on skyrmion or antiskyrmion as studied in §6.1.2. While there are various different skyrmion racetrack designs [145], moving skyrmions have this skyrmion Hall effect unless the spin density of the magnetic layer vanishes as described below (4.52) in §4.4.2.1. We note that there is another universal transport component due to the Hall viscosity, which exists regardless of the underlying spin density. Here we explain the Hall viscosity direction in the skyrmion motion and model it in the skyrmion linear motion by generalizing the Thiele equation. We start with recent advancements of domain-wall velocity in anti-ferromagnets and ferrimagnets.

7.1 Ferro-, antiferro- and ferrimagnets

We briefly consider the development of the domain-wall velocity in anti-ferromagnets and ferrimagnets as they are beneficial for

---

Skyrmions and Hall Transport
Bom Soo Kim
Copyright © 2023 Jenny Stanford Publishing Pte. Ltd.
ISBN 978-981-4968-34-8 (Hardcover), 978-1-003-37253-0 (eBook)
www.jennystanford.com

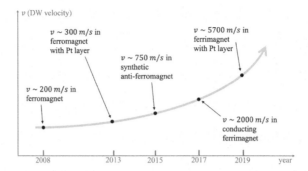

Figure 7.1 Recent advancements of domain wall velocity using different materials and techniques.

understanding the magnetic materials where skyrmions exist. This section is liberally borrowed from the inspiring talk given by S. K. Kim during the 2021 APS March meeting [146].

7.1.1 Domain wall velocity in ferrimagnets

As can be seen in Fig. 7.1, the domain wall (DW) velocity has increased by introducing new materials and new techniques that manipulate the corresponding spin motions. Around 2008, it was demonstrated that domain wall can achieve $v \sim 200 \, m/s$ in a thin ferromagnetic permalloy (NiFe) [147] in the context of building domain wall racetrack. The deriving mechanism of spins is the spin transfer torque (STT) in the ferromagnetic material that is introduced in §4.2.1. One disadvantage of the spin transfer torque is the joule heating that increases with electric current as $P \propto I^2 R$ with power P, current I, and resistance R. Thus we cannot inject a large electric current without losing a significant amount of energy. To overcome this drawback, a heavy metal layer, such as Platinum (Pt) or Tungsten (W), is attached to the magnetic layer. The domain wall velocity was demonstrated to be $v \sim 300 \, m/s$ [59][148] by injecting electric currents through the heavy metal layer and by controlling the domain wall motion using the spin orbit torque (SOT) introduced in §4.2.3 or the spin Hall torque (SHT) in §4.2.5.

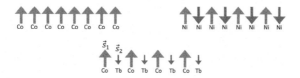

Figure 7.2 Schematic illustrations of magnetic materials. Top left: ferromagnet (Co, Fe, $\cdots$). Top right: anti-ferromagnet (NiO, Cr$_2$O$_3$, $\cdots$). Bottom: ferrimagnet (CoTb, GdFeCo, $\cdots$). The two different atoms in ferrimagnet have different magnetic moments $\vec{s}_1$ and $\vec{s}_2$.

As an alternative for ferromagnet, anti-ferromagnetic materials have been considered due to its fast intrinsic frequency at THz (terahertz =$10^{12}/s$) range and faster domain wall velocity as a result of the suppression of the angular precession known as Walker breakdown of domain wall motion in ferromagnets. In the context of experimental verification of domain wall motion [149], a multi-layer structure with two magnetic layers separated by an ultra-thin anti-ferromagnetic coupling layer was used. This synthetic anti-ferromagnet is different from the naturally occurring anti-ferromagnetic layer with anti-parallel spins for the same atoms illustrated in the top right panel of Fig. 7.2. By coupling another heavy metal layer to the synthetic anti-ferromagnet and injecting current through the heavy metal layer, the domain wall velocity was increased to $v \sim 750\ m/s$. While achieving faster motion is exciting, it is difficult to observe or manipulate the magnetic signals in anti-ferromagnets as the overall magnetic moments vanishes due to anti-parallel spin alignments. This immunity of anti-ferromagnets to magnetic fields presents difficulties in creating, manipulating, and detecting anti-ferromagnetic domain walls.

An exciting recent progresses have been realized with yet an another different type of magnetic materials, ferrimagnets illustrated in the bottom panel of Fig. 7.2. By overcoming the above-mentioned difficulties, joule heating and field immunity, ferrimagnetic materials offer fast dynamics and easy detection of magnetic signals. They have offered much faster domain wall velocity $v \sim 2000\ m/s$ when certain specific conditions meet, *i.e.* at temperature $T = 310\ K$ and driving field $\mu_0 H = 100\ mT$. This particular temperature is the so-called angular momentum compensation point. Away from the temperature and smaller driving field, the domain wall velocity drops quickly [150]. This experimental results have led to generalize the Landau-Lifshitz-Gilbert (LLG) equation, which is explained along with the dynamics of the ferrimagnets in §7.1.2. Moreover, by combining the ferrimagnets and heavy metal layer such as Pt, the domain wall velocity has been increased to $v \sim 5700\,m/s$ in cobalt-gadolinium (CoGd/Pt) alloy [150] and $v \sim 4300\,m/s$ in insulating bismuth-substituted yttrium iron garnet (BiYIG/Pt) [151].

7.1.2 Theory for ferrimagnet dynamics

As mentioned in the previous subsection, §7.1, the domain wall velocity in ferrimagnets has a strong temperature dependence. To understand the basic mechanism, we revisit the Landau-Lifshitz-Gilbert (LLG) equation from a slightly different view

$$\partial_t \vec{m} = -\gamma \vec{m} \times \vec{H}_{\text{eff}} + \alpha \vec{m} \times \partial_t \vec{m}\,, \tag{7.1}$$

where α is a damping coefficient and the gyromagnetic ratio γ is given by the magnetic moment M divided by angular momentum s

$$\gamma = \frac{M}{s}\,. \tag{7.2}$$

For ferromagnets depicted in the top left panel of Fig. 7.2, the the damping coefficient and the gyromagnetic ratio are constants as M and s are proportional to each other. Thus, the temperature dependence of domain wall motion cannot be explained by the LLG equation given in (7.1).

We note that the effective gyroscopic ratio can be changed with temperature, which happens in the rare-earth 2d transition metal compounds (illustrated in the bottom panel of Fig. 7.3). We rewrite the LLG equation by separating γ into M and s as

$$s(\partial_t \vec{m} - \alpha \vec{m} \times \partial_t \vec{m}) = -M\vec{m} \times \vec{H}_{\text{eff}}\,. \tag{7.3}$$

This form allows us to consider the spin density s and magnetization M as independent parameters, in particular with the separate temperature dependence. This is crucial for the ferrimagnetic

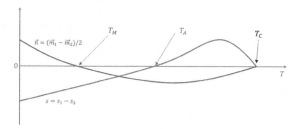

Figure 7.3 Schematic illustrations of the temperature dependence of the magnetic moment $\vec{n} = (\vec{m}_1 - \vec{m}_2)/2$ and the spin density $s = s_1 - s_2$. Two temperatures T_M and T_A are the magnetization and angular momentum compensation points. T_c is Curie temperature, above which the magnetic properties of material vanishes.

material such as GdFeCo, as its parameters s and M reveal non-trivial temperature dependence as illustrated in Fig. 7.3. The corresponding gyromagnetic ratio γ is highly non-trivial as it vanishes at $T = T_M$ and diverges at $T = T_A$. In particular, the LLG equation in the form (7.1) cannot be used near the angular momentum compensation point. Furthermore, the first order time derivative term (that comes from Berry phase and leads slower dynamics) in (7.3) vanishes at the compensation point. This indicates that we need second time derivative term to properly describe its associated dynamics.

Two different spins $\vec{s}_1 = s_1 \vec{m}_1$ and $\vec{s}_2 = s_2 \vec{m}_2$ in ferrimagnets, as in the bottom panel of Fig. 7.2, satisfy the LLG equations in the form (7.3) with an additional contribution from the exchange interaction between $\vec{m}_1$ and $\vec{m}_2$.

$$s_1(\partial_t \vec{m}_1 - \alpha_1 \vec{m}_1 \times \partial_t \vec{m}_1) = -M_1 \vec{m}_1 \times \vec{H}_{\text{eff},1} - h M_1 M_2 \vec{m}_1 \times \vec{m}_2,$$
$$s_2(\partial_t \vec{m}_2 - \alpha_2 \vec{m}_2 \times \partial_t \vec{m}_2) = -M_2 \vec{m}_2 \times \vec{H}_{\text{eff},2} - h M_1 M_2 \vec{m}_2 \times \vec{m}_1,$$
$$(7.4)$$

where h is the strength of the exchange interaction. More appropriately, the combined system can be described by the following combinations of the magnetization $\vec{n}$ and the spin density s [150][152]

$$\vec{n} = \frac{\vec{m}_1 + \vec{m}_2}{2}, \quad s = s_1 + s_2,$$
$$\vec{m} = \vec{m}_1 - \vec{m}_2, \quad s_m = \frac{s_1 - s_2}{2}. \quad (7.5)$$

The vector $\vec{n}$ is the order parameter that captures the collinear structure in equilibrium. $\vec{m}$ corresponds to the relative canting of the two sub-lattices that vanishes in equilibrium. s and s_m are the net and staggered spin densities in equilibrium, respectively. The resulting equation is given by

$$s\partial_t \vec{n} + \rho \vec{n} \times \partial_t^2 \vec{n} = M\vec{n} \times \vec{H}_{\text{eff}} - \alpha_{\text{eff}} s_m \vec{n} \times \partial_t \vec{n}. \quad (7.6)$$

We note that the second derivative term arises by substituting the equation of motion of the field, $\vec{m} = s_m \chi \dot{\vec{n}} \times \vec{n}$ which has a clear derivation below in the form of the Lagrangian given in §7.1.2.1. This mass density has the form $\rho = s_m^2 \chi$ with the magnetic susceptibility χ. We come back to the derivation for this equation below when considering the corresponding action. One can also add other spin torque terms such as spin transfer torque or spin orbit torque to (7.6).

The equation (7.6) is interesting. It interpolates the dynamics between the ferromagnetic and anti-ferromagnetic systems by changing the temperature or other turning parameter such as chemical compositions. Away from the angular momentum compensation point, $T \neq T_A$, the first order time derivative dominates the second time derivative term as discussed in §1.2. Thus,

$$s\partial_t \vec{n} = M\vec{n} \times \vec{H}_{\text{eff}}, \quad (7.7)$$

where we simply omit the damping term. This equation is nothing but (7.3) for a net magnetic moment $\vec{n}$.

On the other hand, the spin density vanishes, $s = 0$, at the angular momentum compensation point $T = T_A$. As the spin density vanishes, it describes anti-ferromagnets. In general, their magnetic structures are more numerous and varied compared to ferromagnets. The total magnetic moment within the magnetic unit cell sums up to vanish, while individual atoms have non-zero magnetic moments in the microscopic level. Here, $s = s_1 + s_2 = 0$. Thus, the equation

$$\rho \vec{n}_A \times \partial_t^2 \vec{n}_A = M \vec{n}_A \times \vec{H}_{\text{eff}} \tag{7.8}$$

describes the dynamics of anti-ferromagnet with $\vec{n}_A = \vec{m}_1 = -\vec{m}_2$. Note that the spins in anti-ferromagnetic materials are described by the second time derivative term. Typical dynamical time scale given by the frequency is THz (Terahertz). This is much faster than GHz (Gigahertz), typical frequency of dynamics of ferromagnet with the first time derivative.

We describe the Thiele equation for the ferrimagnets. We can derive it from the generalized LLG equation (7.6) following our discussion in §4.4. Here we use the collective coordinate $\vec{R}(t)$ for the center of rigid skyrmion as

$$\vec{n}(\vec{x}, t) = \vec{n}_0(\vec{x} - \vec{R}(t)) . \tag{7.9}$$

The corresponding Thiele equation can be evaluated following §4.4 and has the form

$$\mathcal{M}\ddot{\vec{R}} = Q\dot{\vec{R}} \times \vec{B} - \mathcal{D}\dot{\vec{R}} + \vec{F} , \tag{7.10}$$

where $\dot{}$ refers to the time derivative. Here we use slightly different notations compared to (4.51) in §4.4.2.1. The emergent magnetic field is given by $\vec{B} = B\hat{z} = -s\hat{z}$. The drag coefficient $\mathcal{D}$ and mass $\mathcal{M}$ have the forms

$$\begin{aligned} \mathcal{D} &= \alpha_{\text{eff}} s_m \int_{UC} dxdy \, \frac{\partial \vec{n}_0}{\partial x} \cdot \frac{\partial \vec{n}_0}{\partial x} , \\ \mathcal{M} &= \rho \int_{UC} dxdy \, \frac{\partial \vec{n}_0}{\partial x} \cdot \frac{\partial \vec{n}_0}{\partial x} , \end{aligned} \tag{7.11}$$

where UC refers unit cell. To avoid confusion, we explicitly use the two dimensional coordinates (x,y) for volume element. The same result can be obtained by taking the derivative with y coordinate instead of x coordinate. The last term contains various forces $\vec{F} = \vec{F}_U + \vec{F}_{STT} + \cdots$. The contribution from the spin transfer torque F_{STT} can be written as a part of the force term, while it is explicitly written as $\vec{v}_s$ in (4.51). There is also the internal force, that can be derived by taking the derivative of the internal energy $U(\vec{R})$ as

$$\vec{F}_U = -\frac{dU}{d\vec{R}}. \tag{7.12}$$

7.1.2.1 Lagrangian for ferrimagnets

The corresponding Lagrangian for the spin dynamics of ferrimagnets has been developed in [152][153]. We briefly mention its derivation.

Let us consider a two-dimensional collinear magnet that consists of two inequivalent sublattices with the local spin densities $\vec{s}_1 = s_1\vec{m}_1$ and $\vec{s}_2 = s_2\vec{m}_2$ with slowly varying unit vectors $|\vec{m}_1|^2 = |\vec{m}_2|^2 = 1$. The spin densities s_1 and s_2 can be positive or negative. We use the new spin vectors $\vec{n} = (\vec{m}_1 + \vec{m}_2)/2$, $\vec{m} = \vec{m}_1 - \vec{m}_2$, $s = s_1 + s_2$, and $s_m = (s_1 - s_2)/2$ as in (7.5). We can easily check $\vec{n} \cdot \vec{m} = 0$. In this notation, a ferromagnetic exchange can be described by $s_1, s_2 > 0$, and the corresponding $\vec{n}$ describes the net spin density in equilibrium. For anti-ferromagnet, $s_1 \cdot s_2 < 0$, and $\vec{n}$ is the staggered spin density. In equilibrium the spin vectors are collinear $\vec{m}_1 = \vec{m}_2$. For a small deviation from the equilibrium, $|\vec{n}| = 1$ and $|\vec{m}| \ll 1$.

The Lagrangian density for the Berry phase is given by

$$\mathcal{L}_B = -s_1 \mathbf{a}(\vec{m}_1) \cdot \dot{\vec{m}}_1 - s_2 \mathbf{a}(\vec{m}_2) \cdot \dot{\vec{m}}_2. \tag{7.13}$$

The vector potential for magnetic monopoles satisfies the equation

$$\vec{\nabla}_{\vec{m}} \times \mathbf{a}(\vec{m}) = \vec{m}. \tag{7.14}$$

This equation makes sense when we impose $\vec{m} = \hat{r}$ for the magnetic monopole that has the magnetic field $\vec{B} = \hat{r}/r^2$ with $r^2 = 1$ as discussed in §1.2.3.

We expand the Lagrangian up to the second order in $\vec{m}$ and $\dot{\vec{n}}$, which are small near the equilibrium. We evaluate one term explicitly as

$$\begin{aligned}\mathbf{a}(\vec{m}_1) \cdot \dot{\vec{m}}_1 &= \mathbf{a}_j\left[\vec{n} + \frac{\vec{m}}{2}\right]\left(\dot{n}_j + \frac{\dot{m}_j}{2}\right) \\ &= \left(\dot{n}_j + \frac{\dot{m}_j}{2}\right)\left[\mathbf{a}_j(\vec{n}) + \left(\frac{m_i}{2}\partial_i\right)\mathbf{a}_j(\vec{n})\right] + \cdots.\end{aligned} \tag{7.15}$$

Similar expression for $\vec{m}_2$ has relative signs. Thus the Lagrangian $\mathcal{L}_B$ can be written as

$$\mathcal{L}_B = -s\big[\mathbf{a}(\vec{n}) \cdot \dot{\vec{n}} + \dot{m}_j m_i \partial_i \mathbf{a}_j(\vec{n})/4\big] \\ - s_m\big[\mathbf{a}(\vec{n}) \cdot \dot{\vec{m}} + \dot{n}_j m_i \partial_i \mathbf{a}_j(\vec{n})\big] + \cdots . \quad (7.16)$$

We can set $\partial_i \mathbf{a}_j = A\delta_{ij} + \epsilon_{ijk}\beta_k$. We can check $\beta_k = n_k$ from the equation of vector potential. The last term gives $\dot{n}_j m_i \partial_i \mathbf{a}_j = \dot{n}_j m_i(A\delta_{ij} + \epsilon_{ijk}n_k) = -A\dot{m}_i n_i - \vec{n} \cdot (\dot{\vec{n}} \times \vec{m}) = -\vec{n} \cdot (\dot{\vec{n}} \times \vec{m})$, where we use $\dot{n}_i m_i = -\dot{m}_i n_i$ as $n_i m_i = 0$. To describe the slow dynamics, we ignore the terms with $\dot{\vec{m}}$. Then we get

$$\mathcal{L}_B = -s\mathbf{a}(\vec{n}) \cdot \dot{\vec{n}} + s_m \vec{n} \cdot (\dot{\vec{n}} \times \vec{m}) , \quad (7.17)$$

where the first term is Lagrangian for the ferromagnets with net spin Berry phase, while the second term is the Lagrangian for the anti-ferromagnets that can describe the system when the spin Berry phase of the two sub-lattices cancel. There is also a potential energy term $U(\vec{n}, \vec{m})$ that can be also expanded for small $\vec{m}$ as

$$\mathcal{U}(\vec{n}, \vec{m}) = \mathcal{U}(\vec{n}) + \frac{|\vec{m}|^2}{2\chi} + \cdots , \quad (7.18)$$

where χ is proportional to the magnetic susceptibility. Terms linear in $\vec{m}$ do not usually appear in Lagrangian as they can be absorbed into the square term by shifting the vector unless there are physical reasons that require the term such as the time-dependent Berry phase term.

The total Lagrangian without other elements such as magnetic fields or currents, we get $\mathcal{L} = \mathcal{L}_B - \mathcal{U}$.

$$L(\vec{n}, \vec{m}) = \int d^2x \Big[- s\mathbf{a}(\vec{n}) \cdot \dot{\vec{n}} + s_m \vec{n} \cdot (\dot{\vec{n}} \times \vec{m}) \\ - \frac{|\vec{m}|^2}{2\chi} - \mathcal{U}(\vec{n}) \Big] . \quad (7.19)$$

The equation of motion of $\vec{m}$ gives

$$\vec{m} = s\chi \dot{\vec{n}} \times \vec{n} . \quad (7.20)$$

By using this, we get the Lagrangian with only vector $\vec{n}$.

$$L(\vec{n}) = \int d^2x \left[-s\mathbf{a}(\vec{n}) \cdot \dot{\vec{n}} + \frac{\rho}{2}\dot{\vec{n}}^2 - \mathcal{U}(\vec{n}) \right] . \tag{7.21}$$

Thus, we achieve our goal to derive the spin Lagrangian that contains the terms with both first and second time derivatives.

This Lagrangian can be extended to include an external field $\vec{B}$ as [153]

$$\mathcal{L}(\vec{n}) = -s\mathbf{a}(\vec{n}) \cdot \dot{\vec{n}} + \frac{\rho}{2}\left(\dot{\vec{n}} - g_t \vec{n} \times \vec{B}\right)^2 - \mathcal{U}(\vec{n}) , \tag{7.22}$$

where g_t is the gyromagnetic ratio for the transverse component of the spin density with respect to the spin direction $\vec{n}$.

With the digression of recent experimental and theoretical advancements for anti-ferromagnetic and ferrimagnetic materials in the context of domain wall motion, we return to the transverse motion of the skyrmions on these materials. The connection comes with an interesting experimental observation of vanishing skyrmion Hall effect at the angular momentum compensation point reported in [154]. This is interesting from the skyrmion racetrack applications as we can have better controls on skyrmion motion.

7.2 Hall angle data of skyrmion and antiskyrmion

In this section, we consider the difference between Hall effects acting on skyrmion and antiskyrmion as illustrated in Fig. 7.4. We already discussed the skyrmion Hall effect and the corresponding skyrmion Hall angle in §6.1.2. While there have been lots of skyrmion Hall experiments, there are not many data that contain both the skyrmion and antiskyrmion Hall angles in systematic ways. Here we list two different data points for Néel-type skyrmions, one for a ferromagnetic material given in [123] and another for a ferrimagnet in [154].

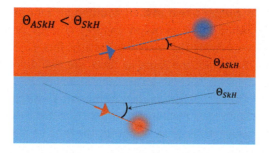

Figure 7.4 Illustrations of skyrmion and antiskyrmion motion in their appropriate ferromagnetic backgrounds. It turns out that their Hall angles are not the same. The Hall viscosity is estimated from the difference as the Hall viscosity does not depend on the skyrmion charge, while skyrmion Hall angle does.

7.2.1 For a ferromagnet

In [123], a systematic study for skyrmion and antiskyrmion motion with two opposite moving directions was performed in the Ta/CoFeB/TaO$_x$ trilayer material with a ferromagnetic cobalt-iron-boron layer sandwiched with tantalum and tantalum-oxide layers. The Néel-type skyrmions in the middle layer are controlled by the spin Hall spin transfer torque by the electric current passing through the heavy metal tantalum (Ta) layer, which has a large spin Hall angle.

The material has strong pinning potential due to randomly distributed defects, and thus transverse motion was not observed at low current densities. The transverse motion saturates with a constant values of skyrmion Hall angles at pulsed high current density, which was observed by current-driven imaging data acquired by a polar magneto-optical Kerr effect (MOKE) microscope at room temperature.

	(J_e, Q)	$(B, \Theta_{(A)SkH})$	$(B, \Theta_{(A)SkH})$	$(B, \Theta_{(A)SkH})$
I	$(+,-)$	$(4.8, 28°)$	$(5.4, 32°)$	$(5.0, 29.3°)$
II	$(-,+)$	$(-4.6, 28°)$	$(-5.2, 34°)$	$(-5.0, 32°)$
III	$(-,-)$	$(4.8, -29°)$	$(5.4, -33°)$	$(5.0, -30.3°)$
IV	$(+,+)$	$(-4.6, -29°)$	$(-5.2, -33°)$	$(-5.0, -31.6°)$

The table contains two sets of 4 different data points in the middle two columns for skrmion charges $Q = \pm 1$ and currents directions $J_e = \pm$ with the electron particle flux J_e. The applied magnetic fields are not the same as there are two different values either with $B = (4.8, 5.4)$ Oe for antiskyrmions or $B = (-4.5, -5.2)$ Oe (oersted = $(1/4\pi) \times 10^2 A/m$) for skyrmions. These data are from Fig. 3c of [123]. To have meaningful comparison, we normalized the effects of the magnetic field by obtaining the data for $B = 5.0$ Oe by interpolating these data. These are listed in the right column.

From these data, we note $|\Theta_{SkH}| = 31.6°$ (IV) and $\Theta_{ASkH} = 29.3°$ (I) are significantly different ($\Delta|\Theta| = 2.3°$) for the positive current $+J_e$ and $B = \pm 5.0$ Oe. Similarly, $|\Theta_{SkH}| = 32°$ (II) and $\Theta_{ASkH} = 30.3°$ (III) are also different ($\Delta|\Theta| = 1.7°$) for the negative current $-J_e$ and $B = \pm 5.0$ Oe. Fig. 7.4 illustrates this for $+J_e$. For both directions of current, the skyrmion Hall angel is different from that of the antiskyrmion. Below in §7.4, we come back to these data after developing a generalized Thiele equation, that can take these differences into account.

We also note that the skyrmion Hall angles change when current is reversed. For the skyrmion, $|\Theta_{SkH}| = 31.6°$ (IV) for $+J_e$ and $\Theta_{SkH} = 32°$ (II) for $-J_e$. Thus, the corresponding Hall angle changes, $\Delta|\Theta_{SkH}| = 0.4°$ for Q. This is not significant compared to the difference originated by skyrmion and antiskyrmion. For antiskyrmion, the change under the current reversal is $\Delta|\Theta_{ASkH}| = 1°$ for $-Q$ as $|\Theta_{ASkH}| = 30.3°$ (III) for $-J_e$ and $\Theta_{ASkH} = 29.3°$ (I) for J_e. Again, this is not as significant as the difference between the skyrmion and antiskyrmion. This is consistent with the experimental results in [72], where skyrmion Hall angles driven by opposite currents are shown to be the same.

We provide one more comment on the experimental results given in [123]: various 'not-saturated' skyrmion Hall angle data are presented in Fig. 3c along with the data sets in Figs. 2r and 2x. With similar investigations as above, these data show more significant differences between the skyrmion Hall angle and antiskyrmion Hall angle.

7.2.2 For a ferrimagnet

Here we return to the motion of a ferrimagnetic domain wall, which can be interpreted as half skyrmion. This allows us to discuss the skyrmion and antiskyrmion Hall angles as they are directly measured in [154].

The ferrimagnetic material used is GdFeCo/Pt film with the Pt layer. The latter provides the spin current that can control the domain wall or skyrmion using the spin orbit torque. By measuring the sharp peak of the domain wall velocity as a function of temperature, the angular momentum compensation point is determined as $T_A = 287 \pm 5K$. This sharp peak signifies that the spin dynamics is that of the antiferromagnet, which can be described by the zero spin density $s = 0$. At this compensation point $T = T_A$, the skyrmion Hall angle vanishes [154].

The rare earth Gd and transition metal FeCo have different temperature-dependent spin densities due to different intra-atomic exchanges. The net spin density changes gradually with temperature. Away from the compensation point $T = T_A$, the magnet GdFeCo behaves as ferromagnet as the compound's anti-aligned moments between Gd and FeCo are not completely canceled out. Thus, at $T = 343K > T_A$, the material behaves as a ferromagnet. This particular experiment observed that the bubble domian was elongated in response to the applied current, instead of moving along. This unexpected behavior was interpreted to happen due to a strong pinning potential near the half-ring shaped writing line (lithography-induced damage).

With the setup at $T = 343K$, Hall angle measurements were performed on Néel-type magnetization up domain wall (identified as half skyrmions $Q = 1/2$) in spin-down ferrimagnetic GdFeCo film. The corresponding skyrmion Hall angle is measured as $\Theta_{SkH} = -35^o$. On the other hand, with the same setup, the antiskyrmion Hall angles is measured as $\Theta_{ASkH} = 31^o$ for the magnetization down domain wall (identified as half antiskyrmions $Q = -1/2$) in spin-up film. This set of data have significant difference between the skyrmion Hall angle and the antiskyrmion Hall angle. We return to this data set after developing a generalized Hall angle with the Hall viscosity as the conventional skyrmion Hall angle cannot explain the difference.

7.3 Direction of Hall viscosity

The significant difference between skyrmion and antiskyrmio Hall angles cannot be explained by the known skyrmion Hall effect. They have the same magnitudes with opposite directions as the Hall angle is directly proportional to skyrmion charge. Here we investigate the direction of Hall viscosity and see that the transverse motion due to the Hall viscosity is the same for skyrmion and antiskyrmion.

To figure out the direction of Hall viscosity η_H on skyrmions, we start with its explicit formula developed in §2.3. Let us consider a small deformation ξ_i of a fluid. Under a strain rate, $\dot{\xi}_{ij} = \partial_i \dot{\xi}_j + \partial_j \dot{\xi}_i$ with a time derivative ˙, the viscosity tensor contains the antisymmetric component in the system with broken parity as

$$T_{ij} = -(\eta^S_{ijkl} + \eta^A_{ijkl})\dot{\xi}_{kl} . \tag{7.23}$$

T^{ij} is stress energy tensor. For a 2 dimensional fluid with rotational invariance,

$$\begin{aligned}\eta^S_{ijkl} &= \eta(\delta_{ik}\delta_{jl} + \delta_{il}\delta_{jk}) + (\zeta - \eta)\delta_{ij}\delta_{kl} , \\ \eta^A_{ijkl} &= -\eta^A_{klij} = -\frac{\eta_H}{2}(\epsilon_{ik}\delta_{jl} + \epsilon_{jl}\delta_{ik} + \epsilon_{il}\delta_{jk} + \epsilon_{jk}\delta_{il}) ,\end{aligned} \tag{7.24}$$

where η, ζ, and η_H are shear, bulk, and Hall viscosities and $\delta_{ij}, \epsilon_{ij}$ are symmetric and antisymmetric unit tensors. Here, i, j, k, l refer to coordinates x, y [29].

First, the Hall viscosity is transverse to the direction of the shear viscosity. We explicitly compute $T^{ij} da_j$, that is the force acting on the surface da_j due to the momentum along the i-th direction. Let us consider a radially outward flow depicted in the left panel of Fig. 7.5. Then, the force due to x momentum acting on small area element $\Delta a_y = \Delta x$ facing y coordinate along the horizontal direction is given by $T^{xy} \Delta a_y$. For the radial flow and focusing on the small region near the horizontal axis, we have $\dot{\xi}_x = v_x(x)$ and $\dot{\xi}_y = v_y = 0$. Then,

$$\begin{aligned}T^{xy}\Delta a_y &= -2\left[\eta(\partial_x v_y + \partial_y v_x) - 2\eta_H(\epsilon_{xy}\partial_y v_y + \epsilon_{yx}\partial_x v_x)\right]\Delta a_y \\ &= -2\eta_H(\partial_x v_x)\Delta x .\end{aligned} \tag{7.25}$$

Skyrmions and Hall Transport | 337

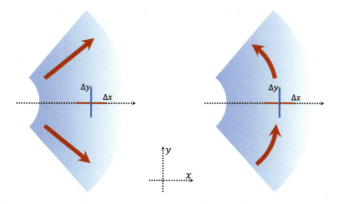

Figure 7.5 Illustrations of two quadrants for rotationally invariant flows as thick red arrows. Left: A radial flow is depicted with $\dot{\xi}_x = v_x(x)$ and $\dot{\xi}_y = v_y = 0$ along the x-coordinate. Right: A circular flow is depicted with $\dot{\xi}_x = v_x = 0$ and $\dot{\xi}_y = v_y(x)$ along the x-coordinate. Solid horizontal red and vertical blue lines are small area elements $\Delta a_y = \Delta x$ facing toward y direction and $\Delta a_x = \Delta y$ toward x direction, respectively.

The direction is perpendicular to the area element, which is y direction. Note that change in x momentum along x direction produces force in y direction, which is from the Hall viscosity.

It is instructive to compute the force due to x momentum acting on an area $\Delta a_x = \Delta y$ facing x coordinate (in the same left panel of the figure). Then,

$$\begin{aligned}T^{xx}\Delta a_x &= -2\left[2\eta\partial_x v_x - (\xi - \eta)(\partial_x v_x + \partial_y v_y)\right]\Delta a_x \\ &\quad + 2\eta_H \epsilon_{xy}(\partial_y v_x + \partial_x v_y)\Delta a_x \\ &= 2(\xi - 3\eta)(\partial_x v_x)\Delta y\,.\end{aligned} \quad (7.26)$$

The contribution comes from the shear and bulk viscosities and its direction is along the horizontal direction, along the flow. This is transverse to the contribution of the Hall viscosity. Thus the force due to Hall viscosity acts transverse to the fluid motion. This fact signifies that the Hall viscosity does not slow motion and is dissipationless. This radial flow is relevant for device applications where skyrmions are pushed with pulse-like forces.

We can also consider the circulating flow, depicted in the right panel of Fig. 7.5. We consider two different cases where the flow along the y direction at $y = 0$ near the horizontal direction. First, we take the flow motion to be $\dot{\xi}_y = v_y(y)$ and $\dot{\xi}_x = v_x = 0$. Similar to the previous computations, we obtain

$$\begin{aligned}
T^{xx}\Delta a_x &= 2(\xi - \eta)(\partial_y v_y)\Delta y\,, \\
T^{yy}\Delta a_y &= 2(\xi - 3\eta)(\partial_y v_y)\Delta x\,, \\
T^{xy}\Delta a_y &= 2\eta_H(\partial_y v_y)\Delta x\,, \\
T^{yx}\Delta a_x &= 2\eta_H(\partial_y v_y)\Delta y\,.
\end{aligned} \quad (7.27)$$

This flow contains the bulk viscosity that has expansion of the flow. Still the Hall viscosity is separated from the shear and bulk viscosity. Second, we consider the flow described by $\dot{\xi}_y = v_y(x)$ and $\dot{\xi}_x = v_x = 0$. Repeating the same computations, we obtain

$$\begin{aligned}
T^{xx}\Delta a_x &= 2\eta_H(\partial_x v_y)\Delta y\,, \\
T^{yy}\Delta a_y &= -2\eta_H(\partial_x v_y)\Delta x\,, \\
T^{xy}\Delta a_y &= -2\eta(\partial_x v_y)\Delta x\,, \\
T^{yx}\Delta a_x &= -2\eta(\partial_x v_y)\Delta y\,.
\end{aligned} \quad (7.28)$$

The forms are different from the previous cases as the derivative of the velocity already picks up asymmetric contribution $\partial_x v_y \neq 0$. In this case, the Hall viscosity is parallel to the direction of fluid motion. The result can be confirmed in cylindrical coordinates with a finite area element. In all these cases, we see that Hall viscosity is transverse to the shear viscosity.

7.3.1 Hall viscosity is independent of skyrmion charge

While the above computations confirm that the Hall viscosity is transverse to shear viscosity and dissipationless, it does not directly tell us the directions of the skyrmion motion. Thus, we look for other ways to figure out the direction for skyrmions.

Here we demonstrate that the Hall viscosity and the corresponding transverse force only depend on the direction of flow and are independent of skyrmion charge. One decisive way comes from the Kubo formula for Hall viscosity [34]. The expression is given by

$$\eta_H = \lim_{\omega \to 0} \frac{\epsilon_{ik}\delta_{jl}}{4i\omega} \tilde{G}_R^{ij,kl}(\omega, \vec{0}), \tag{7.29}$$

where ω is a frequency, repeated indices are summed over, and $\tilde{G}_R^{ij,kl}$ is a momentum space representation of the retarded Green's function for the stress energy tensor T^{ij},

$$G_R^{ij,kl}(t,\vec{x};t',\vec{x}') = -i\theta(t-t')\langle [T^{ij}(t,\vec{x}), T^{kl}(t',\vec{x}')]\rangle. \tag{7.30}$$

This is an even function of the tensor T^{ij}.

Let us consider a simple Lagrangian

$$\mathcal{L} = \dot{\Phi}(\cos\Theta - 1) - \frac{J}{2}\partial_i\vec{n}\cdot\partial_i\vec{n}, \tag{7.31}$$

which accommodates skyrmions. In cylindrical coordinate (ρ, ϕ, z), the magnetization $\vec{n}$ ($\vec{n}^2 = 1$) is parameterized as

$$\vec{n} = (\sin\Theta(\rho)\cos\Phi(\phi), \sin\Theta(\rho)\sin\Phi(\phi), \cos\Theta(\rho)). \tag{7.32}$$

The corresponding stress energy tensor, $T_{ij} \propto \partial_i\vec{n}\cdot\partial_j\vec{n}$, is a quadratic function of $\vec{n}$. The Green's function should have even number of vector $\vec{n}$. Thus, we confirm that Hall viscosity is invariant under $\vec{n} \to -\vec{n}$.

On the other hand, the skyrmion charge

$$Q = \int d^2x\, \vec{n}\cdot(\partial_x\vec{n}\times\partial_y\vec{n}) \tag{7.33}$$

is an odd function of $\vec{n}$ and changes its sign under $\vec{n} \to -\vec{n}$. Thus, we conclude that both skyrmion and antiskyrmion experience the same transverse force due to the Hall viscosity. This is depicted in the right panel of Fig. 7.6. We also illustrate the skyrmion Hall effect in the left panel of Fig. 7.6 to see their differences clearly.

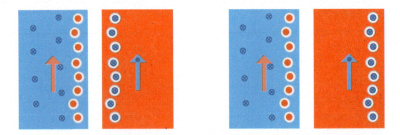

Figure 7.6 Left: Illustration of skyrmion Hall effect. Skyrmions (middle up-spin, red dots) in down-spin ferromagnet (blue background with cross circles) move toward right, while antiskyrmions (middle down-spin, blue cross circles) in up-spin ferromagnet (red background) toward left. Right: Illustration of the skyrmion motion due to the Hall viscosity. Both skyrmions and antiskyrmions move toward right. Thick arrow is the direction of a driving force, such as a positive electron current. Reversing the force results in turning around all figures by 180°.

7.4 Modeling Hall viscosity in Thiele equation

Thiele described the steady-state motion of a skyrmion (originally a domain wall) by parameterizing the magnetization vector $\vec{n}$ as

$$n_i = \frac{1}{M_s} M_i(x_j - X_j) , \qquad (7.34)$$

where $M_s = |\vec{M}|$ is a saturation magnetization, x_j is the field position, and $X_j = v_j t$ represents the linear motion of the skyrmion center moving with a velocity v_i [82].

We note that various spin torques that we have introduced in this book or in the literature cannot take into account of the effects of the Hall viscosity, which is independent of skyrmion charge. Thus, here we offer another generalization for the linear motion of the skrmion center. Motivated by the fact that Hall viscosity is actually a fundamental and universal transverse motion, we generalize the center of motion of skyrmion with the transverse component as

$$X_i = v_i t + R_{ij} v_j t , \qquad (7.35)$$

where $R_{ij} = R \epsilon_{ij}$.

It is important to note that the generalization does not alter the Thiele's initial setup: (i) conserved magnetization, $M_i \partial_t M_i = 0$ and thus $|\vec{M}| = const.$, and (ii) spatially constant saturation magnetization, $M_i \partial_j M_i = 0$. The former is a consequence of Landau-Lifshitz-Gilbert (LLG) equation, while the latter is followed by the parameterization $M_i(x_j - X_j)$ and (7.35) for $\partial_t M_i = (v_j + R_{jk} v_k) \partial_j M_i$.

From the fact that M_i and $\partial_t M_i$ are orthogonal, Thiele showed the equation

$$-\frac{\epsilon_{jkl} M_k \partial_t M_l}{\gamma_0 M_s^2} - \alpha \frac{\partial_t M_j}{\gamma_0 M_s} + \tilde{\beta} M_j + H_j = 0 \qquad (7.36)$$

is equivalent to LLG equation. One can check it by multiplying $-\epsilon_{jik} M_k$ to (7.36), summing over j, and renaming the indices. Multiplying M_j with (7.36) fixes the value $\tilde{\beta} = -M_j H_j / M_s^2$, which does not contribute below. γ_0 is the gyromagnetic ratio and α is a damping parameter. We also provided the detailed derivation in §4.4.1.

By multiplying $-\partial M_j / \partial x_i$ on (7.36) followed by integrating over unit skyrmion volume, one arrives at the generalized Thiele equation.

$$\mathcal{G}_{ij}(v_j + R_{jk} v_k) + \alpha \mathcal{D}_{ij}(v_j + R_{jk} v_k) + F_i = 0, \qquad (7.37)$$

here $i, j, k = x, y$ in (7.37),

$$\begin{aligned} \mathcal{G}_{ij} &= \epsilon_{ij} \frac{M_s}{\gamma_0} Q, \\ \mathcal{D}_{ij} &= -\frac{M_s}{\gamma_0} \int d^2 x (\partial_i n_k)(\partial_j n_k), \end{aligned} \qquad (7.38)$$

where $\mathcal{G}_{ij} v_j$ is the Magnus force with the skyrmion charge Q, $\mathcal{D}_{ij} v_j$ is total dissipative drag force, and $F_i = -M_s \int d^2 x (\partial_i n_j) H_j$ is total external force that can include forces due to various spin torques. Internal forces due to anisotropy and exchange energies, internal demagnetizing fields, and magnetostriction do not contribute [82]. The third term in (7.36) does not contribute either.

We consider a skyrmion configuration $\vec{n}$ parameterized by

$$\Phi(\phi) = m\phi + \delta_0, \qquad (7.39)$$

where the integer m characterizing the topological skyrmion charge. Then, $Q = 4\pi m$ and

$$\mathcal{D}_{xx} = \mathcal{D}_{yy} = -\frac{M_s}{2\gamma_0} \int \rho d\rho \left(\Theta'(\rho)^2 + \frac{m^2}{\rho^2} \sin^2 \Theta \right),$$

$$\mathcal{D}_{xy} = \mathcal{D}_{yx} \propto \int_0^{2\pi} d\phi \sin(2\phi) = 0.$$

(7.40)

We check the drag term $\mathcal{D}_{ij}v_j \equiv (M_s/\gamma_0)\mathcal{D}v_i$ is parallel to a force direction F_i. A new contribution,

$$\mathcal{D}_{ij}R_{jk}v_k = \frac{M_s}{\gamma_0}\mathcal{D}R\epsilon_{ik}v_k,$$

(7.41)

is transverse to F_i. It is quadratic in $\vec{n}$ and independent of the skyrmion charge. The physical origin of this *universal transverse drag force* can be identified as the Hall viscosity discussed above. There exist other possible transverse forces surveyed and studied below, but they are from various spin torques depending on applied currents and can be distinguished easily.

Generalization (7.35) adds another contribution in (7.37),

$$\mathcal{G}_{ij}R_{jk}v_k = -\frac{M_s}{\gamma_0}QRv_i,$$

(7.42)

that depends on the skyrmion charge. Note it is parallel to the skyrmion moving direction. The combination QR from the skyrmion Hall effect and the Hall viscosity accelerates skyrmion and decelerates antiskyrmions, or vice versa, depending on the sign of R. This is in contrast to the transverse force (7.41). Thus, generalized Thiele equation (7.37), that includes (7.41) and (7.42), describes drastically different Hall transport phenomena.

7.4.1 Generalized skyrmion Hall angle

We consider the generalized Thiele equation (7.37) and set $F' = (\gamma_0/M_s)F_x$ and $F_y = 0$ for simplicity.

$$(\alpha \mathcal{D} - QR)v_x + (Q + \alpha \mathcal{D}R)v_y + F' = 0 ,$$
$$-(Q + \alpha \mathcal{D}R)v_x + (\alpha \mathcal{D} - QR)v_y = 0 . \tag{7.43}$$

From these two equations, we can compute the velocity components as

$$v_x = \frac{(QR - \alpha \mathcal{D})F'}{(Q^2 + \alpha^2 \mathcal{D}^2)(1 + R^2)} ,$$
$$v_y = -\frac{(Q + \alpha \mathcal{D}R)F'}{(Q^2 + \alpha^2 \mathcal{D}^2)(1 + R^2)} . \tag{7.44}$$

Velocity along the forcing direction is determined by $\alpha \mathcal{D} - QR$, which depends on the combination QR mentioned in (7.42), while the transverse velocity determined by $Q + \alpha \mathcal{D}R$. Note that we use slightly different normalization for Q without 4π to simplify the expression.

From the velocities, we compute the generalized skyrmion Hall angle with the Hall viscosity contribution as

$$\tan \Theta_{SkH} = \frac{v_y}{v_x} = \frac{Q + \alpha \mathcal{D}R}{\alpha \mathcal{D} - QR} . \tag{7.45}$$

Note the generalized Hall angle Θ_{SkH} depends on R as well as Q. Θ_{SkH} reduces to the usual skyrmion Hall angle $\tan \theta_{SkH} = Q/\alpha \mathcal{D}$ when $R = 0$ in the absence of transverse drag term, and antiskyrmion has the same Hall angle as skyrmion, $\theta_{ASkH} = -\theta_{SkH}$. Equation (7.45) is useful for insulating and conducting magnets with (anti-)skyrmions with combined applied forces in one direction, chosen to be x coordinate here.

7.4.2 Hall viscosity from data

Here we first estimate the Hall viscosity contribution using R in (7.45) from experimental data given in §7.2.2 [154]. Hall angle for magnetization-up states (half skyrmion $Q = 1/2 \times 4\pi$ for a slightly different normalization) is measured as $\Theta_{SkH} = -35°$, while that for magnetization-down states (half antiskyrmion $Q = -1/2 \times 4\pi$) is $\Theta_{ASkH} = 31°$. Using (7.45) twice, we obtain that $R = -0.035$ and $\alpha\mathcal{D} = -9.68$.

We compute the ratio of transverse force due to Hall viscosity $\alpha\mathcal{D}_{ij}R_{jk}v_k$ to that due to skyrmion Hall effect $\mathcal{G}_{ij}v_j$.

$$\frac{\text{Hall viscosity}}{\text{Skyrmion Hall effect}} = \frac{\alpha\mathcal{D}R}{Q} = 5.4\% . \tag{7.46}$$

It is a significant contribution. For this, we assume that the universal transverse force (independent of skyrmion charge) is entirely from Hall viscosity. We also assume that possible pinning effects due to material imperfection, excitations of internal degrees of freedom, and shape distortions for skyrmions are the same for antiskyrmions, along with the system-specific forces such as tension presented in [154]. We also recount the result (7.46) by surveying various other spin torques, especially SOT below.

Next, we use (7.45) to estimate the Hall viscosity contribution from the experimental data given in [123]. They are summarized in §7.2.1. Here the angles are measured with respect to the force direction. For the data set, $|\Theta_{SkH}| = 31.6°$ (IV) and $\Theta_{ASkH} = 29.3°$ (I), we get $R = -0.020$, $\alpha\mathcal{D} = -1.70$ and the Hall viscosity to skyrmion Hall effect ratio $\alpha\mathcal{D}R/Q = 3.5\%$. The other data sets, $|\Theta_{SkH}| = 32°$ (II) and $\Theta_{ASkH} = 30.3°$ (III), give $R = -0.015$, $\alpha\mathcal{D} = -1.65$, and $\alpha\mathcal{D}R/Q = 2.5\%$. By taking the average of 3.5% and 2.5%, we get

$$\frac{\text{Hall viscosity}}{\text{Skyrmion Hall effect}} = \frac{\alpha\mathcal{D}R}{Q} = 3\% . \tag{7.47}$$

Thus, from these two experimental data, we conclude that the Hall viscosity amounts a significant part $3\% - 5.4\%$ of the transverse motion of skyrmion motion. For fast moving skyrmions, this will add significant transverse motion.

7.4.3 Contributions from spin torques?

The original Thiele equation (4.48) in §4.4.1 has been generalized with forces from different spin torques, such as spin orbit torque, spin Hall torque, spin transfer torque, and emergent electromagnetic fields in §4.4.2. One or a combination of them can play a dominant role on skyrmion dynamics depending on physical situations.

In this section, we also include these spin torques to the generalized Thiele equation (7.37) with the transverse component (7.35). The different forces stemming from these spin torques, $F_i = \int d^2x f_i$, are integrals of force densities f_i over the skyrmion unit volume. In particular, we examine that whether the spin torques modify the estimates of the Hall viscosity presented in the previous section §7.4.2. Here we focus on the ferromagnetic materials and expect similar stores hold for the ferrimagnets.

7.4.3.1 Spin orbit torque (SOT)

Let us start with the spin orbit torque (SOT). Local magnetization in uniform ferromagnets experiences a torque through indirect s-d exchange interactions with conduction electron spins, which feel the so-called Rashba magnetic field in their rest frames due to local electric field [70][72].

There are two independent SOT force densities, damping-like $f_{i,DL}^{SOT}$ and field-like $f_{i,FL}^{SOT}$ ones with corresponding parameters a and $a\eta$. These force densities have been discussed in (4.26) and (4.27) of §4.2.3 and in the context of Thiele equation in §4.4.2.2. We list these two SOT terms as

$$f_{i,DL}^{SOT} = a\epsilon_{zlj}\epsilon_{lpq}(\partial_i n_p)n_q J_j^e ,$$
$$f_{i,FL}^{SOT} = -a\eta \left[\epsilon_{zpq}n_j - \epsilon_{jpq}n_z\right] n_p(\partial_i n_q) J_j^e .$$
(7.48)

Note they depend linearly on the electric current $\vec{J}^e$. We choose $J_x^e = J^e, J_y^e = 0$ to make expressions simple.

For detailed computations, we adapt the parametrization

$$\vec{n} = \sin\Theta(\rho)\cos\Phi(\varphi)\hat{x} + \sin\Theta(\rho)\sin\Phi(\varphi)\hat{y} + \cos\Theta(\rho)\hat{z} , \quad (7.49)$$

where ρ, φ are the two dimensional coordinates. We use the parametrization $\Phi(\phi) = m\phi + \delta_0$ as in (7.39) with a skyrmion topological number m and a phase δ_0 that is related to different skyrmion species. For example, $\delta_0 = 0$ for the Néel-type skyrmion.

Damping-like $F_{i,DL}^{SOT}$ has its force components as

$$\begin{aligned} f_{i,DL}^{SOT} &= a\epsilon_{zlx}\epsilon_{lpq}(\partial_i n_p)n_q J^e \\ &= -a\big[(\partial_i n_z)n_x - (\partial_i n_x)n_z\big]J^e \\ &= a\big[(\partial_i \Theta(\rho))\cos\Phi - (\partial_i \Phi(\varphi))\sin\Theta\cos\Theta\sin\Phi\big]J^e \, . \end{aligned} \qquad (7.50)$$

Now we can get the transverse force by integrating the density along y direction as the driving force is parallel to the current. Thus,

$$\begin{aligned} F_{y,DL}^{SOT} &= a\int \rho d\rho d\varphi \big[(\partial_y \Theta(\rho))\cos\Phi - (\partial_y \Phi(\varphi))\sin\Theta\cos\Theta\sin\Phi\big]J^e \\ &= a\int \rho d\rho d\varphi \Big[\Theta'(\rho)\sin\varphi\cos(m\varphi + \delta_0) \\ &\quad - \frac{\sin 2\Theta}{2}\frac{m}{\rho}\cos\varphi\sin(m\varphi + \delta_0)\Big]J^e \\ &= -a\frac{2\sin(m\pi)\sin(m\pi + \delta_0)}{m^2 - 1}\int \rho d\rho\Big[\Theta'(\rho) + \frac{m^2}{\rho}\frac{\sin 2\Theta}{2}\Big]J^e \, , \end{aligned} \qquad (7.51)$$

where we use the following to derive the second line

$$\begin{aligned} \partial_y \Theta(\rho) &= \frac{\partial \rho}{\partial y}\Theta'(\rho) = \sin\varphi\, \Theta'(\rho) \, , \\ \partial_y \Phi(\varphi) &= \frac{\partial(m\varphi + \delta_0)}{\partial y} = \frac{m}{\rho}\cos\varphi \, . \end{aligned} \qquad (7.52)$$

For stable skyrmions with $m = \pm 1$, the integral in (7.51) gives a simpler result as

$$F_{y,DL}^{SOT} = -am\pi(\sin\delta_0)\delta_{m,\pm 1}\int \rho d\rho\Big[\Theta'(\rho) + \frac{1}{\rho}\frac{\sin 2\Theta}{2}\Big]J^e \, . \qquad (7.53)$$

The same result can be obtained by taking the limit $m \to \pm 1$ from the bottom line of (7.51) using the L'Hospital's rule. This computation tells us interesting results that the Néel-type skyrmions with $\delta_0 = 0, \pi$ do not receive any transverse force component from the damping-like SOT. Before considering the field-like SOT, we mention that the Bloch-type skyrmion with $\delta_0 = \pm\pi/2$ receives transverse force due to damping-like SOT.

We can also compute the longitudinal force component as

$$F^{SOT}_{x,DL} = a \int \rho d\rho d\varphi \left[(\partial_x \Theta(\rho)) \cos\Phi - (\partial_x \Phi(\varphi)) \sin\Theta \cos\Theta \sin\Phi \right] J^e$$

$$= a \frac{2m \sin(m\pi) \cos(m\pi + \delta_0)}{m^2 - 1} \int \rho d\rho \left[\Theta'(\rho) + \frac{\sin 2\Theta}{2\rho} \right] J^e \quad (7.54)$$

$$\to a\pi (\cos\delta_0) \delta_{m,\pm 1} \int \rho d\rho \left[\Theta'(\rho) + \frac{\sin 2\Theta}{2\rho} \right] J^e,$$

where the last line is for $m = \pm 1$. Thus the Néel-type skyrmions have longitudinal force contributions, while the Bloch-type skyrmions do not receive contributions. We expect that this is not a major determining factor even for the longitudinal motion.

We turn to the field-like SOT force for the spin (7.49) for the current along x direction for $J^e = J^e_x$. Thus,

$$f^{SOT}_{i,FL} = -a\tilde{\eta} \left[\epsilon_{zpq} n_x - \epsilon_{xpq} n_z \right] n_p (\partial_i n_q) J^e$$

$$= -a\tilde{\eta} \left[(\partial_i \rho) \Theta'(\rho) \cos\Theta \sin\Phi + (\partial_i \Phi(\varphi)) \sin\Theta \cos\Phi \right] J^e. \quad (7.55)$$

Similar computations lead the integrated force along the transverse direction $i = y$ as

$$F^{SOT}_{y,FL} = -a\tilde{\eta} \int \rho d\rho d\varphi \left[(\partial_y \rho) \Theta'(\rho) \cos\Theta \sin\Phi + (\partial_y \Phi(\varphi)) \sin\Theta \cos\Phi \right] J^e$$

$$= -a\tilde{\eta} \frac{2 \sin(m\pi) \cos(m\pi + \delta_0)}{m^2 - 1} \int \rho d\rho \left[\Theta'(\rho) \cos\Theta + \frac{m^2 \sin\Theta}{\rho} \right] J^e$$

$$\to -a\tilde{\eta} \, m\pi (\cos\delta_0) \delta_{m,\pm 1} J^e \left[\rho \sin\Theta(\rho) \right]_{\rho=0}^{\rho=\infty} \quad (7.56)$$

$$= 0,$$

where we use $\rho d\rho (\Theta'(\rho) \cos\Theta + \sin\Theta/\rho) = d(\rho \sin\Theta(\rho))$ for $m = \pm 1$. This result is interesting as the field-like SOT force integrated over the skyrmion volume vanishes for ideal skyrmions as $\sin\Theta(\rho = 0) = \sin\Theta(\rho = \infty) = 0$ (as a physical skyrmion does not extend to strictly $\rho = \infty$). This is also the case for the force along the longitudinal direction $F^{SOT}_{x,FL} = 0$ as the integral along the radial coordinate is identical. While field-like SOT can play important roles in Hall transport when the skyrmion shape is deformed [72], numerical analysis in [154] shows that it does not alter the main results significantly.

These analysis confirms that the estimate (7.46) is reasonable after taking into account of the full SOT forces, both damping-like and field-like SOTs.

7.4.3.2 Spin Hall torque (SHT)

Let us move on to discuss the spin Hall torque (SHT). When a ferromagnetic layer is placed on top of a heavy metal layer, polarized electric currents along the heavy metal layer $\vec{J}^{HM}$ can be used to pump polarized spins into the ferromagnetic layer through the spin Hall effect [74][123].

The corresponding Thiele equation has the force term

$$f_i^{SHT} = -b\epsilon_{zlj}\epsilon_{lpq}(\partial_i n_p) n_q J_j^{HM} , \qquad (7.57)$$

where b parametrizes the strength of the SHT that depends on spin Hall angle. We notice (7.57) has the same structure as the damping-like SOT force given in (7.48). Thus, the Néel-type skyrmions do not receive transverse forces according to the analysis related to (7.53).

Thus we can conclude that the SHT does not significantly change the estimates of the skyrmion Hall angle given in (7.46).

7.4.3.3 Spin transfer torque (STT) and emergent fields

Here we discuss the effects of spin transfer torque (STT) along with the effects of the emergent electromagnetic fields, which have been discussed in (4.51) in the section §4.4.2.1. We note that this analysis is not directly related to the estimates of the Hall viscosity done in (7.46) as the data does not use STT. Nevertheless we carry out the analysis as it is important to gain deeper understanding of the effects of the transverse motion introduced in (7.35) in the context of the Thiele equation.

Spins of the conduction electrons moving with velocity $\vec{v}_s$ interact with local magnetization. This spin transfer torque is sensitive to the spatial variation of the magnetization [63][64]. The spatial variation of magnetization, $(v_i^s \partial_i)\vec{M}$, fits nicely together with $\partial_t \vec{M} \propto (v_i \partial_i)\vec{M}$ in LLG equation. This gives two additional contributions to Thiele equation (7.37) as

$$\mathcal{G}_{ij}(v_j - v_j^s + R_{jk}v_k) + \alpha \mathcal{D}_{ij}(v_j - \frac{\beta}{\alpha}v_j^s + R_{jk}v_k) = -F_i, \quad (7.58)$$

where β is the non-adiabatic parameter.

We set $F' = (\gamma_0/M_s)F_x$, $v_x^s = v^s$, $F_y = v_y^s = 0$ to get

$$(\alpha \mathcal{D} - Q R)v_x + (Q + \alpha \mathcal{D} R)v_y = (\beta \mathcal{D} - Q R)v^s - F',$$
$$(Q + \alpha \mathcal{D} R)v_x - (\alpha \mathcal{D} - Q R)v_y = (Q + \beta \mathcal{D} R)v^s. \quad (7.59)$$

This reduces to (7.43) for $v^s = 0$. We can solve the equations for v_x and v_y as

$$v_x = \frac{(QR - \alpha \mathcal{D})F'}{(Q^2 + \alpha\beta \mathcal{D}^2)(1 + R^2)} + \frac{Q^2 + \alpha\beta \mathcal{D}^2}{Q^2 + \alpha\beta \mathcal{D}^2}v^s,$$
$$v_y = \frac{-(Q + \alpha \mathcal{D} R)F'}{(Q^2 + \alpha\beta \mathcal{D}^2)(1 + R^2)} + \frac{(\beta - \alpha)Q\mathcal{D}}{Q^2 + \alpha\beta \mathcal{D}^2}v^s. \quad (7.60)$$

By taking the ratio between the transverse velocity and the longitudinal velocity, we get the generalized skyrmion Hall angle

$$\tan \Theta_{SkH} = \frac{(Q + \alpha \mathcal{D} R)F' + (\alpha - \beta)Q\mathcal{D}(1 + R^2)v^s}{(\alpha \mathcal{D} - QR)F' - (Q^2 + \alpha\beta \mathcal{D}^2)(1 + R^2)v^s}. \quad (7.61)$$

Note that Θ_{SkH} depends on F' and v^s and is useful to model skyrmion Hall angle when there are offsets between them.

If the velocity of conduction electrons is proportional to the applied force F' as $v^s = kF'$ with a dimensionful parameter k, we can get the skyrmion Hall angle as

$$\tan \Theta_{SkH} = \frac{(Q + \alpha \mathcal{D} R) + (\alpha - \beta)Q\mathcal{D}(1 + R^2)k}{(\alpha \mathcal{D} - QR) - (Q^2 + \alpha\beta \mathcal{D}^2)(1 + R^2)k}. \quad (7.62)$$

Thus, Θ_{SkH} is independent of F'. More general cases are easy to work out.

Finally, we discuss the effects of the emergent electromagnetic fields introduced in §4.2.7 and §4.4.2.1 to the Thiele equation (7.37). One way to understand the contributions is to add the following force density

$$f_i^{EM} = Q_2(\beta v_i^s - \alpha v_i - \alpha R_{ij}v_j), \tag{7.63}$$

where $Q_2 = \int d^2x\, Q^2/\gamma_0$ with the topological charge density Q defined around (7.33). Q_2 is independent of the sign of the topological skyrmion charge. For $v_x^s = v^s$ and $v_y^s = 0$, Thiele equation (7.37) becomes (7.43) with $F' = -\beta Q_2 v^s$ and $\mathcal{D} \to \mathcal{D} - Q_2$. The Hall angle $\tan\Theta_{SkH}$ in (7.45) is modified accordingly. As the size of damping is expected to increase with this new contribution, the Hall angle will decrease, while the ratio between the transverse force due to the Hall viscosity and the skyrmion Hall effect, given in (7.46), increases. Thus, while the specific estimates done in (7.46) and (7.47) are not directly related, in general the Hall viscosity contributions become more significant with the emergent gauge fields.

Another way to understand the emergent gauge fields is replacing the parameters α and β with $\tilde{\alpha}$ and $\tilde{\beta}$, respectively. These parameters are defined around the equation (4.53) in §4.4.2.1.

7.5 Outlook

The generalized Thiele equation can be used to investigate the existence of the Hall viscosity more effectively in different settings. First, skyrmions in anti-ferromagnetic materials are expected to have vanishing skyrmion Hall effects especially in synthetic materials as the net skyrmion topological number vanishes $Q = 0$ as skyrmion is paired with antiskyrmion. In a different language, the net spin density of the anti-ferromagnetic material vanishes as described in §7.1.2.

Nevertheless, the skyrmion and antiskyrmion pair has non-vanishing Hall viscosity as the two contributions add up. This can be described by taking $Q = 0$ in (7.45). Thus, the Hall viscosity angle is non-zero,

$$\tan\Theta_R = R, \tag{7.64}$$

moving toward right with respect to force direction. Note this angle should be the same for skyrmions and antiskyrmions for it is independent of the topological skyrmion charge. Experimental verifications would be straightforward.

Second, in the presence of Hall viscosity, the precise location of vanishing skyrmion Hall effect in various parameter spaces, if any, would be deviated from the expected one. Resolving the mismatch will provide more effective and precise experimental verifications of the Hall viscosity.

Bibliography

[1] E. Witten, Global aspects of current algebra. Nucl. Phys. B **223**, 422 (1983).

[2] L. D. Landau and E. M. Lifshitz, On the theory of the dispersion of magnetic permeability in ferromagnetic bodies. Physik. Zeits. Sowjetunion **8**, 153 (1935).

[3] J. H. Han, Skyrmions in Condensed Matter. Springer International Publishing (2017).

[4] V. Borshch, et al., Nematic twist-bend phase with nanoscale modulation of molecular orientation. Nature Communications **4**, 2635 (2013).

[5] A. Bogdanov and D. Yablonskii, Thermodynamically stable "vortices" in magnetically ordered crystals. The mixed state of magnets. Zh. Eksp. Teor. Fiz. **95**, 178 (1989) [Sov. Phys. JETP **68**, 101 (1989)].

[6] A. N. Bogdanov, U. K. Rössler, M. Wolf, and K.-H. Müller, Magnetic structures and reorientation transitions in noncentrosymmetric uniaxial antiferromagnets. Phys. Rev. B **66**, 214410 (2002).

[7] U. Rössler, A. Bogdanov, and C. Pfleiderer, Spontaneous skyrmion ground states in magnetic metals. Nature (London) **442**, 797 (2006).

[8] L. D. Landau and E. M. Lifshitz, Statistical Mechanics. 3rd Edition. Pergamon Press LTD (1980).

[9] T. Jeong and W. E. Pickett, Implications of the B20 crystal structure for the magnetoelectronic structure of MnSi. Phys. Rev. **B 70**, 075114 (2004).

[10] P. Bak and M. H. Jensen, Theory of helical magnetic structures and phase transitions in MnSi and FeGe. J. Phys. C: Solid State Phys. **13**, L881 (1980).

[11] O. Nakanishia, A. Yanase, A. Hasegawa, and M. Kataokaa, The origin of the helical spin density wave in MnSi. Solid State Commun. **35**, 995 (1980).

[12] L. D. Landau and E. M. Lifshitz, Electrodynamics of Continuous Media. Pergamon Press LTD (1984).

[13] I. E. Dzyaloshinskii, A thermodynamic theory of "weak" ferromagnetism of antiferromagnetics. J. Phys. Chem. Solids. **4**, 241 (1958).

[14] T. Moriya, Anisotropic superexchange interaction and weak ferromagnetism. Phys. Rev. **120**, 91 (1960).

[15] G. H. Derrick, Comments on nonlinear wave equations as models for elementary particles. J. Math. Phys. **5**, 1252 (1964).

[16] J. H. Han, J. Zang, Z. Yang, J.-H. Park, and N. Nagaosa, Skyrmion lattice in a two-dimensional chiral magnet. Phys. Rev. B **82**, 094429 (2010).

[17] A. Tonomura, et al., Real-space observation of skyrmion lattice in helimagnet MnSi thin samples. Nano Lett. **12**, 1673 (2012).

[18] C. Schütte and M. Garst, Magnon-skyrmion scattering in chiral magnets. Phys. Rev. B **90**, 094423 (2014).

[19] I. E. Dzyaloshinskii, Theory of helicoidal structures in antiferromagnets. I. Nonmetals. Sov. Phys. JETP **19**, 960 (1964).

[20] N. Papanicolaou and T. N. Tomaras, Dynamics of magnetic vortices. Nucl. Phys. B **360**, 425 (1991).

[21] H. Liu, H. Ooguri, B. Stoica, and N. Yunes, Spontaneous generation of angular momentum in holographic theories. Phys. Rev. Lett. **110**, 211601 (2013).

[22] C. Hoyos, B. S. Kim, and Y. Oz, Ward identities for transport in 2+1 Dimensions. JHEP **1503**, 164 (2015).

[23] L. D. Landau and E. M. Lifshitz, Fluid Mechanics, Third Impression. Pergamon Press Ltd. (1966).

[24] K. Jensen, et al., Parity-violating hydrodynamics in 2+1 dimensions. JHEP **1205**, 102 (2012).

[25] J. Bhattacharya, S. Bhattacharyya, S. Minwalla, and A. Yarom, A theory of first order dissipative superfluid dynamics. JHEP **1405**, 147 (2014).

[26] C. Hoyos, B. S. Kim, and Y. Oz, Lifshitz hydrodynamics. JHEP **1311**, 145 (2013).

[27] C. Hoyos, B. S. Kim, and Y. Oz, Lifshitz field theories at non-zero temperature, hydrodynamics and gravity. JHEP **1403**, 029 (2014).

[28] L. D. Landau and E. M. Lifshitz, Theory of Elasticity, 3rd Edition. Pergamon Press, Oxford (1986).

[29] J. E. Avron, R. Seiler, and P. G. Zograf, Viscosity of quantum Hall fluids. Phys. Rev. Lett. **75**, 697 (1995).

[30] J. E. Avron, Odd viscosity. J. Stat. Phys. **92**, 543 (1998).

[31] R. Kubo, Statistical-mechanical theory of irreversible processes. I. General theory and simple applications to magnetic and conduction problems. J. Phys. Soc. Jpn. **12**, 570 (1957).

[32] R. Kubo, M. Yokota, and S. Nakajima, Statistical-mechanical theory of irreversible processes. II. Response to thermal disturbance. J. Phys. Soc. Jpn. **12**, 1203 (1957).

[33] D. Evans and G. Morriss, Statistical Mechanics of Nonequilibrium Liquids. Australian National University Press, Australia (2007).

[34] O. Saremi and D. T. Son, Hall viscosity from gauge/gravity duality. JHEP **1204**, 091 (2012).

[35] L. Onsager, Reciprocal relations in irreversible processes. I. Phys. Rev. **37**, 405 (1931).

[36] L. Onsager, Reciprocal relations in irreversible processes. II. Phys. Rev. **38**, 2265 (1931).

[37] H. B. Callen, The application of Onsager's reciprocal relations to thermoelectric, thermomagnetic, and galvanomagnetic effects. Phys. Rev. **73**, 1349 (1948).

[38] H. B. Callen, Thermodynamics: An introduction to the physical theories of equilibrium thermostatics and

irreversible thermodynamics. John Wiley and Sons, Inc. (1960).

[39] S. Maekawa, et al., Physics of Transition Metal Oxides. Springer-Verlag Berlin Heidelberg (2004).

[40] B. S. Kim, Skyrmions and Hall transport, J. Phys. Condens. Matter **31**, 383001 (2019).

[41] M. V. Berry, Quantal phase factors accompanying adiabatic changes. Proc. R. Soc. Lond. **A 392**, 45 (1984).

[42] N. Read, Non-Abelian adiabatic statistics and Hall viscosity in quantum Hall states and $p(x) + ip(y)$ paired superfluids. Phys. Rev. B **79**, 045308 (2009).

[43] C. Hoyos and D. T. Son, Hall viscosity and electromagnetic response. Phys. Rev. Lett. **108**, 066805 (2012).

[44] D. Tong, Lectures on the Quantum Hall Effect, [arXiv:1606.06687 [hep-th]].

[45] P. Levay, Berry phases for Landau Hamiltonians on deformed tori. J. Math. Phys. **36**, 2792 (1995).

[46] D. Son and M. Wingate, General coordinate invariance and conformal invariance in nonrelativistic physics: Unitary Fermi gas. Annals Phys. **321**, 197 (2006).

[47] X. G. Wen and A. Zee, Shift and spin vector: New topological quantum numbers for the Hall fluids. Phys. Rev. Lett. **69**, 953 (1992).

[48] N. Read and E. H. Rezayi, Hall viscosity, orbital spin, and geometry: paired superfluids and quantum Hall systems. Phys. Rev. B **84**, 085316 (2011).

[49] I. V. Tokatly and G. Vignale, Lorentz shear modulus of a two-dimensional electron gas at high magnetic field. Phys. Rev. B **76**, 161305(R) (2007).

[50] I. V. Tokatly and G. Vignale, Lorentz shear modulus of fractional quantum Hall states. J. Phys: Condensed Matter, **21**, 275603 (2009).

[51] K. v. Klitzing, G. Dorda, and M. Pepper, New method for high-accuracy determination of the fine-structure constant based on quantized Hall resistance, Phys. Rev. Lett. **45**, 494 (1980).

[52] D. C. Tsui, H. L. Stormer, and A. C. Gossard, Two-dimensional magnetotransport in the extreme quantum limit, Phys. Rev. Lett. **48**, 1559 (1982).

[53] R. Willett, J. P. Eisenstein, H. L. Störmer, D. C. Tsui, A. C. Gossard, and J. H. English, Observation of an even-denominator quantum number in the fractional quantum Hall effect, Phys. Rev. Lett. **59**, 1776 (1987).

[54] D. Tong, Lectures on Kinetic Theory, http://www.damtp.cam.ac.uk/user/tong/kinetic.html

[55] D. J. Thouless, M. Kohmoto, M. P. Nightingale, and M. den Nijs, Quantized Hall conductance in a two-dimensional periodic potential, Phys. Rev. Lett. **49**, 405 (1982).

[56] Y. Chen, F. Wilczex, E. Witten, and B. Halperin, On anyon superconductivity, Int. J. Mod. Phys. B **3**, 1001 (1989).

[57] D. J. Griffiths, Introduction to Quantum Mechanics. Prentice Hall (1995).

[58] T. L. Gilbert, A phenomenological theory of damping in ferromagnetic materials. IEEE Trans. Magn. **40**, 3443 (2004).

[59] S. Emori, et al., Current-driven dynamics of chiral ferromagnetic domain walls. Nature Materials **12**, 611 (2013).

[60] M. N. Baibich, et al., Giant Magnetoresistance of (001)Fe/(001)Cr Magnetic Superlattices. Phys. Rev. Lett. **61**, 2472 (1988).

[61] M. Johnson and R. H. Silsbee, Thermodynamic analysis of interfacial transport and of the thermomagnetoelectric system. Phys. Rev. B **35**, 4959 (1987).

[62] T. Valet and A. Fert, Theory of the perpendicular magnetoresistance in magnetic multilayers. Phys. Rev. B **48**, 7099 (1993).

[63] S. Zhang and Z. Li, Roles of nonequilibrium conduction electrons on the magnetization dynamics of ferromagnets. Phys. Rev. Lett. **93**, 127204 (2004).

[64] A. Thiaville, Y. Nakatani, J. Miltat, and Y. Suzuki, Micromagnetic understanding of current-driven domain

wall motion in patterned nanowires. Europhys. Lett. **69**, 990 (2005).

[65] G. S. D. Beach, M. Tsoi, and J. L. Erskine, Current-induced domain wall motion. J. Magn. Magn. Mater. **320**, 1272 (2008).

[66] V. B. Berestetskii, E. M. Lifshitz, and L. P. Pitaevskii, Quamtum Electrodynamics, 2nd Ed. Pergamon Press LTD (1982).

[67] Yu. A. Bychkov and E. I. Rashba, Properties of a 2D electron gas with lifted spectral degeneracy. Pis'ma Zh. Eksp. Teor. Fiz **39**, 66 (1983) [JETP Lett. **39**, 78 (1984).]

[68] G. Dresselhaus, Spin–orbit coupling effects in zinc blende structures. Phys. Rev. **100**, 580 (1955).

[69] A. Manchon and S. Zhang, Theory of nonequilibrium intrinsic spin torque in a single nanomagnet. Phys. Rev. B **78**, 212405 (2008).

[70] I. Miron, et al., Current-driven spin torque induced by the Rashba effect in a ferromagnetic metal layer. Nature Materials **9**, 230 (2010).

[71] I. Miron, et al., Fast current-induced domain-wall motion controlled by the Rashba effect. Nature Materials. **10**, 419 (2011).

[72] K. Litzius, et al., Skyrmion Hall effect revealed by direct time-resolved X-ray microscopy. Nature Physics **13**, 170 (2017).

[73] M. I. Dyakonov and V. I. Perel, Possibility of orienting electron spins with current. JETP Lett. **13**, 467 (1971).

[74] J. E. Hirsch, Spin Hall effect. Phys. Rev. Lett. **83**, 1834 (1999).

[75] J. C. Slonczewski, Current-driven excitation of magnetic multilayers. J. Magn. Magn. Mater. **159**, L1 (1996).

[76] L. Berger, Emission of spin waves by a magnetic multilayer traversed by a current. Phys. Rev. B **54**, 9353 (1996).

[77] L. Liu, et al., Current-induced switching of perpendicularly magnetized magnetic layers using spin torque from the spin Hall effect. Phys. Rev. Lett. **109**, 096602 (2012).

[78] L. Liu, et al., Spin-torque switching with the giant spin Hall effect of tantalum. Science **336**, 555 (2012).

[79] G. E. Volovik, Linear momentum in ferromagnets. J. Phys. C **20**, L83 (1987).

[80] J. Zang, et al., Dynamics of skyrmion crystals in metallic thin films. Phys. Rev. Lett. **107**, 136804 (2011).

[81] S. Zhang and S. S.-L. Zhang, Generalization of the Landau-Lifshitz-Gilbert equation for conducting ferromagnets. Phys. Rev. Lett. **102**, 086601 (2009).

[82] A. A. Thiele, Steady-state motion of magnetic domains. Phys. Rev. Lett. **30**, 230 (1973).

[83] K. Everschor, et al., Rotating skyrmion lattices by spin torques and field or temperature gradients. Phys. Rev. B **86**, 054432 (2012).

[84] J. Iwasaki, M. Mochizuki and N. Nagaosa, Universal current-velocity relation of skyrmion motion in chiral magnets. Nature Communications **4**, 1463 (2013).

[85] B. S. Kim, Modeling Hall viscosity in magnetic-skyrmion systems. Phys. Rev. Res. **2**, 013268 (2020).

[86] L. Gravier, et al., Spin-dependent Peltier effect of perpendicular currents in multilayered nanowires. Phys. Rev. B **73**, 052410 (2006).

[87] G. E. W. Bauer, E. Saitoh, and B. J. van Wees, Spin caloritronics. Nature Materials **11**, 391 (2012).

[88] P. Atkins and J. de Paula, Physical Chemistry, Eighth Edition. W. H. Freeman and Company (2006).

[89] K. Uchida, et al., Observation of the spin Seebeck effect. Nature **455**, 778 (2008).

[90] K. Uchida, et al., Spin Seebeck insulator. Nature Materials **9**, 894 (2010).

[91] C. M. Jaworski, et al., Observation of the spin-Seebeck effect in a ferromagnetic semiconductor. Nature Materials **9**, 898 (2010).

[92] H. Adachi, K. Uchida, E. Saitoh, and S. Maekawa, Observation of the spin Seebeck effect. Rep. Prog. Phys. **76**, 036501 (2013).

[93] J. Xiao, G. E. W. Bauer, K. Uchida, E. Saitoh, and S. Maekawa, Theory of magnon-driven spin Seebeck effect. Phys. Rev. B **81**, 214418 (2010).

[94] A. Slachter, et al., Thermally driven spin injection from a ferromagnet into a non-magnetic metal. Nature Physics **6**, 879 (2010).

[95] J. Flipse, et al., Direct observation of the spin-dependent Peltier effect. Nature Nanotechnology **7**, 166 (2012).

[96] J. C. Ward, An identity in quantum electrodynamics. Phys. Rev. **78**, 182 (1950).

[97] Y. Takahashi, On the generalized ward identity. Nuovo Cimento, **6**, 370 (1957).

[98] B. Bradlyn, M. Goldstein, and N. Read, Kubo formulas for viscosity: Hall viscosity, Ward identities, and the relation with conductivity. Phys. Rev. B **86**, 245309 (2012).

[99] C. Hoyos, B. S. Kim, and Y. Oz, Ward identities for Hall transport. JHEP **1407**, 054 (2014).

[100] S. Weinberg, The Quantum Theory of Fields Volume II: Modern Application. Section 22. Cambridge University Press (1996).

[101] J. Schwinger, Field theory commutators. Phys. Rev. Lett. **3**, 296 (1959).

[102] J. Schwinger, Energy and momentum density in field theory. Phys. Rev. **130**, 800 (1963).

[103] D. G. Boulware and S. Deser, Stress-tensor commutators and Schwinger terms. J. Math. Phys. **8**, 1468 (1967).

[104] S. A. Hartnoll and P. Kovtun, Hall conductivity from dyonic black holes. Phys. Rev. D **76**, 066001 (2007).

[105] S. A. Hartnoll and C. P. Herzog, Ohm's Law at strong coupling: S duality and the cyclotron resonance. Phys. Rev. D **76**, 106012 (2007).

[106] C. P. Herzog, Lectures on Holographic Superfluidity and Superconductivity. J. Phys. A **42**, 343001 (2009).

[107] M. Greiter, F. Wilczek, and E. Witten, Hydrodynamic relations in superconductivity. Mod. Phys. Lett. B **03**, 903 (1989).

[108] I. V. Tokatly, Magnetoelasticity theory of incompressible quantum Hall liquids. Phys. Rev. B **73**, 205340 (2006).

[109] M. Geracie, D.T. Son, C. Wu, and S. F. Wu, Spacetime symmetries of the quantum Hall effect. Phys. Rev. D **91**, 045030 (2015).

[110] S. Weinberg, The Quantum Theory of Fields, Vol. I. Cambridge University Press, Cambridge, England (1995).

[111] H. Watanabe and H. Murayama, Noncommuting momenta of topological solitons. Phys. Rev. Lett. **112**, 191804 (2014).

[112] N. Nagaosa and Y. Tokura, Topological properties and dynamics of magnetic skyrmions. Nature Nanotechnology **8**, 899 (2013).

[113] S. Mühlbauer, et al., Skyrmion lattice in a chiral magnet. Science **323**, 915 (2009).

[114] X. Z. Yu et al., Real-space observation of a two-dimensional skyrmion crystal. Nature **465**, 901 (2010).

[115] S. Heinze, et al., Spontaneous atomic-scale magnetic skyrmion lattice in two dimensions. Nature Phys. **7**, 713 (2011).

[116] B. S. Kim and A. D. Shapere, Skyrmions and Hall transport. Phys. Rev. Lett. **117**, 116805 (2016).

[117] F. Jonietz, et al., Spin transfer torque in MnSi at ultralow current densities. Science **330**, 1648 (2010).

[118] T. Schulz, et al., Emergent electrodynamics of skyrmions in a chiral magnet. Nature Physics **8**, 301 (2012).

[119] M. Lee, et al., Unusual Hall anomaly in MnSi under pressure. Phys. Rev. Lett. **102**, 186601 (2009).

[120] A. Neubauer, et al., Topological Hall effect in the A phase of MnSi. Phys. Rev. Lett. **102**, 186602 (2009).

[121] N. Kanazawa, et al., Large topological Hall effect in a short-period helimagnet MnGe. Phys. Rev. Lett. **106**, 156603 (2011).

[122] Y. Li, et al. Robust formation of skyrmions and topological Hall effect anomaly in epitaxial thin films of MnSi. Phys. Rev. Lett. **110**, 117202 (2013).

[123] W. Jiang, et al., Direct observation of the skyrmion Hall effect. Nature Physics **13**, 162 (2017).

[124] Y. Kajiwara, et al., Transmission of electrical signals by spin-wave interconversion in a magnetic insulator. Nature **464**, 262 (2010).

[125] K.-i. Uchida, et al., Spin-wave spin current in magnetic insulators. Solid State Physics **64**, 1 (2013).

[126] M. J. Stevens, et al., Quantum interference control of ballistic pure spin currents in semiconductors. Phys. Rev. Lett. **90**, 136603 (2003).

[127] J. Hüber, et al., Direct observation of optically injected spin-polarized currents in semiconductors. Phys. Rev. Lett. **90**, 216601 (2003).

[128] T. Schneider, et al., Realization of spin-wave logic gates. Appl. Phys. Lett. **92**, 022505 (2008).

[129] Y. Onose, et al., Observation of the magnon Hall effect. Science **329**, 297 (2010).

[130] H. Katsura, N. Nagaosa, and P. A. Lee, Theory of the thermal Hall effect in quantum magnets. Phys. Rev. Lett. **104**, 066403 (2010).

[131] T. Ideue, et al., Effect of lattice geometry on magnon Hall effect in ferromagnetic insulators. Phys. Rev. B **85**, 134411 (2012).

[132] C. Strohm, G. L. J. A. Rikken, and P. Wyder, Phenomenological evidence for the phonon Hall effect. Phys. Rev. Lett. **95**, 155901 (2005).

[133] L. Kong and J. Zang, Dynamics of an insulating skyrmion under a temperature gradient. Phys. Rev. Lett. **111**, 067203 (2013).

[134] Shi-Zeng Lin, Cristian D. Batista, Charles Reichhardt, and Avadh Saxena, AC current generation in chiral magnetic insulators and skyrmion motion induced by the spin Seebeck effect. Phys. Rev. Lett. **112**, 187203 (2014).

[135] A. A. Kovalev, Skyrmionic spin Seebeck effect via dissipative thermomagnonic torques. Phys. Rev. B **89**, 241101(R) (2015).

[136] S. Seki, X. Z. Yu, S. Ishiwata, and Y. Tokura, Observation of skyrmions in a multiferroic material. Science **336**, 198 (2012).

[137] S. Seki, et al., Formation and rotation of skyrmion crystal in the chiral-lattice insulator Cu_2OSeO_3. Phys. Rev. B **85**, 220406(R) (2012).

[138] X. Z. Yu, et al., Magnetic stripes and skyrmions with helicity reversals. PNAS **109**, 8856 (2012).

[139] A. A. Kovalev and Y. Tserkovnyak, Thermomagnonic spin transfer and Peltier effects in insulating magnets. Europhysics Letters **97**, 67002 (2012).

[140] S. K. Kim et al., Tunable magnonic thermal Hall effect in skyrmion crystal phases of ferrimagnets. Phys. Rev. Lett. **122**, 057204 (2019).

[141] M. Mochizuki, et al., Thermally driven ratchet motion of a skyrmion microcrystal and topological magnon Hall effect. Nature Materials **13**, 241 (2014).

[142] S. L. Zhang, et al., Manipulation of skyrmion motion by magnetic field gradients. Nat. Commun. **9**, 2115 (2018).

[143] K. Everschor, M. Garst, R. A. Duine, and A. Rosch, Current-induced rotational torques in the skyrmion lattice phase of chiral magnets. Phys. Rev. B **84**, 064401 (2011).

[144] K. A. van Hoogdalem, Y. Tserkovnyak, and D. Loss, Magnetic texture-induced thermal Hall effects. Phys. Rev. B **87**, 024402 (2013).

[145] R. Tomasello, et al., A strategy for the design of skyrmion racetrack memories. Sci. Rep. **4**, 6784 (2014).

[146] S. K. Kim, Unconventional dynamics of ferrimagnets in the vicinity of compensation points. APS March Meeting 2021, March 15-19, 2021.

[147] S. S. P. Parkin, M. Hayashi, and L. Thomas, Magnetic domain-wall racetrack memory. Science **320**, 190 (2008).

[148] K.-S. Ryu, L. Thomas, S.-H. Yang, and S. Parkin, Chiral spin torque at magnetic domain walls. Nature Nanotech. **8**, 527 (2013).

[149] S.-H. Yang, K.-S. Ryu, and S. Parkin, Domain-wall velocities of up to 750 m s^{-1} driven by exchange-coupling torque

in synthetic antiferromagnets. Nature Nanotech. **10**, 221 (2015).

[150] K.-J. Kim, et al., Fast domain wall motion in the vicinity of the angular momentum compensation temperature of ferrimagnets. Nature Mater. **16**, 1187 (2017).

[151] L. Caretta, et al., Relativistic kinematics of a magnetic soliton. Science **370**, 1438 (2020).

[152] S. K. Kim, Y. Tserkovnyak, and O. Tchernyshyov, Propulsion of a domain wall in an antiferromagnet by magnons. Phys. Rev. B **90**, 104406 (2014).

[153] S. K. Kim, K.-J. Lee, and Y. Tserkovnyak, Self-focusing skyrmion racetracks in ferrimagnets. Phys. Rev. B **95**, 140404(R) (2017).

[154] Y. Hirata, et al., Vanishing skyrmion Hall effect at the angular momentum compensation temperature of a ferrimagnet. Nature Nanotech. **14**, 232 (2019).

Index

adiabatic approximation 129–130, 166
adiabatic process 106, 108, 128, 146, 166
affinities 98
anisotropy 161, 341
anomaly 215, 217, 220
angular momentum compensation point 312, 325, 327–328, 332, 335
anti-de Sitter space conformal field theory (AdS/CFT) correspondence 241, 291
anti-ferromagnet 6, 156, 323, 325, 328–331
antiskyrmion 276, 311–312, 323, 332, 334–336, 340, 342–344, 350
Atiyah-Singer index theorem 122

background field method 83, 88, 90, 151
background fields 93, 130
Berry connection 128, 130, 146, 275
Berry curvature 128, 133–134, 146–147
Berry phase 6, 109, 113, 128–130, 146–147, 230, 316, 327, 330
black hole 93, 241, 291
Bohr magneton 11, 160, 166
broken parity 19, 22, 62, 66, 71, 78–80, 82, 135, 270, 319, 336
Brownian diffusion 309

bulk viscosity 61–62, 64–65, 74, 90–91, 224, 228–229, 248, 337–338

CCD *see* charge-coupled device
charge carrier 97, 102, 153, 197–198, 301
charge-coupled device (CCD) 313
charge density 88, 208, 240, 266, 271, 290, 321
  electric 289
  quantized topological 319
  topological skyrmion 18, 210
charged particle 109, 111, 120, 133, 148–150, 180
Chern number 146–147
Chiral magnetic skyrmion 19, 155
chiral superfluid 261–262
coefficient 22, 25, 59, 61, 64, 81, 83, 98, 100, 108, 199, 250, 252
  damping 326
  dissipative 83
  drag 329
  dynamical 198
  isothermal 106
  Peltier 102, 104, 203–204
  Seebeck 99–100, 102, 106, 200–201, 302
  Thompson 103–104
  transport 57–58, 61–62, 64–66, 77, 80, 82–83, 90, 92, 94, 98–100, 104–106, 108, 130–132, 205, 207

commutation relation 12, 259–260, 263–264, 266, 268
  canonical 114
  equal time 222
  momentum-momentum 263
commutator 12, 122, 131, 209–211, 213, 216, 221–222, 230–231, 250, 257–258, 262
conduction electrons 164, 166, 168–170, 172–173, 175, 177, 187–188, 194, 202, 274–275, 292–294, 301–304, 310–311, 317–318, 348–349
conductivity 81, 83, 99–100, 137, 139–141, 151, 153, 183, 199, 204, 207, 210, 245–246, 289–291, 319–321
conductor 95, 100, 103–104, 221, 298
conservation equation 37, 40, 44, 46, 51–53, 57–58, 74–75, 208, 210, 213–214, 216, 221–222, 234–236, 266–267, 288–289
conservation law 18, 40–41, 207
constraint 2, 7, 13, 21, 23–24, 37, 53, 70, 80, 105
contact terms 88–90, 207, 209–210, 222–223, 229, 237–240, 243, 245, 249–251, 260–261, 268–271
contribution 10–11, 26–27, 31–33, 153–154, 164, 176–177, 187–188, 193–196, 260, 270–271, 278–280, 283, 293, 317–319, 337, 342, 344–345, 349–350
  anomaly 215, 217, 220
  asymmetric 338
  derivative 275
  dissipative 319
  Hall viscosity 154, 251, 343–344, 350
  magnon 197, 200, 310–311, 319, 321–322

Cooper pair 262
coupling 173, 187, 248, 275, 292, 325
Curie point 159
current algebra 214, 216
current density 97, 99, 144, 165–166, 208, 212, 248, 274, 287, 333
cyclotron frequency 96, 137, 244, 262

damping term 159–160, 162, 164, 169, 171, 187, 328
deformation 20, 66–68, 70, 73, 121, 126, 130–131, 189, 192, 310, 313–315, 317
density matrix 84–85, 89
derivative 20, 22, 25–26, 42–43, 61, 63, 81, 110–111, 133–134, 168, 187–188, 221–222, 225, 270–271, 327–329
Derrick's argument 29–31
Derrick's theorem 29–30
diamagnetic 89, 153–154, 260
Dirac fermions 256
Dirac's quantization condition 5
divergence 22, 60, 65, 67, 80, 98, 208, 254, 260
DM interaction *see* Dzyaloshinskii-Moriya interaction
domain wall (DW) 69, 162–164, 167–173, 176–180, 185–186, 189, 191–192, 274, 282–284, 303–304, 324–326, 332, 335
Drude model 96, 137, 249
DW *see* domain wall
Dzyaloshinskii-Moriya interaction (DM interaction) 26–28, 30–33, 47, 49–50

Eckart frame 58, 76
edge current 54, 239

effective field 10, 12–14, 160–162, 169–170, 172, 174, 176–178, 184–187, 189
effective field approach 169, 171–172, 178
eigenstates 126, 128–129, 143, 253, 256
electric conductivity 80, 88, 106, 142, 224, 242, 289, 295
electric field 94–96, 99–102, 106–107, 137, 139, 141–142, 148–153, 174–175, 177, 274, 276–278, 280, 298, 301
electrochemical potential 97, 197–198, 200, 204, 303
emergent electric field 18, 187, 193, 277–278
emergent electromagnetic field 155, 187–188, 192–193, 275–276, 345, 348–349
emergent magnetic field 187, 192–193, 275–279, 292, 318, 329
energy density 7–9, 41, 45, 59, 76, 97, 99, 154
energy spectrum 115–117, 259
entropy 58, 60–61, 65–66, 74–75, 77, 80, 82, 92, 97–98, 102, 197
entropy current see entropy
broken boost system 66
broken parity 80–82
charged hydrodynamics 77
first derivative order 65
hydrodynamics 58
spin contribution 197
equal time commutator 213, 222, 236–238, 244–245, 257–258
equilibrium 9, 20, 58, 67–68, 70, 79, 90, 93, 328, 330
Ettingshausen effect 107–108
Euclidean space 38–39
Euclidean time 4, 93
Euler equation 3, 8

Euler-Lagrange equation 4, 42–44, 47–50, 86, 111
exchange interaction 10, 25, 28, 156–158, 161, 167, 327, 345
expectation value 53, 55, 84–85, 88–89, 140, 142, 209–210, 215, 227, 239–240, 250, 254

Faraday's law 18
fermions 156, 257, 259, 262
Fermi sea 165
ferrimagnet 6, 156, 312, 323–327, 329–330, 332, 335, 345
  angular momentum compensation temperature (T_A) 312, 325, 327, 328, 332, 335
ferromagnet 166, 169, 176, 180, 183, 309–310, 322, 325–326, 328–329, 331, 333, 335, 340
Feynman path integral 4
field-like spin orbit torque (FL SOT) 176, 178–179, 184–185
filling factor 134–135, 154
FL SOT see field-like spin orbit torque
fluctuation-dissipation theorem 304
force 66–68, 73, 83, 86, 152–153, 155, 181, 318, 329–330, 336–337, 340–341, 343–347, 349
  Coriolis 153
  dissipative 70, 313
  dissipative drag 191, 341
  electromotive 101, 201–202
  gauge invariance 209
  Lorentz 4, 107, 151, 153, 277, 281
  Magnus 278, 281, 313, 341
  molecular 67
  transverse 338–339, 342, 344, 346–348, 350
Fourier modes 220

Fourier transform 87, 89, 217, 219–220, 222–224, 231–232, 237, 241, 246, 249, 254, 268–269
free energy 20, 22, 24–31, 34–36, 69–71
function 41, 43, 101, 103, 105, 134–136, 150, 159, 224, 226, 229, 286–287, 300, 303, 339
 anti-holomorphic 134
 arbitrary 87
 current-current 89
 delta 216, 218–219, 222, 226, 228, 233, 252–255
 dissipative 70
 Green's 83, 89–90, 94, 220, 222–224, 236–237, 239–241, 244, 268, 289, 339
 Heaviside step 215, 221
 holomorphic 118, 123, 125–126
 Maxwell's distribution 255
 one point 88, 93, 217, 220, 222–223, 227, 245, 268–270
 partition 88–89, 92–93, 228–229
 probability 119
 retarded Green's 83, 87–90, 93, 131–132, 220–221, 223, 235, 254, 267, 269, 339
 spectral 255–256
 step 221, 232, 236, 254
 theta 86, 124–125
 time-ordered 88
 two-point 254
 wave 83–84, 115, 117–120, 125, 134, 137–139, 144, 148, 156–158, 166

Galilean invariance 154, 244, 260
gauge field 17–18, 30, 74–75, 77, 79, 88–89, 93, 125, 130, 350
gauge transformation 111, 114, 125–126, 144, 235

geometric deformation 130, 133
Gilbert damping 52, 161, 177
Gilbert damping parameter 188, 285, 304
Goldstone bosons 226, 256, 319
guiding center 112, 150–151
gyromagnetic ratio 11–12, 160, 188, 248, 304, 326–327, 332, 341

Hall angle 311–312, 332–336, 343, 349–350
 experimental data 333, 335
 skyrmion 285–287, 332–335, 343, 348–349
 spin 181, 184, 281, 283, 285, 348
Hall conductivity 80, 90, 135, 141–145, 147–148, 151, 153–154, 207, 214, 291, 293–295
 electric 292–293, 295
 momentum dependent 148
 static 260
 thermal 293–294, 300–301, 309, 318–321
 zero-momentum 148
Hall effect 107, 147, 180, 274, 287, 300, 332
 electron 274
 inverse spin 182, 201–202, 298–299, 302–303, 322
 phonon 301
 spin 164, 180, 182–183, 297, 348
 skyrmion 274, 277–278, 280, 285–287, 308, 323, 332, 336, 339–340, 342, 344, 350
 thermal 108, 300, 312, 320
 topological 274–275, 277–280
 topological magnon 318
Hall resistivity 96, 137, 278–280, 292

Hall viscosity 64–66, 71, 73–74, 91, 109–154, 207, 214, 224–225, 228–231, 253, 255–256, 261–262, 291, 293–295, 320–323, 333, 335–340, 342–345, 350–351
  direction 336–340
  hydrodynamics 64–65, 80, 83, 91
  Kubo formula 91, 130–132
  relation to angular momentum 134–135
  relation to Hall conductivity 148–154, 320
  skyrmion Hall angle 344
  tensor structure 224
  v.s. skyrmion Hall effect 344
  Ward identity 225, 228, 243, 250–262, 291–294
Hamiltonian 4, 7, 12, 83, 85–86, 114–116, 120–122, 128–130, 132, 138–139, 141, 144–146, 165, 173–175, 259–260
  charged particle 111
  equation of motion 12
  fermions in magnetic field 259–260
  Landau (on torus) 120–126
  quantum Hall fluid 113–115
  Rashba 174–175
  s-d 165
  skyrmion 304
  spin-orbit coupling 173–174
harmonic oscillator 114–115, 121, 138, 140
heavy metal (HM) 165, 180, 184–186, 195, 280–281, 283, 285, 324–326, 333, 348
Heisenberg equation 84
helical state 24, 28–29, 34–37
HM *see* heavy metal
Hund's rule coupling 156, 167, 169, 172–173, 176–177, 187, 274–275, 277, 291–293, 297

index notation 12, 21, 27, 227
isothermal Nernst effect 106–107

Jacobi identity 263–265
Jacobi theta function 123–125
joule heating 104, 204, 324–325

Kelvin relation 102, 104
  first 104
  second 102
Kubo formula 83, 90–92, 128, 131–132, 141–142, 144, 220
  bulk viscosity 91
  conductivity 90
  Hall conductivity 90, 144
  Hall viscosity 91, 132, 338
  shear viscosity 91

Lagrangian 2–4, 10, 17, 24, 40, 45–46, 110–112, 132, 328, 330–332, 339
  charged particle 110–112
  CP^1 action 17
  fermions in magnetic field 256
  ferrimagnet 330–332
  skyrmion 266, 339
  spin 15–17
Landau frame condition 58–59, 61, 65, 75–76, 79, 81
Landau-Ginsberg theory 25
Landau Hamiltonian 120–122
Landau levels 115, 117–118, 137, 140, 259
Landau-Lifshitz equation 6, 10–11, 13–15, 17, 62, 191
Landau-Lifshitz-Gilbert equation (LLG equation) 156, 159–160, 162–164, 167–169, 176–177, 179, 187–193, 195, 303–307, 311, 314–317, 326–327, 341
Leduk-Righi effect 108, 301
L'Hospital's rule 346
linear response theory 83, 85, 88, 132, 141–142, 151

LLG equation *see*
    Landau-Lifshitz-Gilbert
    equation
Lorentz force law 111, 151, 180,
    274, 277
Lorentz invariant 58, 75, 78, 221
Lorentz transformation 42, 64, 174

magnet 35, 179, 200, 273–322,
    330, 343
  conducting 273, 288, 295–296,
    301, 307, 314, 317, 343,
  insulating 296, 300, 303, 308,
    311, 314, 317–318,
    321
magnetic field 34–37, 79–80,
    96–98, 100–101, 104–109,
    111–113, 120, 133–134,
    148–151, 153–155, 174–175,
    180–181, 201, 234–235,
    237–238, 255–256, 259–262,
    285–290, 292–294, 299–301
magnetic flux 120, 135, 262
magnetic moment 6–11, 15, 19, 25,
    35, 156, 158, 183, 186, 325,
    328–329
magnetic translation operator 123,
    144
magnetization 30–31, 153–154,
    156, 158–161, 163–168,
    178–179, 182–184, 239, 304,
    312–313, 326–327, 339–341,
    348
  anti-parallel 34, 162
  conserved 161, 341
  in-plane 162, 173
  local 160, 163–164, 166–167,
    169, 172–173, 175–179, 185,
    345, 348
  non-zero 239
  spontaneous 24, 156, 159, 280,
    300
  up-direction 183
magnon 108, 197, 200, 202,
    296–303, 305–312, 317–318,
    321

and thermal Hall effect 312
magnon current 305–307,
    310–311
magnon Hall effect 296–301,
    318
Magnus force 191, 278, 281, 313,
    341
majority spin 182–184, 198, 204
material 9–10, 26, 101–102, 156,
    194, 263, 288, 295, 298, 324,
    327, 332–333, 335
  anti-ferromagnetic 193, 325,
    329, 350
  conducing 313
  conducting 100, 102–103, 194,
    200, 275, 293, 301
  ferrimagnetic 325, 332, 335
  ferromagnetic 6, 9, 11, 27, 52,
    195, 286, 296–297, 322, 324,
    332
  insulating 301, 308, 313
  magnetic 159, 168, 273, 296,
    324–325
  metallic 187, 308
  nonmagnetic 204
  synthetic 350
  thin 181
Miller index 24
minority spin 182–184, 198,
    204
moduli 121, 126, 128–129, 133
momentum 46, 51, 54–55,
    138–139, 148, 217, 220,
    248–249, 277, 288, 290–293,
    295, 318–321, 336–337
  angular 11, 49–55, 134–135,
    153, 159–160, 174, 213–214,
    223, 227, 229–230, 234–235,
    250–253, 256, 259–262,
    270–271, 297–298
  canonical 110–111, 113–114,
    139
  linear 50–51, 174, 319
  non-zero 291
  zero 241, 288, 319
  zero spatial 91, 131

momentum density 41, 45–46, 221–222, 224, 227–228, 230–231, 239, 245, 248, 250, 253, 255–256, 258
momentum space 87, 89–90, 208, 217–218, 223–225, 229, 245–246, 269–270, 288–289
motion 3, 11–13, 15–17, 70, 73, 84–85, 109, 111–112, 148–151, 171, 196–197, 274–278, 304, 308, 310
  antiskyrmion 333
  ballistic 297
  Brownian 311–312
  cycloid 150–151
  cyclotron 134
  DW 155, 162, 168–173, 176–179, 184–186, 189, 304, 324–326, 332
  helical 23
  linear 148, 150, 323, 340
  longitudinal 196, 274, 277, 300, 308, 310, 312, 347
  rotation 70, 159, 197, 313–314, 317
  rotational 313–314, 316–317
  skyrmion 187, 274, 285, 287–289, 296, 308–310, 313–314, 317, 319, 322–323, 338, 340, 344
  steady-state 96, 190, 340
  translational 192
  transverse 274, 277, 280–281, 308, 310, 312, 332–333, 336, 340, 344, 348

Néel-type skyrmions 32–33, 39, 177, 281, 286, 332–333, 346–348
Nernst effect 106–107, 201
Noether's theorem 40–41, 209
non-linear sigma model 25, 29, 30
normalization factor 123, 125–126, 134

Ohm's equation 98

Ohm's law 80, 94
Onsager's reciprocal relation 94, 96, 100
operator 12, 85–86, 117–118, 122, 126, 141–142, 145, 209, 230, 238, 240, 253, 259
  annihilation 121, 126, 140
  composite 257
  conserved current 291
  creation 117–118, 123, 126, 309
  current 257
  differential 30, 126
  finite translation 122
  gauge-invariant 259
  hermitian 87
  ladder 116–117
  magnetic translation 123
  mechanical 86, 165
  momentum 53, 55, 210–211, 230, 235, 259–260, 263–265

parity invariance 20, 22, 80
parity transformation 3, 26, 62–63
Pauli exclusion principle 156
Pauli matrices 12, 17, 71–72, 122, 166, 174, 211–212, 230, 234, 251
Peltier effect 100, 102, 107
perturbation theory 86, 141, 145, 208
phase 19–20, 37, 128–130, 133, 266, 279, 296, 300, 346
phonon 300, 302–303
Planck constant 11, 114, 160, 276
polarization 101, 166, 176, 182–183, 209, 297, 312
pressure 59, 61, 68, 70–71, 78–79, 88, 132, 152, 227–229, 271, 279–280
projection 59–60, 66, 78, 193

quantum Hall fluid 109, 113, 153, 210, 214

Rashba effect 164

Rashba field 175–176
Rashba Hamiltonian 174–175
Rashba magnetic field 177, 345
relativistic Dirac equation 173
resistivity 100, 106, 136–137, 140, 279–280
retarded correlator 231–232, 254
retarded Green's function 87–94, 131–132, 220–226, 237–241, 254, 267–269, 289, 339
right-handed Bloch DW 162–163, 172, 185
right-handed Néel DW 162–163, 172, 185
rotational invariance 74, 143, 224–227, 239–241, 245, 315, 319, 336

saturation magnetization 6, 160, 165, 190, 340–341
Schrödinger equation 129
Schrödinger picture 84–85
Schwinger term 215, 217, 219, 222
Seebeck effect 100–102, 107–108, 200, 301–302, 309–310, 312
 Seebeck coefficient ϵ 99, 100, 106, 201–202
 skyrmion Seebeck effect 308–310
 spin Seebeck effect 200–203, 301–303
shear 60, 71, 210–211, 241, 336, 338
shear transformation 210–212, 230, 258–259
shift 135, 259–262
SHT *see* spin Hall torque
skyrmion 29–53, 274–321
 action 30
 angular momentum 51–53
 Bloch-type 32–33, 39, 275–276
 center 188, 191
 conducting magnet 274–296
 continuous skyrmion model 281
 drift velocity 188, 281, 308
 emergent electromagnetic fields 17, 187, 193, 275–278, 292
 Hamiltonian 304
 half 335, 344
 Hall angle 285–286, 332–335, 343, 348–349
 Hall effect 274, 280–288
 insulating magnet 296–321
 Néel-type 32–33, 39, 193, 277, 281, 286
 phase diagram 36
 rotation motion 313–318
 rotation symmetry 47–51, 268
 Seebeck effect 308–312
 skyrmion charge 37, 51, 264, 266, 268, 281, 312, 333, 336, 338–342, 344
 skyrmion crystal 29–31, 35–37, 40, 308, 312, 318
 skyrmion lattice 191, 313–314, 316
 skyrmion spin 155, 275, 289, 291–292
spin transfer torque 192–194, 349–350
spin Hall torque 195–196, 280, 283, 348
spin orbit torque 194–195, 345–348
Thiele equation 190–192, 196–197
topological charge 18, 39, 210, 267, 270, 276, 286, 294, 319
topological Hall effect 275–280
translation symmetry 44–47, 268
transverse motion 281, 285
Ward identity 262–271
Slonczewski mechanism 180, 182–183
Slonczewski torque 183–184
SOC *see* spin-orbit coupling
SOT *see* spin orbit torque
spatial translations 46, 235

spectrum 115, 146, 255–256
spin configuration 26, 155, 187, 192, 265–266, 275, 292
spin currents 94, 176, 202, 296–298, 302–303, 316
spin density 165–167, 175–176, 323, 326–328, 330, 332, 335
spin imbalance 180–181, 322
spin mixing 198–199
spin-orbit coupling (SOC) 25–26, 173–174, 180, 315
spin polarization 101, 166, 168, 180–185, 200, 202
spin pumping 298–299, 302, 310, 312
spin Seebeck effect 200–202, 301, 308
spin torque 155–156, 159, 162, 164, 175–176, 182–183, 185, 187, 189, 192, 280–281, 340–342, 344–345
  emergent electromagnetic field 187
  spin Hall torque (SHT) 155, 161, 180, 183–186, 192, 280–281, 285, 324, 345, 348
  spin orbit torque (SOT) 176–178, 184–185, 192, 194–196, 280, 287–288, 324, 328, 335, 344–345, 347–348
  spin transfer torque (STT) 161, 164–165, 167–173, 175, 177, 187–188, 192, 195, 274–275, 298–299, 304, 307, 324, 345, 348
spin transport 182, 184, 198
state 20, 28, 35–37, 60, 67, 84–85, 114–115, 117, 119, 129, 133–134, 157, 254–255
  antisymetrized 158
  coherent 17
  deformed 19–20
  ferromagnetic 36
  ground 19, 36–37, 114, 117, 120, 122–123, 125–127, 129, 141, 143, 145, 253–254, 296–297
  helical spin 23, 28–30, 34
  isotropic 213
  macroscopic 20
  magnetization-down 344
  many-body 141
  quantum Hall 132, 148, 154
  ring-shaped 119
  stationary 129
  steady 306
  symmetric 158
stereographic projection 38–39
Stokes's theorem 5, 130
structure constant 215
STT *see* spin transfer torque
symmetry 1–3, 6–7, 20, 22, 24, 40–41, 62, 74–75, 77–78, 115–116, 207–208, 213–214, 220–221, 223, 267–269
  boost 64–66, 264–266
  broken 62
  compatible 55, 109
  cubic 24–25, 174
  discrete 2, 7, 24–25
  inversion 174
  Lorentz 44, 59, 64, 77
  non-relativistic 130
  parity 19–20, 22, 25–26, 54, 62–63, 71, 94, 105, 109, 224, 248, 262, 269, 274, 319
  Poincaré 2, 264
  rotation 2, 47, 55, 64, 66, 68, 223–226, 265, 270, 288–289, 294
  rotational 2, 55, 64, 71, 115, 213, 220, 227, 250, 265, 281
  spontaneously broken 7, 256
  translation 2, 55, 210, 214, 217, 223, 225, 227, 229, 231, 245–246, 249, 268–271
  uniaxial 174

U(1) 74–75, 208
system 3–4, 54, 58, 66, 71, 78–80, 84–85, 87, 112–115, 128–130, 135–136, 144–145, 223, 225, 228, 231, 240, 249–250, 252–254, 256, 262, 265, 268, 270, 320–321
  anti-ferromagnetic 328
  finite 119
  gapped 135
  homogeneous 227, 239
  hydrodynamic 60
  infinite 231
  insulating 221
  isothermal 99
  many-body 141
  microscopic 87
  neutral 66
  quantum Hall 55, 74, 109, 113–114, 120, 135–136, 214, 256, 261, 294
  relativistic 241
  rotational invariant 70, 132
  skyrmion 266, 321–322

tensor 12, 20–21, 41, 47, 59–64, 66, 72, 244, 251, 281, 284
  antisymmetric 3, 63, 99, 248
  antisymmetric unit 336
  conductivity 95–96, 136, 241–243, 247, 250, 253, 295
  damping torque 311
  delta symbol 12
  dissipative force 281–282
  elasticity 240
  elastic modulus 69, 71, 132
  epsilon 53, 63
  field strength 5, 18, 215
  resistivity 95, 137
  second rank 60, 67, 78, 191–192, 195–196, 281
  shear 60, 71–73, 106, 321
  strain 67–69, 71
  stress 66, 68–71, 76, 152, 211, 240
  stress energy 40–41, 44, 46, 57–59, 63–64, 74–77, 80–81, 92, 131–132, 210, 220–223, 229–231, 234–235, 240–241, 339
  symmetric stress energy 259
  symmetric traceless 77
  transverse 76
  viscosity 71, 73, 225, 245, 336
temperature gradient 57, 80, 94, 99–103, 105–108, 200–202, 204–205, 299–302, 308–309, 311, 313–314, 318, 321
tensor structure 76, 78, 223–225, 227, 241–242, 246–247, 250, 252, 269, 271, 290
thermal conductivity 77, 99, 224–226, 270, 289, 319–320
thermocouple 101, 104, 200
thermoelectric power 99–102, 106–107, 199, 201
Thiele equation 188–189, 191–196, 280, 283, 285–286, 310–312, 314, 317, 329, 334, 340–343, 345, 348–350
  emergent electromagnetic field 192–194, 348
  spin Hall torque 195–196, 280, 283, 348
  spin orbit torque 194–195, 345–348
  spin transfer torque 192–194, 349–350
  transverse velocity 196, 340–343
thin film 36, 181, 267, 280
Thompson effect 100, 103–104
TKNN invariant 147
topological charge 38–39, 44, 192–193, 220, 225, 262–263, 267–268, 270, 276, 286, 288, 290, 294
topological charge density 191, 210, 266, 268, 276, 292–293, 295, 350

topological Hall effect 275–280
topological skyrmion charge 187, 266, 276–277, 319–320, 342, 350–351
torque 11, 159–161, 163–164, 167, 169, 171–172, 176, 179, 183–184, 187, 315
  damping 161, 281
  negative transfer 310
  spin-orbit 173, 189, 280
torus 55, 120–122, 124–126, 128, 133–134, 144–145
transformation 22, 42–44, 93, 95, 111, 124–125, 235
  canonical 43
  guage 209
  linear 230
  mirror 62
  orthogonal 64
  strain 128
  time reversal 3, 87
translation invariance 51, 53, 138, 143, 227, 231, 240, 243, 250, 252, 271

vector 4–5, 13–15, 19–23, 26, 28, 32, 39–40, 47–49, 78–80, 159, 197–198, 209, 304, 315–316, 330–332
  basis 16
  bra 130
  dimensional column 198
  displacement 6, 66, 130
  electromagnetic 289
  gyromagnetic coupling 191, 281
  inhomogeneous 20
  magnetization 34, 159, 164–165, 171–172, 185–186, 190, 197, 298, 304, 307, 317
  pseudo 79

spin 24–25, 28, 31, 38, 47, 159, 277–278, 330
spin-down 282
spin-up 282
surface element 68
  unit 10, 15, 20, 174, 192–193, 330
  wave 24, 26–28, 90, 140, 152–153
velocity 58–60, 63, 65, 70, 139, 151, 153, 188–189, 193–194, 277–278, 285, 288, 309–311, 343, 348–349
  angular 70
  atomic 10
  domain wall 323–326, 335
  drift 188, 191, 193, 281, 285, 315
  dynamic 287
  in-plane 285
  relative 70, 188, 278
  relativistic 62

Walker breakdown 325
Ward identity 53, 55, 154, 207–271, 288–293, 295, 318–321
  conducting magnet 242–243, 247, 290–291, 294–295
  conserved current 214
  continuity equation 208
  Hall viscosity 243, 256, 261, 270–271, 291, 294, 320
  insulting magnet 225, 228, 270–271, 319–320
  topological 267–270, 288–289, 292, 318
Wess-Zumino action 17
winding number 31, 38, 266

Zeeman Hamiltonian 11